Hydrology and Water Resource Systems Analysis

Hydrology and Water Resource Systems Analysis

MARIA A. MIMIKOU
National Technical University of Athens, Greece
EVANGELOS A. BALTAS
National Technical University of Athens, Greece
VASSILIOS A. TSIHRINTZIS
National Technical University of Athens, Greece

CRC Press
Taylor & Francis Group
Boca Raton London New York

CRC Press is an imprint of the
Taylor & Francis Group, an **informa** business

CRC Press
Taylor & Francis Group
6000 Broken Sound Parkway NW, Suite 300
Boca Raton, FL 33487-2742

First issued in paperback 2018

© 2016 by Taylor & Francis Group, LLC
CRC Press is an imprint of Taylor & Francis Group, an Informa business

No claim to original U.S. Government works

ISBN-13: 978-1-4665-8130-2 (hbk)
ISBN-13: 978-0-367-02801-5 (pbk)

Library of Congress Cataloging-in-Publication Data

Names: Mimikou, Maria A., author. | Baltas, Evangelos A., author. |
Tsihrintzis, Vassilios A., 1959- author.
Title: Hydrology and water resource systems analysis / authors, Maria A
Mimikou, Evangelos A Baltas, Vassilios A Tsihrintzis.
Description: Boca Raton : Taylor & Francis, 2016. | Includes bibliographical
references and index.
Identifiers: LCCN 2015039103 | ISBN 9781466581302 (alk. paper)
Subjects: LCSH: Water resources development. | Hydrology. | System analysis.
Classification: LCC TC405 .M49 2016 | DDC 551.48--dc23
LC record available at http://lccn.loc.gov/2015039103

Visit the Taylor & Francis Web site at
http://www.taylorandfrancis.com

and the CRC Press Web site at
http://www.crcpress.com

To my husband Yiannis and my daughter Alexandra for their support

Maria Mimikou

To my family

Evangelos Baltas

In memory of my parents Andreas and Evangelia to whom I owe everything.

To my wife Alexandra and my two sons Andreas and Konstantinos
for their support, encouragement and patience

Vassilios A. Tsihrintzis

Contents

Preface

The overall goal is to provide students and practitioners with a complete and comprehensive guidebook on hydrological and water resources issues by facilitating their understanding of basic and theoretical knowledge and concepts through a significant number of examples. We believe that this book will contribute to the education of undergraduate and graduate students who attend classes related to hydrology and water resources and will provide better insights to scientists, technicians, practitioners and professional engineers regarding integrated approaches in hydrological processes.

Authors

Dr. Maria A. Mimikou is a professor in the School of Civil Engineering at the National Technical University of Athens (NTUA) and director of the Laboratory of Hydrology and Water Resources Management, Athens, Greece. She established the Center of Hydrology and Informatics (CHI) (www.chi.civil.ntua.gr). She has vast experience (35 years) in the areas of water resources management; water resources systems planning and operation; urban, rural and coastal hydrology; stochastic hydrology; hydrological and water quality modelling; soil erosion and sediment transport; climate change and land use change at the catchment scale; flood forecasting; risk assessment and mapping and water scarcity and drought analysis. She has written several books and has authored or co-authored more than 80 papers in peer-reviewed scientific journals and has more than 100 peer-reviewed publications in national and international conferences, in addition to several scientific technical reports. Dr. Mimikou coordinates several undergraduate and postgraduate courses in the School of Civil Engineering (NTUA) and has supervised 90 graduate diplomas and postgraduate theses. She has vast academic and non-academic administrative experience. She has served as director of the Hydrology Section of the Public Power Corporation. She has served as dean of the School of Civil Engineering (NTUA) and director of the Department of Water Resources and Environmental Engineering. She is experienced in the management of research programs and has been scientifically responsible for and administrator of many European competitive research programs and has been a coordinator of and scientifically responsible for other International competitive programs (http://mimikou.chi.civil.ntua.gr/). Also, she has served as a member of the External Advisory Groups in DG Research, as chairman and member of steering and organizing committees in several conferences, as a member of different European scientific networks like EurAqua, EXCIFF, etc. and has contributed to position papers for the Water Framework Directive, Horizon 2020, Water Science – Policy Interfacing. She established the national academic network on hydrology and water resources 'HYDROMEDON'.

Dr. Evangelos A. Baltas is a professor in the School of Civil Engineering at the National Technical University of Athens (NTUA), Athens, Greece. He actively participated in the establishment of the Center of Hydrology and Informatics (CHI) in Athens, which comprises the NTUA meteorological network, the database of the hydrological information and the experimental basin. He has more than 25 years experience in the areas of water resources management, water resources systems planning and operation, hydrometeorology, hydrological modelling, climate change and land use change at the catchment scale, flood forecasting, risk assessment and mapping analysis. He has written several books and has authored or co-authored more than 70 papers in peer-reviewed scientific journals and more than 100 peer-reviewed publications in national and international conferences, in addition to a number of technical reports. He has supervised more than 60 undergraduate and postgraduate theses and has extensive academic and non-academic administrative experience.

He has also offered engineering consultation services in the fields of his expertise to the EU, Greek ministries, public organizations and private companies in the United States and Europe. He has been the principal investigator or researcher in competitive EU (more than 30) and nationally funded programs related to integrated water resources management. He also served as a secretary general for the Ministry of Environment, Physical Planning and Public Works from 2006 to 2009. During that period, he was appointed as a Greek delegate to a number of councils at a ministerial level in the European Union, UNESCO, OECD and the United Nation for issues concerning environmental legislation, climate change, renewable energy water resources, etc.

Dr. Vassilios A. Tsihrintzis is a professor of ecological engineering and technology at the School of Rural and Surveying Engineering, National Technical University of Athens, Greece. His research interests concentrate, among others, on water resources engineering and management with an emphasis on urban and agricultural drainage and non-point source pollution, water quality of aquatic systems and pollution control, ecohydrology and ecohydraulics and the use of natural treatment systems (i.e. constructed wetlands and stabilization ponds) for runoff and wastewater treatment. His published research work includes more than 130 papers in peer-reviewed scientific journals, more than 250 papers in conference proceedings and more than 100 technical reports. He has also authored or co-authored books and book chapters on operations research, urban hydrology and runoff quality management and natural systems for wastewater and runoff treatment, among others. He has participated as a principal investigator/coordinator or a team member in various research projects in the United States, the EU and Greece. Dr. Tsihrintzis has supervised more than 80 undergraduate and postgraduate theses and 12 doctoral dissertations. He regularly teaches engineering hydrology, urban water management, fluid mechanics, groundwater, environmental engineering and natural wastewater treatment systems. He has also served as a professor and the head of the Department of Environmental Engineering, Democritus University of Thrace, Greece, for several years, and previously was an associate professor of water resources engineering at Florida International University, Miami, Florida. Dr. Tsihrintzis has extensive professional experience as a practicing civil and environmental engineer in leading engineering: consulting firms both in the United States (he was a registered professional engineer in California and a certified professional hydrologist by the American Institute of Hydrology) and in Greece, having been involved in several projects related to land development, drainage, urban hydrology, sediment transport and channel design, water resources management, wetlands restoration and constructed wetlands.

Chapter 1

Introduction

1.1 GENERAL

This book contains the description and analysis of basic concepts, as well as, a great number of applications and solutions that cover the scientific disciplines of engineering hydrology and water resources management.

More specifically, in Chapter 2, the forms of precipitation, the precipitation-measuring instruments and their function and installation design, the basic preprocessing of the point rainfall data (homogenization, addition of missing data, altitude correction) and the methods of integration for the estimation of the areal rainfall are described. Also, in the same chapter, we explain the terms and the estimation methods of evaporation, transpiration and actual potential evapotranspiration. Finally, the hydrological losses on the ground are presented, such as storage depression, retention and infiltration. Finally, various methods of infiltration evaluation are also presented.

In Chapter 3, we deal with runoff components, the geomorphological and physiographic characteristics of a river basin and the characteristics and the components of a hydrograph. Moreover, the hydrometry, the rainfall–runoff relationships, and selected models are presented.

In Chapter 4, a detailed description of groundwater is given. Particularly, soil parameters, groundwater aquifers and their classification, the mathematical problem and the general solution of groundwater flow, analytical solutions of steady and unsteady flows, the theory of images, pumping tests, aquifer recharge and seawater intrusion are presented.

In Chapter 5, the basic concepts of the probability and statistics theory in hydrology are described. Continuous and discrete probability functions that are appropriate for use in hydrology and water resources, extreme distributions (floods and droughts) and rainfall intensity–duration–frequency curves are presented.

In Chapter 6, the hydrologic design of hydraulic and other engineering works is presented. The issues discussed include hydrologic design (sizing) of reservoirs and safety (flood protection) structures like spillways and river diversion works. Both deterministic and stochastic approaches are presented. In addition, regional models are described for the design of works at ungauged sites and other specific issues related, for example, to the design of irrigation uptakes and small hydroelectric dams.

In Chapter 7, issues of urban hydrology and stormwater management are presented. The chapter starts with the description of the urban drainage system and proceeds with the impacts of urbanization in affecting the shape of the runoff hydrograph. Then, issues related to urban runoff quality, including factors affecting runoff quality, the pollutant generation processes, the types and sources of pollutants and the impacts on receiving waters are discussed. The chapter then describes methods to compute urban runoff quantity, such as the rational and the SCS methods. Then, three methods for runoff quality computations, as well as methods to evaluate pollutant accumulation on the street surface and washoff of pollutants, are described. Finally, the last section of the chapter presents in detail the design of a great number of structural and non-structural best management practices to manage urban runoff quantity and quality. Among these are porous pavements, infiltration trenches, grass filters and dry and wet ponds, as well as methods to deal with combined sewer overflows.

Chapter 8, the last chapter of this book, is an introduction to sediment transport and erosion. The chapter first introduces the sediment properties, and then it addresses flow resistance issues in open channels and alluvial streams and rivers, bed forms, incipient motion and stable channel design and selected sediment transport formulas for bed load, suspended load and total load. It finally presents watershed erosion and sediment yield and control measures.

1.2 THE SCIENCE OF HYDROLOGY

Hydrology is the science that describes the appearance, circulation and distribution of earth's water, as well as the interaction of its physical and chemical properties with the environment and humans.

The objective of hydrology is the scientific evaluation of water in various phases, especially its variation in time and space. In the science of hydrology, various complex processes of primary concern are involved, including evaporation, rainfall, infiltration, transpiration, storage and runoff. Hydrology is closely connected to a wide range of sciences like biology, chemistry, agriculture, geography, geology, oceanography, physics and volcanology, as shown in Figure 1.1. Its connection with these sciences is a physical consequence of the close connection of water with the atmosphere and the ground. This connection states clearly that hydrology is an interdisciplinary science relevant to almost every aspect of life (Singh, 1992).

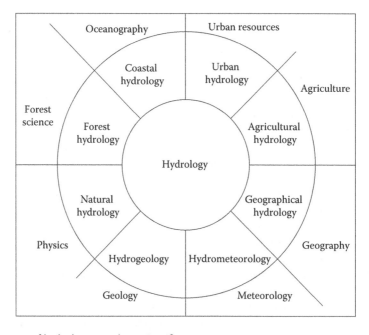

Figure 1.1 Relevance of hydrology to other scientific areas.

Many techniques and methods with origins from other scientific areas like mathematics, statistics, probability theory, operational research, control theory and systems analysis have been used to solve hydrology problems.

The field study of hydrology includes the atmosphere (up to a distance of 15 km), the surface and the interior of the lithosphere (down to a depth of 1 km) and the hydrosphere (oceans). Within the system of these three areas, the hydrological cycle takes place.

1.3 HISTORICAL EVOLUTION OF HYDROLOGY

Mankind before many centuries constructed hydraulic projects worth mentioning, relying mostly on philosophical theories, since the knowledge of hydrological processes was almost nonexistent, if not conceptually incorrect. The hydrological cycle was known from Plato's era; nevertheless, a precise theory was formulated by Marcus Vitruvius during the first century AD (Mimikou and Baltas, 2012).

The period of modern hydrology blossomed in the seventeenth century, during which the knowledge of hydrological phenomena started to be based on measurements. Perrault measured the rainfall and evaporation of the Seine River basin (Perrault, 1966; Nace, 1974), Mariotte (Chisholm, 1911) estimated the discharge of this river and later executed measurements of the flow velocity and the wet cross section, and finally, astronomer Halley (Brutsaert, 2005) measured the evaporation of the Mediterranean Sea, accounting the inputs and outputs from that sea. With these measurements, it became possible to extract relatively precise conclusions about various hydrological phenomena.

The experimental period in hydrology started in the eighteenth century and brought consequent developments in hydraulics. For discharge measurement, the Pitot tube (Saleh, 2002), the mouthpiece of Borda (Chanson, 2014) and Woltman water meter (Arregui et al. 2007) were invented, the equations of flow in open channel conduits and spillways were formulated, and flow in porous media was studied including equations for pumped wells.

The insufficiency of empirical equations for adequate solutions to practical problems became obvious during the beginning of the twentieth century. Due to this, through the foundation of special services and scientific organizations responsible to look into hydrology matters, an effort was made for systematic collection of data, research and study of relevant problems. Thus, the rational management of hydrological phenomena started in the 1930s. Sherman (1932) introduced the concept of the unit hydrograph for surface discharge and statistical and mathematical analyses of discharges.

Finally, in the 1950s, the period of application of theoretical methods to the study of hydrological problems began. The development of electronic computers made feasible the solution of equations arising from the application of mathematical analysis of hydrological phenomena. Additionally, the development of complex instruments of high precision and above-ground measuring potential (radars, satellites, etc.) made possible today the collection and assessment of detailed meteorological and hydrological data, and the verification of applied methods.

1.4 CLASSIFICATION OF HYDROLOGY

Depending on the way and the target of the approach on the subject, the field of hydrology is divided into various basic subdisciplines, as shown in Figure 1.2.

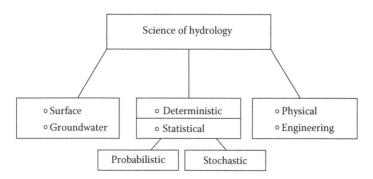

Figure 1.2 Classification of hydrology in disciplines.

A way of dividing the various disciplines of hydrology is based on the space in which various hydrological phenomena occur. So, surface hydrology deals with surface water and groundwater hydrology deals with underground water. This distinction exists due to the different kinetic and dynamic behaviours of water on the ground surface and under the ground.

Another way of distinction is based on the methodological approach of the hydrological procedures. Hence, deterministic hydrology uses methods and models whose parameters are calculated from empirical procedures (e.g. the unit hydrograph method) and statistical hydrology deals with the methods of the theory of probabilities and statistics. The latter is divided into two different sub-categories: probabilistic hydrology, which analyzes and composes the hydrological events without taking into consideration the time continuity of the events, and stochastic hydrology, which takes into consideration the time series in the structure and the occurrence of the hydrological events.

The branch of hydrology that targets the understanding of the physical processes, the causes and mechanisms that evoke them and the natural phenomena that are connected to them is known as physical hydrology, while the discipline that targets the quantitative estimation and the prediction of the hydrological phenomena is known as engineering hydrology. Thus, the engineering hydrology provides the tools for the study, or in other words the hydrologic design for the functioning and sizing of all hydraulic and other engineering works, for example, the hydrologic design of dams for the storage of water, water supply, irrigation and natural land reclamation networks, flood protection structures (embankments, levees), urban hydraulic projects (sewage networks), roads, bridges crossing rivers and projects of groundwater recharge.

1.5 HYDROLOGICAL CYCLE

The hydrological cycle describes the perpetual movement of water between the oceans, the atmosphere and the land, accompanied by changes among the wet, gas and liquid phases of water. Figure 1.3 presents a schematic description of the hydrological cycle.

The start of the hydrological cycle can theoretically be placed in the atmosphere, where the vapour from water evaporation from the land and the oceans, as well as from transpiration from trees and vegetation, is gathered. The vapour is carried by winds and when appropriate conditions develop, clouds are formed. At a later time, the vapour forms atmospheric precipitation (rain, snow, haze) which returns to the earth surface. From the water which reaches the earth surface, a part is detained by the vegetation and evaporates, another part

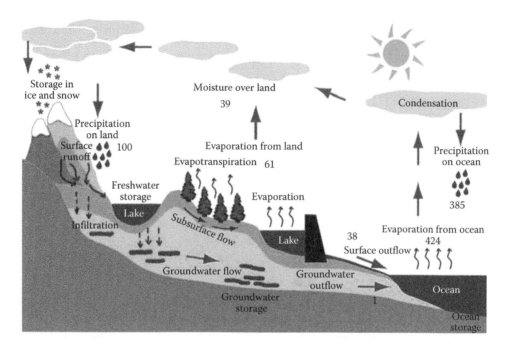

Figure 1.3 The hydrological cycle and the annual world hydrological balance.

infiltrates into the soil, and finally, a last part runs as runoff on the surface and collects in streams or rivers ending in lakes or seas. From the water which infiltrates, a part evaporates or transpires through plants and the rest percolates in deeper layers of the ground, recharging the groundwater aquifers, finding ways out to the surface at a later time at lower altitude, and ending up, finally, into the sea. Then, through evaporation, the water returns to the atmosphere, completing the hydrological cycle.

1.6 HYDROLOGICAL VARIABLES AND THEIR UNITS OF MEASUREMENT

The basic hydrological variables are precipitation, evaporation, interception, retention, infiltration and percolation, and surface runoff. Surface runoff ends into the most important water resources and also creates most hydrological hazards (i.e. floods, droughts). Also underground percolation is connected with the exploitation of water resources and constitutes a major subject of engineering hydrology. Atmospheric precipitations are connected to surface runoff and underground percolation by the cause–effect relation. Evaporation and transpiration, which are also referred together as evapotranspiration, are hydrological losses, meaning the part of precipitation that does not run off and cannot be available for surface or underground storage and exploitation. In conclusion, runoff (surface and groundwater), precipitation and evapotranspiration are the most characteristic processes and variables of the hydrological cycle, and their values quantify the status of the water resources in an area.

Table 1.1 presents the most important units (in the metric system) used in the science of hydrology for all measured variables. These units are determined by the precision of the

Table 1.1 Some commonly used hydrological parameters and their units

Variables	Characteristics	Units
Rainfall	Depth	Millimetres (mm)
	Intensity	Millimetres per hour (mm/h)
	Duration	Hours (h)
Evaporation	Rate	Millimetres per day, month or year (mm/day, mm/month, mm/year)
	Depth	Millimetres (mm)
Infiltration and percolation	Rate	Millimetres per hour (mm/h)
	Depth	Millimetres (mm)
Interception	Equivalent depth	Millimetres per storm duration (mm/time)
Retention	Equivalent depth	Millimetres per storm duration (mm/time)
Runoff	Discharge	Cubic metres per second (m³/s)
	Volume	Cubic metres (m³)
	Equivalent depth	Equivalent millimetres over the river basin (mm)

instruments used and by the physical meaning of the measured value. The most common unit of measurement of discharge is cubic metres per second (m^3/s), while also the equivalent depth of water is used, divided by the surface of the river basin usually measured in km^2. The rainfall depth is measured in mm or cm, indicating the volume of rainfall per unit drainage area. It is unnecessary to express this particular variable in precision of mm, since the rain gauges and rain meters do not have this kind of precision. In any case, in the measurement of the variables and expression of the results, the judgement of the hydrological engineer is incorporated.

1.7 RIVER BASIN

It is very important in hydrology to define the entire area that receives the rainfall, and the area where surface runoff is produced and ends up to a specific location or point of the drainage system. This area is called drainage basin, or watershed, or catchment area of the river, or simply river basin. The boundary line along a topographic ridge separating two adjacent drainage basins is called drainage divide. As shown in Figure 1.4, the single point or location at which all drainage from the basin concentrates as outflow is called concentration or collection point or outlet of the basin (Raghunath, 2006; Figure 1.4).

1.8 SCALE IN HYDROLOGY

Depending on a certain hydrological problem, the hydrological cycle and its components may be considered at different scales of space and time. In the study of hydrological phenomena, it is necessary to know every type of hydrological variables in continuous space and time. It is clearly understandable that the extreme complexity of the hydrological phenomena and the vast extent and depth of time during which they are evolving make this approach impossible. The insertion of scale is connected with the separation of a specific spatial territory and a specific time period in which we study each phenomenon.

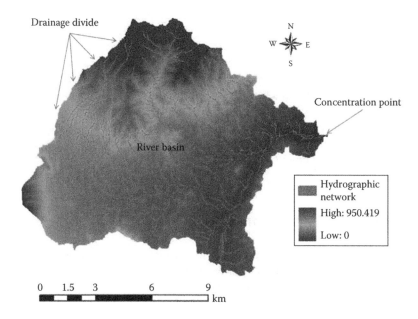

Figure 1.4 A river basin with its concentration point and the drainage divide line.

1.8.1 Spatial scale

The worldwide scale is the largest spatial scale, while the scale of a drainage basin is the smallest as far as the hydrological matters are concerned. It is obvious that there is not a single and unique drainage basin for a stream, but at each point, there is a specific drainage (river sub-basin). This is the most important scale for the science of engineering hydrology, and all the other scales may be composed of summing different river basins. This approach also separates hydrology from hydraulics; in the latter case, the average scale is a pipe/canal or a part of it. It should also be understandable that a drainage basin does not necessarily coincide with the spatial or district limits that are determined for economical or political reasons. Drainage basins may have the size of a small park or even the size of a large river basin. For example, the size of Mississippi river basin occupies approximately 41% of the area of the United States. Usually, large basins are divided into smaller sub-basins so that the hydrological analysis is easier.

The drainage basins which show operational interest have sizes of about tens or thousands of km². However, in research studies regarding mainly the understanding of the physical mechanisms which are connected with the hydrological procedures, the detailed observation and measurement takes place in smaller drainage basins (even smaller than 1 km²), the so-called experimental river basins. Finally, sometimes, the observation of a phenomenon or its measurement takes place in a very small area which is practically represented by one point only, so then we can speak about point observation or measurement.

1.8.2 Timescale

All the hydrological variables introduce time variability. The true knowledge of the time evolution of a hydrological variable demands its continuous time monitoring. However, this is usually impossible, either due to the calculating difficulties or because of the absence of distinct measurements. So, the variables are monitored at different (distinctive) timescales depending on the nature of the problem that is confronted. The timescales which are used in hydrology vary from a fraction of an hour to a year or even many years. The timescale which is used in hydrological analysis depends on the goal of the study and the nature of the examined problem. Usual scales in hydrology are hourly, daily, weekly, monthly and annual scales. The chosen scale is often determined from the time step of the available hydrological data (e.g. measurements of river stage at daily step, rainfall at hourly step).

In conclusion, the timescales of minutes, hours or even days are proper for studies of storms, floods or even for detailed studies or hydrological analyses in an area. In other cases, for example, in water management studies, the efficient standards are timescales of months or years.

1.9 WORLDWIDE DISTRIBUTION OF WATER

Water is one of the biggest resources provided by nature, due to its necessity in the lives of people, animals and plants. The total volume of water on earth is estimated around 1360 million km^3. The distribution of water is given in Table 1.2. We can easily observe that the largest amount of water is stored in the oceans (97.2% of the total volume of earth water). A significant volume of water (2.15% of the total water volume) is stored in the form of ice glaciers and consequently is not exploitable. The remaining amount of water of the planet corresponds to river and lake water, biomass water and groundwater.

Groundwater has a volume of 8.4×10^6 km^3; nevertheless, half of it cannot be exploited because either it is located so deep that pumping is not economically feasible or it is saline. From the exploitable amount of the planet's water, river and lake water comprise only a small fraction of almost 2%. The remaining 98% of the exploitable quantity of water is groundwater. From these arguments, it is clear that a minimal quantity of groundwater is crucial for satisfying human needs. The recharge of a significant part of groundwater is done via the infiltration of the atmospheric precipitation, while on the other hand, substantial amount of runoff ends into the oceans and seas.

Table 1.2 Worldwide distribution of water

	Volume × 1000 km³	Percentage %
Atmospheric water	13	0.001
Surface water		
Ocean water	1,320,000	97.2
Saltwater in lakes	104	0.008
Freshwater in lakes	125	0.009
Freshwater in rivers	1.25	0.0001
Water in ice glaciers	29,000	2.15
Biomass water	50	0.004
Groundwater		
Water in the unsaturated zone	67	0.005
Water at a depth up to 800 m	4,200	0.31
Water at a depth from 800 to 4000 m	4,200	0.31
Total	136,000	100

Source: Bouwer, H., *Groundwater Hydrology*, McGraw-Hill, New York, 1978.

1.10 HYDROLOGICAL BALANCE

The hydrologic or water balance of a drainage basin is the mathematical quantitative expression of its hydrological cycle. It is expressed by equating the difference between inputs and outputs in the drainage basin with the rate of change of the water storage ΔS in a defined time interval Δt. If the drainage basin, or a reservoir, is considered as a system, in which all the inputs and outputs are known and the internal processes are unknown (black box), then the hydrological balance (budget) can be expressed as follows:

$$\frac{\Delta S}{\Delta t} = \bar{I} - \bar{O} \tag{1.1}$$

$$\frac{S_2 - S_1}{\Delta t} = \frac{I_1 + I_2}{2} - \frac{O_1 + O_2}{2} \tag{1.2}$$

where $\bar{I}$ and $\bar{O}$ are, respectively, the average inflow and outflow in the time interval Δt, which should be relatively small for the estimation of average values to be meaningful. The indices 1 and 2 correspond to the start and end of the time interval $\Delta t = t_2 - t_1$. If I and O are continuously changing through time, then Equation 1.2 can be written as

$$\frac{dS(t)}{dt} = I(t) - O(t) \tag{1.3}$$

For a river basin, the rainfall, the snow, the hail and the other forms of precipitation could be considered as inflow. The surface runoff, intermediate runoff, underground runoff, evaporation, transpiration and percolation are the most common forms of outflow. The storage of the drainage basin has also various components like the surface storage (above the ground containing also the storage in water streams and reservoirs), the soil storage (in the voids and the roots of plants), the underground storage (in the aquifers) and intercepted water (by vegetation, buildings). The aforementioned factors are outlined in the next basic equation of hydrological balance:

$$\Delta S = P - R - G - E - T \tag{1.4}$$

According to this equation, the change in a drainage basin storage ΔS is equal to the amount of water falling in the form of precipitation P, minus the amount of water that exits the basin as surface runoff R, percolates underground G, evaporates to the atmosphere E and transpires from the plant leaves T. For individual rain incidents, the constituents of evaporation E and transpiration T are substantially lower than the rest and usually are omitted.

Example 1.1

A rainfall of an intensity of 5 mm/h has dropped on a river basin of an area 5 km^2 in 6 h. At the exit of the river basin, the discharge has been measured during this time period at 100,000 m^3. Estimate the hydrological loss of this 6 h rainfall event.

$I = 0.005 \text{ m/h} \cdot 6 \text{ h} \cdot 5 \cdot 10^6 \text{ m}^2 = 150,000 \text{ m}^3$

The measured outflow is

$O = 100,000 \text{ m}^3$

The difference between outflow and inflow gives the hydrological losses:

$$I - O = 150,000 - 100,000 = 50,000 \text{ m}^3$$

These losses mostly comprise the amount of water infiltrated in the soil after subtracting evaporation and transpiration, which, however, are relatively small for a time interval of 6 h.

The rate of losses, expressed in units of equivalent depth of water per unit area of river basin and time, is as follows:

$$\text{Rate of losses} = \frac{50,000 \text{ m}^3}{5 \cdot 10^6 \text{ m}^2 \cdot 6 \text{ h}} = 0.0016 \text{ m/h} = 1.6 \text{ mm/h}$$

Example 1.2

In a given month, a lake of constant surface area 1,000,000 m² has a mean inflow of 0.40 m³/s, an outflow of 0.34 m³/s and an increase in storage of 20,000 m³. A rain gauge aside the lake measured a total rainfall depth of 30 mm for this month. If we assume that other losses from the lake are insignificant, estimate the monthly evaporation of the lake.

Solution

The general equation of the water balance will be applied which, in this particular case, takes the following form:

$$\Delta S = I + P - O - E$$

where
    I is the inflow in the lake = 0.40 m³/s or 1,036,800 m³ in the month
    O is the outflow from the lake = 0.34 m³/s or 881,280 m³ in the month
    P is the precipitation = 0.030 m · 1,000,000 m² = 30,000 m³
    ΔS is the increase in storage = 20,000 m³
    E is the requested evaporation

By replacement of the values in the water balance equation, we get

$$E = (1,036,800 + 30,000 - 881,280 - 20,000) \text{ m}^3 \Rightarrow E = 165,520 \text{ m}^3$$

Evaporation could be expressed in units of vertical depth of water (cm or mm), by dividing the volume with the area of the lake:

$$E = \frac{165,520 \text{ m}^3}{1,000,000 \text{ m}^2} = 0.165 \text{ m} = 165 \text{ mm}$$

Example 1.3

The drainage basin of a river has a total area of 840 km², and it is covered by four rain stations A, B, C and D. The percentage of the area covered by each station is 30%, 25%, 35% and 10%, respectively. During the hydrological year 1998–1999, these four stations recorded the following monthly rainfall depths in mm (Table 1.3).

Table 1.3 Monthly depths of rain in four rain stations

Month	O	N	D	J	F	M	A	M	J	J	A	S
P(A)	63	105	112	80	110	68	58	55	27	18	11	22
P(B)	69	110	130	88	107	70	62	61	34	22	7	25
P(C)	81	145	122	93	125	79	66	58	29	19	12	18
P(D)	76	137	143	100	119	82	70	67	40	10	5	16

Estimate the following:
1. The point annual rainfall depth in each station
2. The annual rainfall depth and the annual volume of rain on the river basin
3. The annual volume of runoff discharge and the equivalent depth of discharge in mm, if the average annual discharge of runoff at the basin outlet is 8.3 m³/s
4. The annual runoff coefficient
5. The volume of hydrological losses and the equivalent depth of hydrological losses on the river basin
6. The actual evapotranspiration if the hydrological losses from infiltration and underground percolation are 45% of the total rainfall
7. The maximum monthly rainfall depth in the river basin

Solution
1. The point annual depth of rainfall in each station results as the sum of 12 monthly values of rainfall. The results are given in the last column of Table 1.4.
2. For the estimation of the annual rainfall depth, initially we transform the rainfall point measurements on the surface by multiplying the point rain of every month with the respective percentage of area that every station represents. For instance, in March, the average monthly rainfall is equal to

$$P_{March} = 0.30 \cdot 68 + 0.25 \cdot 70 + 0.35 \cdot 79 + 0.1 \cdot 82 = 73.8 \text{ mm} \approx 74 \text{ mm}$$

The results are presented in the last row of Table 1.4
The annual depth of rainfall is produced by the sum of the 12 values calculated and is equal to 798 mm. This amount multiplied by the surface of the river basin gives the annual volume of rainfall:

$$V_{rain} = 0.798 \text{ m} \cdot 840 \cdot 10^6 \text{ m}^2 = 670 \cdot 10^6 \text{ m}^3$$

Table 1.4 Calculation of point annual and average monthly values of rainfall

	O	N	D	J	F	M	A	M	J	J	A	S	Point annual depth
P(A)	63	105	112	80	110	68	58	55	27	18	11	22	729
P(B)	69	110	130	88	107	70	62	61	34	22	7	25	785
P(C)	81	145	122	93	125	79	66	58	29	19	12	18	847
P(D)	76	137	143	100	119	82	70	67	40	10	5	16	865
Surface depth of rain	72.1	123.5	123.1	88.6	115.4	73.8	63.0	58.8	30.8	18.6	9.8	20.8	**798**

3. The runoff volume at the outlet of the river basin is determined by multiplying discharge by the respective time:

$$V_{dis} = 8.3 \text{ m}^3\text{/s} \cdot 365 \text{ day} \cdot 86{,}400 \text{ s/day} = 261.75 \cdot 10^6 \text{ m}^3$$

The equivalent depth of rain is the quotient of this volume divided by the area of the river basin:

$$h_{dis} = V_{dis}/A = 261.75 \cdot 10^6/840 \cdot 10^6 = 0.312 \text{ m} = 312 \text{ mm}$$

4. The annual runoff coefficient is the ratio of the runoff depth (i.e. runoff volume) that exited the river basin to the annual rainfall.

   In this case, $C = 312/798 = 0.391$.

5. The volume of the hydrological losses is

$$V_{los} = V_{rain} - V_{dis} = (670 - 261.75) \cdot 10^6 \text{ m}^3 = 408.25 \cdot 10^6 \text{ m}^3$$

So, the equivalent depth of the hydrological losses is

$$h_{los} = V_{los}/A = 408.25 \cdot 10^6/840 \cdot 10^6 \text{ m} = 0.486 \text{ m} = 486 \text{ mm}$$

6. The water balance of this river basin is expressed in the following equation:

$$P - E - (G + \Delta S) - R = 0$$

where
   P is the rainfall depth = 798 mm
   R is the runoff depth = 312 mm
   $G + \Delta S$ is the infiltration and underground storage = $0.45 \times P$ = 359 mm
   E is the evapotranspiration = 798 − 312 − 359 = 127 mm

7. The maximum monthly depth of rainfall in this river basin is shown in the last row of Table 1.4 and is equal to 123.5 for the month of November.

Example 1.4

In a river location, a dam is to be constructed for the storage of water for irrigation needs of the nearby fields. The river basin upstream of the dam has an area of 818 km² and is considered impermeable. Table 1.5 presents historical data of (a) the annual surface rainfall P of the river basin and (b) the average annual runoff discharge Q at the dam location for 10 hydrological years.

Calculate the following:

1. The volume of runoff discharge in m³ and the respective depth of runoff in mm for every year
2. The actual evapotranspiration (ET) in the river basin in mm and in m³ for every year
3. The runoff coefficient for every year and for the given time period
4. The expected annual volume of water (in mm) that we can exploit if the total losses of the under construction dam (i.e. evaporation, infiltration, overflows) are 18% of the average annual inflow on the reservoir

Table 1.5 Data of average annual surface rainfall and average annual runoff discharge

Hydrological year	P (mm)	Q (m³/s)
1981–1982	1487	23.73
1982–1983	1337	17.01
1983–1984	1357	24.99
1984–1985	1000	15.54
1985–1986	1532	23.10
1986–1987	1182	14.49
1987–1988	1173	13.86
1988–1989	1238	12.39
1989–1990	760	5.67
1990–1991	1477	20.37

5. The area of the arable land that can be irrigated with the given volume of water (in the case of irrigation dam) if the potential evapotranspiration of the selected crop during the irrigation period is 800 mm and the rainfall during the same period is negligible

Solution

1. The discharge volume is given by multiplying the average annual discharge of the river by the respective time (31,536,000 s/yr). It is seen in the fourth column of Table 1.6 in m³ and in the fifth column in mm after division by the area of the river basin.
2. Given that the river basin is impermeable, the only losses are those of the actual evapotranspiration, which is derived by abstracting the equivalent depth of runoff discharge from the surface rainfall:

$$E = P - h_{dis}$$

The calculation of the evaporation for every year in mm and in m³ is shown in columns six and seven of Table 1.6.

Table 1.6 Estimation of discharge, evapotranspiration and runoff coefficient

(1)	(2)	(3)	(4)	(5)	(6)	(7)	(8)
Year	P (mm)	Q (m³/s)	V_{dis} (m³)	h_{dis} (mm)	ET (mm)	ET (m³)	Runoff coefficient
1981–1982	1,487	23.73	748,349,280	915	572	468,016,720	0.615
1982–1983	1,337	17.01	536,427,360	656	681	557,238,640	0.490
1983–1984	1,357	24.99	788,084,640	963	394	321,941,360	0.710
1984–1985	1,000	15.54	490,069,440	599	401	327,930,560	0.599
1985–1986	1,532	23.10	728,481,600	891	641	524,694,400	0.581
1986–1987	1,182	14.49	456,956,640	559	623	509,919,360	0.473
1987–1988	1,173	13.86	437,088,960	534	639	522,425,040	0.456
1988–1989	1,238	12.39	390,731,040	478	760	621,952,960	0.386
1989–1990	760	5.67	178,809,120	219	541	442,870,880	0.288
1990–1991	1,477	20.37	642,388,320	785	692	565,797,680	0.532
Sum	12,543		5,397,386,400	6598	5945		0.526

3. The runoff coefficient for every year is defined as the ratio of the amount of water discharging from the exit of the river basin to the total amount of water that precipitates to the river basin each year, that is,

$$C = h_{dis}/P$$

The values of the runoff coefficient for every year are calculated in the last column of Table 1.6. We remind that the runoff coefficient is a dimensionless parameter.

The runoff coefficient for the whole time period of the sample, i.e. 10 years, is estimated again as the quotient of the total amount of water discharged to the total precipitation depth:

$$C_{tot} = \frac{\Sigma h_{dis}}{\Sigma P} = \frac{6,598\,\text{mm}}{12,543\,\text{mm}} = 0.526$$

It must be underlined that it is incorrect to calculate C_{tot} as the average of 10 annual values of C for 10 years.

4. The average annual inflow in the reservoir is equal to the average value of the amount of water discharging into the river where the dam is to be constructed:

$$\overline{V}_{inflow} = \frac{\Sigma V_{annual}}{10} = \frac{5,397,386,400}{10} = 539,738,640\,\text{m}^3$$

The exploitable amount of water is 82% of the quantity given or

$$V_{exp} = 0.82 \cdot 539,738,640\,\text{m}^3 = 442,585,685\,\text{m}^3$$

or by division with the area of the river basin

$$h_{exp} = 442,585,685/818 \cdot 10^6 = 541\,\text{mm}$$

5. Let us assume, for simplicity, that if the irrigation needs of the plants are equal to the potential evapotranspiration, then the area of the arable land that could be irrigated with the given amount of water is

$$A_{irr} = V_{exp}/PET = 442,585,685/0.800 = 553,232,106\,\text{m}^2 = 553.2\,\text{km}^2$$

Example 1.5

In a river basin, a reservoir has been constructed, serving water supply needs of a major city, as well as irrigation needs of a nearby cultivated land. The area of the river basin, upstream of the dam location, is 600 km², while the average area of the lake of the reservoir is 34 km² (Figure 1.5).

Based on data of rain gauges and meteorological stations, the surface rainfall of the river basin and the rainfall and evaporation of the reservoir are estimated for 10 hydrological years, as given in Table 1.7. In addition, Table 1.7 presents the absolute altitude of the water level of the reservoir at the beginning of the hydrological year and the amount of water consumed for water supply and irrigation needs. Finally, based on measurements, the underground water losses of the reservoir have been estimated in the order of 0.08 m³/s.

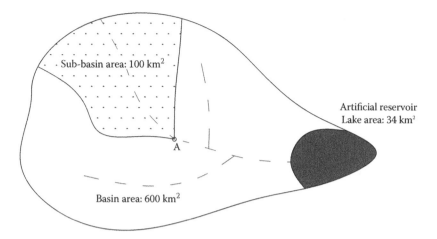

Figure 1.5 The river basin.

Table 1.7 Water balance data of the reservoir

Hydrologic year	Water level H on October 1st (m)	Rainfall on the river basin (mm)	Rainfall on the lake (mm)	Evaporation from lake surface (mm)	Water supply (m³ × 10⁶)	Irrigation (m³ × 10⁶)
2000–2001	50.83	885.6	759.1	1364.0	15	147.0
2001–2002	52.50	857.1	734.7	1397.0	18	157.5
2002–2003	53.15	942.4	807.8	1468.5	17	141.8
2003–2004	54.55	638.4	547.2	1413.5	19	189.0
2004–2005	51.20	715.7	613.4	1402.5	20	186.9
2005–2006	50.10	865.4	741.8	1463.0	19	147.0
2006–2007	51.83	918.6	787.4	1402.5	21	140.7
2007–2008	54.55	625.8	536.4	1364.0	22	180.6
2008–2009	52.42	865.5	741.9	1523.5	22	142.8
2009–2010	54.43	863.9	740.6	1529.0	23	159.6
2010–2011	54.68					

1. Estimate the inflow in the reservoir and the runoff coefficient of the upstream river basin in each hydrological year.
2. Estimate the water level of the reservoir, on 1 October 2011, if it is assumed that during the hydrological years 2010–2011, the inflow, rain and evaporation of the lake, as well as the sum of the uptakes, were equal to the mean average values of the aforementioned variables for the period of 10 hydrological years.
3. Estimate the mean yearly runoff of the river at location A, upstream of the reservoir, in a sub-basin of 100 km². Assume that the surface rainfall, upstream of position A, is increased by 20% in comparison to the whole river basin, while the physiographic characteristics of these two river basins are similar.

Solution

1. From the water balance/equilibrium equation,

$$V_{inflow} + P_{reservoir} - V_{outflow} - L = \Delta V$$

Table 1.8 Calculation of ΔH and ΔV

Hydrologic year	Change in water level ΔH (m)	Change in volume ΔV (×10⁶ m³)
2000–2001	1.67	56.78
2001–2002	0.65	22.1
2002–2003	1.40	47.6
2003–2004	−3.35	−113.9
2004–2005	−1.10	−37.4
2005–2006	1.73	58.82
2006–2007	2.72	92.48
2007–2008	−2.13	−72.42
2008–2009	2.01	68.34
2009–2010	0.25	8.5

Solving for V_{inflow}, we get

$$V_{inflow} = V_{outflow} + L + \Delta V - P_{reservoir} \tag{1.5.1}$$

where

$V_{outflow}$ = Evaporation + Water supply + Irrigation

L(losses) = 0.08 m³/s = 0.08 · 3600 · 24 · 365/10⁶ = 2.52 · 10⁶ m³/year

For the estimation of ΔV, Table 1.8 is formed, where ΔH is the difference in the water level of the lake in comparison to the previous year (i.e. ΔH = (2001–2002) − (2000–2001) = 52.50–50.83 = 1.67) and ΔV is the difference in volume (this value is calculated by multiplying by the mean surface area of the lake, i.e. 1.67 m · 34 · 10⁶ m² = 56.78 · 10⁶ m³).

$P_{reservoir}$ is estimated by multiplying the area of the lake with the rainfall of the lake:

$$P_{reservoir(2000–2001)} = 34 \cdot 10^6 \text{ m}^2 \cdot 759.1/1000 \text{ m} = 25.81 \cdot 10^6 \text{ m}^3$$

In a similar manner, evaporation is estimated by multiplying evaporation with the surface area of the lake, i.e. Evaporation (2000–2001) = 1364/1000 m³ · 34 · 10⁶ m² = 46.4 · 10⁶ m³.

By replacing in Equation 1.5.1, the V_{inflow} is calculated for every hydrological year. If the precipitation volume of the whole river basin is also estimated (by multiplying the area of the river basin with the annual precipitation), the runoff coefficient is calculated as the ratio of V_{inflow} to the annual precipitation P_{river_basin}: $C = V_{inflow}/P_{river\_basin}$. The results are shown in Table 1.9.

2. The average values for the period of 10 hydrological years of the total water supply and irrigation needs are calculated as 19.6 · 10⁶ m³ and 159.3 · 10⁶ m³, respectively. Also from Table 1.9, the average values of $P_{reservoir}$, evaporation and V_{inflow} are estimated.

Substituting the values in Equation 1.5.1 (solving for ΔV), we arrive at the following result:

$$\Delta V = V_{inflow} + P_{reservoir} - \text{Water supply} - \text{Irrigation} - \text{Evaporation} - \text{Losses} \rightarrow$$

$$\Delta V = (219.38 + 23.83 - 19.6 - 159.3 - 48.71 - 2.52) \ 10^6 \text{ m}^3 = 13.9 \cdot 10^6 \text{ m}^3$$

Table 1.9 Estimation of V_{inflow}, P_{river_basin} and C

Hydrologic year	$P_{reservoir}$ (×10⁶ m³)	Evaporation (×10⁶ m³)	V_{inflow} (×10⁶ m³)	P_{river_basin} (×10⁶ m³)	Runoff coefficient
2000–2001	25.81	46.4	241.87	501.25	0.48
2001–2002	24.98	47.5	222.64	485.13	0.46
2002–2003	27.46	49.9	231.39	533.40	0.43
2003–2004	18.60	48.1	126.08	361.33	0.35
2004–2005	20.86	47.7	198.85	405.07	0.49
2005–2006	25.22	49.7	251.87	489.79	0.51
2006–2007	26.77	47.7	277.62	519.90	0.53
2007–2008	18.24	46.4	160.84	354.23	0.45
2008–2009	25.23	51.8	262.24	489.88	0.54
2009–2010	25.18	52.0	220.43	488.98	0.45
Average	**23.83**	**48.71**	**219.38**		

By dividing with the area of the lake, we have

$$\Delta H = 13.9 \cdot 10^6 \text{ m}^3 / 34 \cdot 10^6 \text{ m}^2 = 0.385 \text{ m}$$

Finally, $H_{2010-2011} = 54.68_{(2009-2010)} + 0.385 = 55.07$ m.

3. We assume, according to the exercise, that the runoff coefficient of the sub-river basin is the same as that of the total river basin.

So,

$$C_A = C \rightarrow V_{inflowA}/P_A \cdot (\text{Area } A) = V_{inflow}/P \cdot \text{Area}$$

$$P_A = 1.2P, \quad \text{Area } A = 100 \text{ km}^2, \quad \text{Area} = 600 - 34 = 566 \text{ km}^2.$$

$$V_{inflow} = 219.38 \cdot 10^6 \text{ m}^3$$

Thus,

$$V_{inflowA}/1.2P \cdot 100 = V_{inflow}/P \cdot 566 \geq V_{inflowA} = 46.51 \cdot 10^6 \text{ m}^3$$

REFERENCES

Arregui, F., Cabrera Jr, E., and Cobacho, R., 2007, *Integrated Water Meter Management*, IWA Publishing, UK.

Bouwer, H., 1978, *Groundwater Hydrology*, McGraw-Hill, New York.

Brutsaert, W., 2005, *Hydrology: An Introduction*, Cambridge University Press, New York.

Chanson, H., 2014, *Applied Hydrodynamics: An Introduction to Ideal and Real Fluid Flows*, CRC Press, Taylor & Francis Group, Leiden, the Netherlands.

Chisholm, H., ed., 1911, "Mariotte, Edme". *Encyclopædia Britannica*, 11th edn. Cambridge University Press.

Mimikou, M. and Baltas, E., 2012, *Engineering Hydrology*, Papasotiriou, Athens, Greece.

Nace, R.L., 1974, Pierre Perrault: The man and his contribution to modern hydrology, *Journal of the American Water Resources Association*, 10(4), 633–647. doi:10.1111/j.1752-1688.1974.tb05623.x.

Perrault, P., 1966, *On the Origin of Springs* (Translation of "De l'origine des fontaines" (1674)), Hafner, New York, ASIN B0026MBIHE.

Raghunath, H.M., 2006, *Hydrology: Principles-Analysis-Design*, New Age International Publishers, New Delhi, India.
Saleh, J.M., 2002, *Fluid Flow Handbook*, McGraw-Hill Professional, Cambridge University Press, New York.
Sherman, L.K., 1932, Streamflow from rainfall by the unit hydrograph method, *Engineering News-Record*, 108, 501–505.
Singh, V.P., 1992, *Elementary Hydrology*, Prentice-Hall, Englewood Cliffs, NJ.

Chapter 2

Precipitation and hydrological losses

2.1 GENERAL

This chapter describes the types of precipitation, the measurement instrumentation, the basic analysis processes of the point rainfall information (homogeneity, filling used for measurement gaps, altitude adaptation) and the methods for the estimation of average surface rainfall. Moreover, hydrological losses are discussed thoroughly, which are defined as the part of precipitation which does not end up in a stream after a rainfall. Hydrological losses are mainly divided into evaporation, transpiration, infiltration, percolation and detention.

The term 'precipitation' includes any form of moisture which falls from the atmosphere to the Earth's surface. The humidity in the atmosphere is mainly caused by evaporation from wet surfaces and transpiration. Large amounts of warm water and extensive soil areas covered with lush vegetation are excellent sources for the enrichment of atmospheric moisture.

The amount of water vapour in the atmosphere is less compared to the amounts of other atmospheric gases but is important for human life. The amount of water vapour varies with space and time. The highest concentrations of water vapour are near the surface of the sea and decrease with latitude, altitude and distance from shoreline. Approximately 50% of the atmospheric moisture is found in the first 1500 m from the Earth's surface.

2.2 FORMATION OF ATMOSPHERIC PRECIPITATION

Precipitation is the main component of the hydrological cycle. It feeds the surface receptors, renews the stocks of soil moisture and enriches the underground aquifers. The types of precipitation are rain, snow, hail and variations (dew, mist, fog), which are of minor importance in hydrology. Note that there is no direct relationship between the amount of water vapour over a region and the precipitation of that region (Shaw, 1994). The type and amount of precipitation (product of water vapour in the atmosphere) depend on climatic factors such as wind, temperature and atmospheric pressure, while air humidity is necessary but not sufficient to cause precipitation (Mimikou and Baltas, 2012). Figure 2.1 shows the formation mechanisms of precipitation.

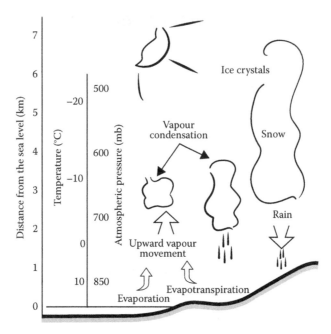

Figure 2.1 Graphic illustration of the formation mechanisms of precipitation.

The precipitated water is the amount of water contained in an atmospheric column of 1 cm² base and height z which extends above the soil surface and is given by

$$W = \int_0^z \rho_w \, dz \qquad (2.1)$$

where
ρ_w is the absolute humidity
W is the depth of water in cm

The vertical structure of the atmosphere is presented in Figure 2.2.

Troposphere: The most important climatic phenomena (clouds, storms, precipitation, winds) occur within the troposphere. The entire quantity of water vapour is in the troposphere. Temperature decreases with height. Strong vertical currents cause severe storms.

Stratosphere: Weak vertical currents are developed. Temperature decreases with height, and the ozone content increases.

Mesosphere: There are ionized molecules, and temperature decreases with altitude.

Thermosphere: The thermosphere constitutes 1% of the mass of the upper atmosphere and is characterized by high temperatures (230°C–1730°C).

Lower atmosphere: It constitutes 99.9% of the mass of the atmosphere and volumetrically is composed of nitrogen 78.08%, oxygen 20.95%, argon 0.93% and water vapour 0%–4%.

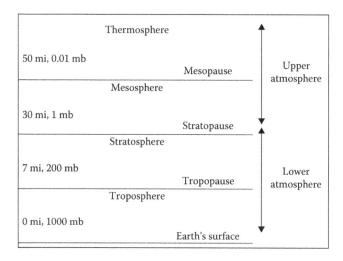

Figure 2.2 Vertical structure of the atmosphere.

2.3 PRECIPITATION TYPES

There are three main types of precipitation as described: rain, snow and hail.

2.3.1 Rain

Rainfall varies geographically, temporally and seasonally. It is obvious that the spatio-temporal variation of rainfall is a key component for all water resources management and hydrological studies. For example, it is important to know that the minimum precipitation coincides with the maximum demand in a region. The amount of rain in a region differs significantly even at very small distances. For example, records have shown deviating measurements greater than 20% at a distance less than 8 m.

The rain consists of water drops with a diameter from 0.5 to 7 mm. A drop of rain as it falls in still air attains a maximum speed called terminal velocity. This speed is obtained when the air resistance offsets the weight of the drop. The terminal velocity increases with the drop size up to a diameter of 5.5 mm and then it decreases for larger sizes because the drops start to widen; thus, resistance in the air increases. Large drops deform easily and break into smaller parts before reaching their terminal velocity. Upward air currents affect the average diameter of the droplets in clouds and thus affect rainfall. Depending on the amount of rain, the event is characterized as light for rainfall depth up to 2.5 mm/h, medium for depth from 2.5 to 7.5 mm/h and intense for depths greater than 7.5 mm/h. A rainfall with drops of diameter smaller than 0.5 mm, falling at a uniform rate, is referred to as a drizzle and has an intensity of less than 1.0 mm/h.

The total rainfall which reaches the ground may undergo various processes. A part is intercepted by obstacles or covers irregularities of the surface, it is temporarily stored there, and eventually evaporates. Another part covers the needs of soil moisture and then recharges the underground reserves, while the final part becomes surface runoff (Viessman and Lewis, 1996). The actual distribution of precipitation depends on the total amount of rainfall, soil moisture conditions, topography, land cover, land use, etc.

2.3.2 Snow

Snow is precipitation in the form of ice crystals. Although individual ice crystals may reach the ground surface, usually many ice crystals coalesce and form snowflakes.

Snow is peculiar when compared to other types of precipitation in that it accumulates and remains on the ground surface for some time before melting and turning into runoff or sublime. Therefore, apart from the study of snow behaviour as precipitation, the hydrologist faces the problem of determining the amount of snow accumulated on the ground and the conditions which determine the rate of melting and sublimation.

The distribution of snowfall over an area is more uniform than the respective distribution of rain, but because of its form, snow concentration and retention on the ground are very heterogeneous. Especially in mountainous regions, the lack of uniformity is more intense due to the combination of strong winds and topography. Avalanches and unusual movements of the snow, caused by local turbulent winds, may cause local accumulation of snow which is much greater than the annual amount of snow in the area. Topographic anomalies, gullies and various structures act as *traps* which accumulate large quantities of snow transported by air from surrounding areas. Additionally, vegetation cover affects the concentration of snow. Within a forest, the concentration of snow on the ground varies, depending on the density and type of trees and the number and type of forest glades.

The mere knowledge of the thickness of the snow in a region is not sufficient to calculate the equivalent water depth; the density of the snow cover is also necessary. The equivalent water depth is the water depth in millimetres resulting from melting snow which has the thickness and characteristics of a particular snow cover.

The snowpack is formed by ice crystals and has different shapes and densities. As time passes, the snow density changes under the influence of the environment. The change in its structure, from fluffy, with a temperature below zero and low density, to coarse, granular and liquid with high density, is called snow maturation. The mature snow cover is ready to give runoff. In a vertical section, the density of the snow cover may not be homogeneous, since it comprises ice parts and snow layers of different densities. The factors affecting the density of a snow cover are as follows:

1. Heat exchange due to transportation, liquefaction, radiation and heat transfer from the ground
2. The wind
3. The temperature of the snow cover
4. The pressure exerted by the overlying layers of snow
5. Variations in the water content of the snow cover
6. Infiltration of the water coming from snowmelt

The properties of the snow layer with respect to reflection and radiation are very interesting in hydrology as they can be used for the estimation of the extent of evaporation from the snowy surface and the melting process.

The snow layer acts as a 'black body' with respect to long-wavelength radiation. It absorbs the majority of the incident long-wavelength radiation from the environment, while transmitting radiation back in accordance with the Stefan–Boltzmann law.

2.3.3 Hail

Hail consists of masses of ice crystals of spherical or irregular shape with a diameter greater than 5 mm. Under certain conditions, the size of hail is very large and hailstones are formed.

Many hailstones consist of alternating layers of solid ice and snow due to successive lifts and falls into the cloud as a consequence of high atmospheric turbulence. Some hailstones reach the ground in the form of pure ice. For the formation of hailstones, it is necessary that strong updrafts hold them in the cloud. Upward velocities in the order of 15 m/s are not uncommon in clouds, and speeds up to 36 m/s have been observed under conditions of extreme instability. Hail occurs only in severe storms, usually in late spring and summer. Another form of hail, which may be characterized as microhail, consists of spherical or conical particles with a diameter of 2–5 mm. It usually appears with rain.

2.4 COOLING MECHANISMS AND TYPES OF PRECIPITATION

Cooling of air sufficient to cause a significant amount of precipitation is achieved when large-scale upward movements take place. The characteristics of precipitation are a function of the characteristics of the ascending air and lifting mechanism (Eagleson, 1970). The supply of air with moisture and the lifting mechanisms change seasonally in a region, so does the type of precipitation falling in this region. The main mechanisms of cooling and the respective types of precipitation are the following.

2.4.1 Cyclonic cooling

This can be divided into non-frontal and frontal cooling. The first is the result of convergence and therefore of the lifting gas mass to lower pressure points. In this case, the rainfall intensity is moderate and quite long. Frontal cyclonic cooling occurs when the movement forces the air to rise and penetrate into a front surface between two air masses of different temperatures. When the front is warm, the rainfall intensity is moderate and the duration is long, while when the front is cold, the rainfall intensity is high, the duration is short and the winds are strong.

2.4.2 Orographic cooling

Air masses fall on mountain slopes, and the air expands and is cooled to a lower pressure level corresponding to the higher elevation. Windy hillsides have many more clouds, more rain and less temperature fluctuations. In non-windy hillsides, the climate is drier with a greater variation in temperature.

2.4.3 Conductive cooling

The heating of the surface of the Earth generates a vertical current transfer, which in turn causes vertical instability of air and hence a lifting mechanism and cooling. The rainfall in this case is intensive and of short duration.

In summary, it could be said that an unsaturated vapour air mass by virtue of one of the three mechanisms rises and cools. The appropriate conditions of pressure and temperature at a certain level cause the saturation of water vapour. As the cooling continues, it causes condensation and the formation of clouds. Heat is released during condensation, which is offered to the mass initially by the process of evaporation and sublimation. Vapour is change of gas to liquid or solid state (for the latter, usually, more cooling of air mass is required). Thus, formed water droplets, snowflakes or icicles gain sufficient size around several cores and precipitate as rain, snow or hail, respectively. These cores are condensing particles such as sea acids, oxides of nitrogen, ammonium salts and soil granules.

Oftentimes, the existing acids, salts, etc., create impurities in the precipitation (characterized as acid rain which damages the natural environment and the monuments).

2.5 MEASUREMENT OF PRECIPITATION

The meteorological variable first measured by humans was probably the rain. The oldest and longest period of record is reported in Egypt in the Nile River in about 980 BC. There are several rainfall measurement and estimation instruments, such as standard rain gauges, the weather radar and satellites. Today, there is a large number of different types of instruments used to measure precipitation; those mostly used are summarized in the following.

2.5.1 Precipitation sensors

They are used in the measurement of cumulative rainfall depth and supplementary snowfall, which are installed at appropriate locations. They measure the total rainfall depth and the equivalent water depth of snowfall at certain time intervals (usually 8, 12 or 24 h). The measurement is recorded from an observer. The classic type of sensor is cylindrical in shape and comprises a collector, a funnel and a recipient, as shown in Figure 2.3. The collector is a cylinder sufficiently tall, with vertical inner walls, which ends up in the funnel. The angle of inclination of the walls of the funnel is at least 45° in order to minimize the water losses which may occur during the collision of raindrops. The funnel leads to the recipient which has a narrow entrance and is protected from sunlight to minimize water loss due to evaporation.

The most common types are the volumetric rain gauge and the 10-fold rain gauge. Volumetric rain gauges with a known section area of the collector are accompanied by specific volumetric tubes. In 10-fold rain gauges, the area of the collector is 10 times greater than

Figure 2.3 Standard rainfall sensor.

the section area of the recipient, which makes the precipitation which reaches the collector 10 times greater in the recipient, leading to the increased accuracy of measurements.

In some cases, the hydrologist is interested only in seasonal rainfall or in the measurement of rainfall in areas not easily accessible, especially during winter. In such cases, the recipient is formed so that it is capable of holding the probable amount of rainfall which is expected to fall within a relatively long period. The rain gauge is supplied with an anticoagulant, with preference to ethyl alcohol, when the instrument is located mainly in mountainous places, where there is a possibility that water can freeze inside the recipient. It is also useful for melting the snow. Additionally, a small amount of oil is added to reduce losses due to evaporation.

2.5.2 Rain gauge recorders

Rain gauge recorders are instruments used for the point measurement of rainfall, installed at appropriate locations, collecting mainly rainfall and supplementary snowfall, recording the change of rainfall depth over time using a simple mechanism, and thus, describing the temporal distribution of point rainfall. Such a rain gauge is shown in Figure 2.4.

Three main types of rain gauges are currently in use: the tipping bucket, the gravimetric and the floating type. Rain gauges give continuous measurements even for very short intervals; thus, they are suitable for the study of variations of rainfall intensity. When quipped with recording paper (old type), their mechanism can be configured so that the drum carrying the recording paper makes a full rotation every day; thus, the rainfall can be measured at 5 min intervals. If the system is configured to change the paper once a week, the rainfall can be measured at 30 min intervals. However, modern recorders have a data logger and are connected to a computer.

The tipping bucket rain gauge, which is the most advanced technology, consists of two small buckets mounted side-by-side on a common horizontal axis, which moves right and left as one of them is filled with rainwater directed at them from the funnel of the instrument.

Figure 2.4 Standard rain gauge.

As one bucket fills, the other empties. Rainfall equal to 0.25 mm is essential for the tip and emptying of a bucket, and the minimum time required for a tip is 2/10 of a second. These properties of the instrument result in errors in cases of very light rain when it takes time to fill the bucket, making the collected water subject to evaporation, but not in cases when the rain is very intense and the bucket fills in less than 2/10 s. The buckets are connected to a recording device, which records the number of tips and the corresponding rainfall depth with time.

The gravimetric type gauge comprises a container which is mounted on a weighing device. As the rainwater is collected in the container, the spring is pressed down and the compression is recorded. The capacity of the instrument may reach 1000 mm of rain.

The floating type gauge consists of a container which collects the rainwater, and it has a float which is used to measure and record variations in the water-level.

To avoid gross errors, a comparison between the amount of water collected and the amount recorded should be made. Instead of the conventional recording mechanism, current technology enables the conversion of the moving mechanism of the instrument into a digital signal. This signal may be stored in an electronic data logger. Furthermore, there is the capability of telemetry, i.e. signal transmission (after signal processing) wirelessly (via radio, GPRS, cell phone network, satellite, etc.) or wired (via land phone line) to a base station.

2.5.3 Snowfall measurement

The snowfall depth is usually measured by snow banks. These are simple horizontal surfaces where snow accumulates and the depth is measured with a common bar. After measuring, the bank is cleared of the snow so that it is ready for the next measurement of snowfall. The equivalent amount of water from snow and the corresponding density can be measured at the snow bank if a simple weighing system is provided, which measures the weight of snow.

The depth of snow cover is measured easily by punching a common bar or reading a permanently installed stage (level staff), in which the zero level corresponds to the ground surface. The equivalent water bar of snow cover is measured by taking a sample of snow through penetration of an appropriate cylindrical snow sampler and then by weighing the collected snow. To obtain representative samples of snow, the average of measurements from about 6 points along a fixed (permanent) route with a standard length of 150–250 m is calculated.

Another class of methods for measuring the snowfall height, which give the extent of snow cover, is based on photogrammetric (with aerial photographs) or terrestrial observation (with ground photographs). The extent of snow cover can also be determined from satellite images.

2.5.4 Meteorological radar

Conventional rainfall data are generally considered inadequate to describe the spatial and temporal distribution of rainfall. One method which has been given great attention recently is the measurement of rainfall with data from a weather radar. The radar outweighs the network of rain gauges based on the following characteristics.

The total information of the distribution of rainfall in time and space is concentrated and processed separately for each element, i.e. problems caused by environmental conditions are overcome by using the radar.

The use of meteorological radar is much easier than using a network of rain gauges.

The rain gauges and the weather radar are two different systems for measuring rainfall. The measuring range of a rain gauge is very small compared to that of the radar. Another reason for the difference in the amount of rainfall measured by the two instruments is that the radar scans from a distance from the surface, while the rain gauge measures rainfall on the ground surface. The temporal scale of the measurements of rain gauges is more than 5 min (1 min for the most modern), while the measurements of the radar are almost instantaneous (20 s/rotation) so that, for the comparison with the measured amounts of rainfall, the radar data should be completed in time.

The meteorological radar utilizes the transmission of electromagnetic waves in the atmosphere and their collision with particles of rain. The beam is concentrated at an angle of 1° or 2° to the antenna, which receives the part of the beam reflected on the rain particles. The amount of returned energy depends on the number of particles of rain pulse inside the volume of the radar beam and their size, composition, relative position, shape and orientation.

The general equation of the meteorological radar is simplified to the following form (Anderson and Burt, 1985):

$$P_r = \frac{C_1 |K|^2 Z}{r^2} = \frac{C_2 Z}{r^2} \tag{2.2}$$

where
$$Z = \sum D_i^6$$

P_r is the returned power
D is the diameter of the rain particles
r is the distance from the target
C_1 and C_2 are the constants of the radar
K is a parameter related to the material of the target, the temperature and the wavelength

The water particles vary in size from 0.2 up to 5 mm. Knowing the size distribution, the reflectivity factor Z from all water particles can be calculated (Collier, 1989):

$$Z = \sum_i N_i D_i^6 = \int_0^\infty N(D)D^6 dD \tag{2.3}$$

where
$N(D)$ is the distribution of the water particles
D is the diameter of the water particles
N_i is the number of water particles with a diameter ranging between D and $D + dD$ per unit volume of the atmosphere

The distribution of water particles becomes 0 when $D < D_{min}$ and $D > D_{max}$, since the limits of integration are from 0 to ∞. The units of the radar reflectivity are mm^6/m^3. The reflectivities start from 0.001 mm^6/m^3 for fog or sparse clouds and reach up to

50,000,000 mm⁶/m³ for intense hail. This difference led to the definition of a new parameter which measures reflectivity:

$$\zeta = 10 \log_{10}\left(\frac{Z}{1\ \mathrm{mm}^6/\mathrm{m}^3}\right) \tag{2.4}$$

where ζ is the logarithmic parameter measured in dBZ and Z in mm⁶/m³.

This conversion gives values of ζ from −30 dBZ for sparse clouds up to 75 dBZ for intense hail. The measurement of the returned power (distance, azimuth, altitude, power, speed and other features) at predefined points is important information for their description.

2.5.5 Joss–Waldvogel disdrometers

The Joss–Waldvogel RD-69 and RD-80 disdrometers are instruments for measuring raindrop size distributions (DSDs) continuously and automatically. The RD-80 is an enhanced version of RD-69 disdrometer. They were developed because statistically meaningful samples of raindrops could not be measured without a prohibitive amount of work (Distromet Ltd., 2011). These instruments usually are combined with weather radars and provide data, such as radar reflectivity values. Moreover, they can be used as stand-alone instruments measuring rainfall intensity, water liquid content, kinetic energy flux and other hydro-meteorological parameters, as will be described in the following.

Both instruments work under the same principle, that they transform the vertical momentum of an impacting drop into an electric pulse, whose amplitude is a function of the drop diameter. A conventional pulse height analysis yields the size distribution of raindrops.

An example of a system is shown in Figure 2.5. It consists of a disdrometer RD-80 and a personal computer.

The sensor is exposed to the raindrops to be measured, and along with the processor, produces an electric pulse for every drop hitting. The amplitude of the pulse is directly related to the diameter of the raindrop. In the processor, pulses are divided into 127 classes of drop diameter. A computer program, developed for the disdrometer system, can be used to transform the data into a suitable format. In order to get statistically meaningful samples and to reduce the amount of data, the program reduces the number of classes to 20.

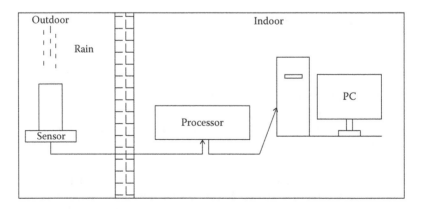

Figure 2.5 System for measuring drop size data.

2.5.5.1 Description of the disdrometer RD-80

The disdrometer for raindrops consists of two units: the sensor, which is exposed to the rain, and the processor for analog processing and digitizing of the sensor signal. A cable, 4 m long, is used to connect the two units.

The sensor transforms the mechanical momentum of drop into an electric pulse, whose amplitude is roughly proportional to the mechanical momentum.

The processor contains circuits to eliminate false signals, mainly due to acoustic noise, to reduce the 90 dB dynamic range of the sensor signal.

The sensor consists of an electromechanical unit and a feedback amplifier. A conical styrofoam body is used to transmit the mechanical impulse of an impacting drop to a set of two moving coil systems in magnetic fields. At the impact of a drop, the styrofoam body together with the two coils moves downwards and a voltage is induced in the sensing coil. This voltage is amplified and applied to the driving coil, producing a force which counteracts the movement. Therefore, it takes some time for the system to return to its original resting position, and therefore, to get ready for the next drop. The amplitude of the pulse at the amplifier output is a measure of the size of the drop that caused it (Distromet Ltd., 2011).

The processor has three main functions:

1. It supplies power to the sensor.
2. It processes the signal from the sensor.
3. It contains circuits for testing the performance of the instrument.

The signal processing circuit consists of four parts:

1. A noise rejection filter
2. A dynamic range compressor
3. A signal recognition circuit
4. A non-linear analog to digital (A–D) converter

The noise rejection filter is an active band pass filter, whose frequency response is designed to give an optimum ratio between signals from raindrops and signals due to acoustic noise affecting the sensor.

The signal recognition circuit can separate the signal pulses caused by drops hitting the sensor from the more uniform oscillations caused by acoustic noise. If a pulse caused by a raindrop exceeds the oscillations caused by noise, a gate passes it to the pulse standardizer, which produces a constant pulse duration without changing the amplitude of the original pulse.

The RD-80 has the specifications listed in Table 2.1.

2.5.5.2 Hydro-meteorological parameters which can be derived by the RD-69 and RD-80 disdrometers

Disdrometers are useful in estimating rain DSDs. A DSD is commonly represented by the function $N(D)$, the concentration of raindrops with the diameter D in a given volume of air. Because of the complicated processes involved in the formation of precipitation, the function $N(D)$ varies and cannot be described by a simple function. In many cases, however, a DSD

Table 2.1 Specifications of the RD-80 disdrometer

Range of drop diameter	0.3–5 mm
Sampling area	50 cm²
Accuracy	±5% of measured drop diameter
Resolution	127 size classes distributed more or less exponentially over the range of drop diameters
Output format	According to RS-232-C standard, 7 data bits, even parity, 1 stop bit
Baud rate	9600 Baud
Handshake	DCD and DTR signals
Display	8 LEDs for 8 groups of 16 channels each
Power requirements	Plug-in power supply included in delivery: 115/230 V AC, 5.5 VA, 50/60 Hz (9–18 V DC; 3.3 W, also possible)
Operating temperature range	0°C–40°C for processor
	0°C–50°C for sensor
Dimensions of the sensor	10 × 10 × 17 cm height
Dimensions of the processor	12 × 26 × 27 cm deep
Weight	Sensor, 2.9 kg; processor, 2.2 kg
Standard length of sensor cable	4 m

can be approximated fairly well by an exponential law and the following parameterization can be used to characterize it:

$$N(D) = N_0 \exp(-\Lambda D) \tag{2.5}$$

where

N_0 is the number concentration of drops with diameter 0 on the exponential approximation
Λ is its slope

Table 2.2 presents the hydro-meteorological parameters that can be derived.

2.5.5.3 Derivation of rain characteristics with the use of a Joss–Waldvogel RD-69 disdrometer

A Joss–Waldvogel RD-69 disdrometer recorded the data for a rainfall event on 22 February 2013 (Table 2.3). This type of disdrometer can record the number of raindrops per diameter class per time step (in this example, the time step is 2 min), classifying the measured diameters in 20 classes. The lower and upper limits of the diameters of each class are shown in Table 2.3. The sampling area of the disdrometer is 50 cm². The estimation of the terminal velocity using four empirical relationships in correlation with the mean raindrop diameter of each class is presented in Table 2.4.

Example 2.1

Find for every relationship of terminal velocity the following:

1. The mean diameter and the range of each class.
2. The number density of drops of the diameter corresponding to size class *i* per unit volume $N(D_i)$ in 1/m³ mm.
3. The intensity of the rain *R* in mm/h.

Table 2.2 Hydro-meteorological parameters which can be derived by the RD-69 and RD-80 disdrometers

$N_A(D_i)$	Number of drops measured in every drop size class i during the time interval t
R	Rainfall rate (mm/h)
	$R = \dfrac{\pi}{6} \dfrac{3.6}{10^3} \sum_{i=1}^{20} N(D_i) D_i^3 V(D_i) \Delta D_i \ \text{(mm/h)}$
RT	Total rain amount since the start of the measurement (mm)
	$RA = \Sigma(R \times t)$ (mm)
Wg	Liquid water content (g/m^3)
	$W = \dfrac{\pi}{6} \dfrac{1}{A * t} \sum_{i=1}^{20} \dfrac{N_A(D_i)}{V(D_i)} D_i^3$
$Z(\zeta)$	Radar reflectivity factor (mm^6/m^3, dBZ)
	$Z = \sum_{i=1}^{20} N(D_i) D_i^6 \Delta D_i \ \text{(mm}^6/\text{m}^3)$
	$\zeta = 10 \log(Z)$ (where Z in mm^6/m^3 and ζ in dBZ)
EF	Energy flux (J/m^2 h)
	$E = \dfrac{3\pi}{10^4} \sum_{i=1}^{20} N(D_i) D_i^3 V(D_i)^3 \Delta D_i$
D_{max}	Largest drop collected (mm)
N_0	$1/(\text{m}^3 \text{ mm}) \ N_0 = \dfrac{1}{\pi} \left(\dfrac{6!}{\pi} \right)^{4/3} \left(\dfrac{W}{Z} \right)^{4/3} W$
Λ	Slope (1/mm) $\Lambda = \left(\dfrac{6!}{\pi} \dfrac{W}{Z} \right)^{1/3}$
$N(D_i)$	The density of drops of the diameter corresponding to size class i per unit volume (1/m^3 mm),
	$N(D_i) = \dfrac{N_A(D_i)}{AtV(D_i)\Delta D_i}$

4. The total amount of rain and the respective cumulative curve for this rainfall event.
5. The water liquid content W_g in g/m^3.
6. The kinetic energy flux E in J/m^2 h.
7. The radar reflectivity factor Z in mm^6/m^3 and dBZ.
8. Let us assume that the DSD $N(D_i)$ is expressed approximately by the exponential relationship $N(D) = N_0 \exp(-\Lambda D)$. Estimate the parameters N_0 (1/m^3 mm) and Λ (1/mm) of this distribution.
9. Find the parameters a and b in a Z–R relationship of the form $Z = \alpha R^\beta$ with units of Z (mm^6/m^3) and of R (mm/h), using the method of linear regression.

Solution

1. The relationships for the calculation of the mean diameter and the range of each class are the following:

$$D_i = \frac{(D_{i\max} + D_{i\min})}{2} \tag{2.6}$$

$$RD_i = D_{i\max} - D_{(i-1)\max} \tag{2.7}$$

where
D_i is the mean diameter class
$D_{i\max}$ is the upper limit of each class and $D_{i\min}$ is the lower limit, respectively

Table 2.3 Recorded data of drop numbers and lower and upper limits of the respective 20 classes of diameters

Limits of diameter classes (mm)			22/2/2013 7:48:00 p.m.	22/2/2013 7:50:00 p.m.	22/2/2013 7:52:00 p.m.	22/2/2013 7:54:00 p.m.	22/2/2013 7:56:00 p.m.	22/2/2013 7:58:00 p.m.	22/2/2013 8:00:00 p.m.	22/2/2013 8:02:00 p.m.
d1 (min)	0.313	Class								
d1 (max)	0.405	1	2	16	19	5	0	0	0	0
d2 (max)	0.506	2	10	11	15	19	0	0	1	3
d3 (max)	0.597	3	1	3	6	10	0	11	8	4
d4 (max)	0.715	4	4	19	25	11	0	19	34	29
d5 (max)	0.827	5	4	37	19	20	0	23	54	66
d6 (max)	1.000	6	33	59	29	32	0	75	229	223
d7 (max)	1.232	7	27	52	48	48	20	210	349	381
d8 (max)	1.43	8	3	15	21	28	64	205	247	343
d9 (max)	1.582	9	2	5	12	29	120	155	161	218
d10 (max)	1.747	10	0	1	8	33	134	132	102	141
d11 (max)	2.077	11	0	1	6	31	243	133	87	103
d12 (max)	2.441	12	0	0	1	23	153	76	16	30
d13 (max)	2.727	13	0	0	0	15	78	28	4	5
d14 (max)	3.011	14	0	0	0	7	72	20	0	0
d15 (max)	3.385	15	0	0	0	8	46	16	0	0
d16 (max)	3.705	16	0	0	0	1	16	10	0	0
d17 (max)	4.127	17	0	0	0	0	12	1	0	0
d18 (max)	4.573	18	0	0	0	0	9	2	0	0
d19 (max)	5.101	19	0	0	0	0	2	0	0	0
d20 (max)	5.645	20	0	0	0	0	0	0	0	0

Table 2.4 Empirical relationships of final raindrop speed in correlation with the mean diameter of each class (V_i in m/s, D in cm)

$V_1 = 14.2D^{1/2}$
$V_2 = 9.65-10.3e^{(-6D)}$
$V_3 = 17.67D^{0.67}$
$V_4 = 48.54De^{(-1.95D)}$

It is valid that $D_{imax} = D_{(i+1)min}$. By substituting the values of the upper and lower limits, the respective mean diameter and diameter range for each class are derived. The results are shown in Table 2.5.

2. We calculate the DSD using the following relationship:

$$N(D_i) = \frac{N_A(D_i)}{AtV(D_i)\Delta D_i} \tag{2.8}$$

where

$N_A(D_i)$ is the number of hits per class i
A is the hit area of the disdrometer cone = 50 cm²
$V(D_i)$ is the final drop speed of each class i
ΔD_i is the range of each class i

We substitute the values of $A = 50$ cm² $= 50 \times 10^{-4}$ m², $t = 120$ s, $V(D_i) => V_1(D_i)$ (the same procedure is followed for the other three final drop speed relationships which are omitted), resulting in the values given in Table 2.5. The values have been multiplied by the factor 10^{-3} for a better presentation. $N(D_i)$ is expressed in 1/m³ mm.

3. The rain intensity R is given using the following relationship:

$$R = \frac{\pi}{6}\frac{3.6}{10^3}\sum_{i=1}^{20}N(D_i)D_i^3V(D_i)\Delta D_i \tag{2.9}$$

All components have been explained previously. The calculations are shown in Table 2.6. In the last row, the sum of each column is calculated. Figure 2.6 shows the graphical representation of the rain intensity with time.

4. The total rainfall amount is estimated by multiplying the rainfall intensity with the respective time interval (2 min) and summing: $RA = \Sigma(R \times t)$ in mm. Figure 2.7 shows the cumulative plot.

5. The liquid water content is given by the following relationship:

$$W = \frac{\pi}{6}\frac{1}{At}\sum_{i=1}^{20}\frac{N_A(D_i)}{V(D_i)}D_i^3 \quad \text{g/m}^3 \tag{2.10}$$

The results are shown in Table 2.7 and Figure 2.8.

6. The kinetic energy flux is given by the following relationship:

$$E = \frac{3\pi}{10^4}\sum_{i=1}^{20}N(D_i)D_i^3V(D_i)^3\Delta D_i \quad \text{J/m}^2 \text{ h} \tag{2.11}$$

The results are shown in Table 2.8 and Figure 2.9.

Table 2.5 Calculation of mean diameter, class range and drop size distribution $N(D_i) \times 10^{-3}$

Mean diameter (mm)	Class range ΔD_i (mm)	Class	$V_i = 14.2D^{1/2}$ (m/s)	22/2/2013 7:48:00 p.m.	22/2/2013 7:50:00 p.m.	22/2/2013 7:52:00 p.m.	22/2/2013 7:54:00 p.m.	22/2/2013 7:56:00 p.m.	22/2/2013 7:58:00 p.m.	22/2/2013 8:00:00 p.m.	22/2/2013 8:02:00 p.m.
0.359	0.092	1	2,691	0.67	0.67	0.00	0.00	0.00	0.00	5.39	2.02
0.4555	0.101	2	3,031	4.90	2.72	5.44	2.18	1.09	4.90	14.70	47.37
0.5515	0.091	3	3,335	26.36	10.44	18.12	14.83	11.53	12.63	26.91	91.17
0.656	0.118	4	3,637	29.13	15.53	26.02	17.48	13.98	18.25	36.89	77.28
0.771	0.112	5	3,943	32.46	17.74	31.33	26.04	15.10	26.42	39.63	49.82
0.9135	0.173	6	4,292	23.34	18.86	39.73	30.08	22.45	33.67	41.30	56.12
1.116	0.232	7	4,744	12.12	19.23	30.14	18.17	18.17	34.07	28.77	32.26
1.331	0.198	8	5,181	8.45	12.67	12.84	13.65	16.74	37.05	24.37	11.05
1.506	0.152	9	5,511	4.38	10.35	11.54	8.76	9.75	29.05	13.33	2.79
1.6645	0.165	10	5,793	4.71	6.80	5.06	5.93	8.19	15.17	4.88	0.17
1.912	0.33	11	6,209	2.93	4.07	2.52	3.09	3.90	6.18	1.30	0.08
2.259	0.364	12	6,749	1.70	0.95	0.95	1.22	1.97	1.49	0.14	0.00
2.584	0.286	13	7,218	0.81	0.16	0.32	0.81	0.73	0.57	0.08	0.00
2.869	0.284	14	7,606	0.15	0.08	0.23	0.31	0.15	0.23	0.00	0.00
3.198	0.374	15	8,030	0.00	0.06	0.00	0.28	0.00	0.00	0.00	0.00
3.545	0.32	16	8,455	0.00	0.06	0.06	0.06	0.00	0.00	0.00	0.00
3.916	0.422	17	8,886	0.00	0.00	0.00	0.00	0.00	0.00	0.00	0.00
4.35	0.446	18	9,366	0.00	0.00	0.00	0.00	0.00	0.00	0.00	0.00
4.837	0.528	19	9,876	0.00	0.00	0.00	0.00	0.00	0.00	0.00	0.00
5.373	0.544	20	10,409	0.00	0.00	0.00	0.00	0.00	0.00	0.00	0.00

Table 2.6 Calculations of rain intensity R

Mean diameter (mm)	Class range ΔD_i (mm)	Class	$V_i = 14.2D^{1/2}$ (m/s)	22/2/2013 7:48:00 p.m.	22/2/2013 7:50:00 p.m.	22/2/2013 7:52:00 p.m.	22/2/2013 7:54:00 p.m.	22/2/2013 7:56:00 p.m.	22/2/2013 7:58:00 p.m.	22/2/2013 8:00:00 p.m.	22/2/2013 8:02:00 p.m.
0.359	0.092	1	2,691	0.00	0.00	0.00	0.00	0.00	0.00	0.00	0.00
0.4555	0.101	2	3,031	0.00	0.00	0.00	0.00	0.00	0.00	0.01	0.03
0.5515	0.091	3	3,335	0.03	0.01	0.02	0.01	0.01	0.01	0.03	0.09
0.656	0.118	4	3,637	0.07	0.04	0.06	0.04	0.03	0.04	0.08	0.18
0.771	0.112	5	3,943	0.12	0.07	0.12	0.10	0.06	0.10	0.15	0.19
0.9135	0.173	6	4,292	0.25	0.20	0.42	0.32	0.24	0.36	0.44	0.60
1.116	0.232	7	4,744	0.35	0.55	0.87	0.52	0.52	0.98	0.83	0.93
1.331	0.198	8	5,181	0.39	0.58	0.58	0.62	0.76	1.69	1.11	0.50
1.506	0.152	9	5,511	0.24	0.56	0.62	0.47	0.53	1.57	0.72	0.15
1.6645	0.165	10	5,793	0.39	0.56	0.42	0.49	0.68	1.26	0.41	0.01
1.912	0.33	11	6,209	0.79	1.10	0.68	0.83	1.05	1.67	0.35	0.02
2.259	0.364	12	6,749	0.90	0.51	0.51	0.65	1.05	0.80	0.07	0.00
2.584	0.286	13	7,218	0.54	0.11	0.22	0.54	0.49	0.38	0.05	0.00
2.869	0.284	14	7,606	0.15	0.07	0.22	0.30	0.15	0.22	0.00	0.00
3.198	0.374	15	8,030	0.00	0.10	0.00	0.51	0.00	0.00	0.00	0.00
3.545	0.32	16	8,455	0.00	0.14	0.14	0.14	0.00	0.00	0.00	0.00
3.916	0.422	17	8,886	0.00	0.00	0.00	0.00	0.00	0.00	0.00	0.00
4.35	0.446	18	9,366	0.00	0.00	0.00	0.00	0.00	0.00	0.00	0.00
4.837	0.528	19	9,876	0.00	0.00	0.00	0.00	0.00	0.00	0.00	0.00
5.373	0.544	20	10,409	0.00	0.00	0.00	0.00	0.00	0.00	0.00	0.00
Rain intensity R in mm/h				4.21	4.60	4.88	5.56	5.57	9.08	4.25	2.70

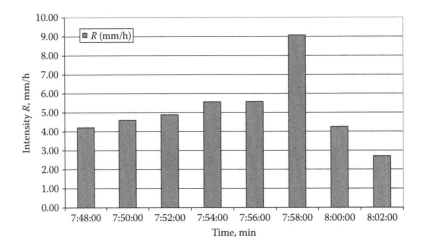

Figure 2.6 Rain intensity.

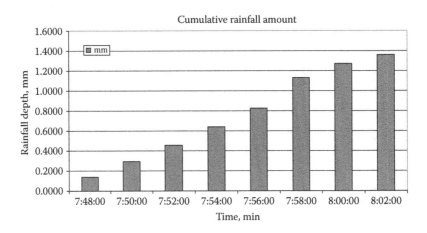

Figure 2.7 Cumulative rainfall depth.

7. The radar reflectivity Z in mm^6/m^3 is calculated by the following relationship:

$$Z = \sum_{i=1}^{20} N(D_i)D_i^6 \Delta D_i \qquad (2.12)$$

To convert into dBZ, the next relationship is used:

$$\zeta = 10 \log(Z) \text{ (where } Z \text{ in } mm^6/m^3 \text{ and } \zeta \text{ in dBZ)} \qquad (2.13)$$

These data were used in preparing Table 2.9 and plotting Figures 2.10 and 2.11.
8. N_0 and Λ are calculated using the following relationships:

$$N_0 = \frac{1}{\pi}\left(\frac{6!}{\pi}\right)^{4/3}\left(\frac{W}{Z}\right)^{4/3} W \qquad (2.14)$$

Table 2.7 Liquid water content W

Class	$V_1 = 14.2D^{\Lambda 1/2}$ (m/s)	22/2/2013 7:48:00 p.m.	22/2/2013 7:50:00 p.m.	22/2/2013 7:52:00 p.m.	22/2/2013 7:54:00 p.m.	22/2/2013 7:56:00 p.m.	22/2/2013 7:58:00 p.m.	22/2/2013 8:00:00 p.m.	22/2/2013 8:02:00 p.m.
1	2,691	0.000	0.000	0.000	0.000	0.000	0.000	0.000	0.000
2	3,031	0.000	0.000	0.000	0.000	0.000	0.000	0.001	0.002
3	3,335	0.002	0.001	0.001	0.001	0.001	0.001	0.002	0.007
4	3,637	0.005	0.003	0.005	0.003	0.002	0.003	0.006	0.013
5	3,943	0.009	0.005	0.008	0.007	0.004	0.007	0.011	0.013
6	4,292	0.016	0.013	0.027	0.021	0.015	0.023	0.029	0.039
7	4,744	0.020	0.032	0.051	0.031	0.031	0.058	0.049	0.054
8	5,181	0.021	0.031	0.031	0.033	0.041	0.091	0.060	0.027
9	5,511	0.012	0.028	0.031	0.024	0.026	0.079	0.036	0.008
10	5,793	0.019	0.027	0.020	0.024	0.033	0.060	0.019	0.001
11	6,209	0.035	0.049	0.030	0.037	0.047	0.075	0.016	0.001
12	6,749	0.037	0.021	0.021	0.027	0.043	0.033	0.003	0.000
13	7,218	0.021	0.004	0.008	0.021	0.019	0.015	0.002	0.000
14	7,606	0.005	0.003	0.008	0.011	0.005	0.008	0.000	0.000
15	8,030	0.000	0.004	0.000	0.018	0.000	0.000	0.000	0.000
16	8,455	0.000	0.005	0.005	0.005	0.000	0.000	0.000	0.000
17	8,886	0.000	0.000	0.000	0.000	0.000	0.000	0.000	0.000
18	9,366	0.000	0.000	0.000	0.000	0.000	0.000	0.000	0.000
19	9,876	0.000	0.000	0.000	0.000	0.000	0.000	0.000	0.000
20	10,409	0.000	0.000	0.000	0.000	0.000	0.000	0.000	0.000
Liquid water content (g/m³)		0.203	0.225	0.248	0.262	0.268	0.452	0.233	0.166

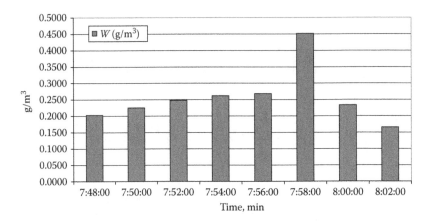

Figure 2.8 Liquid water content W.

$$\Lambda = \left(\frac{6!}{\pi} \frac{W}{Z} \right)^{1/3} \tag{2.15}$$

The results are shown in Table 2.10.
9. The estimation of a relationship of Z versus R of the type $Z = \alpha R^\beta$ is feasible through logarithmic equation: $\ln Z = \ln a + b \ln R$. This relationship is a line of the type $y = C_1 + C_2 x$ in double logarithmic axes ($C_1 = \ln a$, $C_2 = b$). Applying linear regression, the results are $y = 5.4224 + 1.46x$, so $\ln a = 5.4224$, $a = 226.41$, $b = 1.46$ (Figure 2.12).

2.5.6 Installation of rain gauges

The main purpose of the rainfall observations is to obtain values which are representative of a region. This objective is difficult to achieve, since the instrument may affect the measurements, and it is known that the hydro-meteorological parameters are affected by environmental parameters.

Whether they are autographic instruments or not, they must be protected from the air, because the edges of the collector are well above the ground surface. The size of turbulence varies with the velocity of the air and the irregularities of the region. The flow field of the wind during a rainfall or snowfall is the main cause of heterogeneities and discontinuities in the recording of rainfall: the various ground or other abnormalities cause local disturbances in the wind flow lines and corresponding disorders in the movement of raindrops or snowflakes. Regarding non-automatic instruments used for measuring precipitation, the water intake should be less in order to minimize evaporation losses.

Consequently, the basic rules for the installation of rain gauges are summarized as follows:

- Installation of the rain gauge in areas sheltered from winds from all directions, i.e. free from severe turbulence or other disorders.
- The collector should be placed at a height of 1–1.5 m above the ground surface.
- The number of trees or shrubs surrounding it should be limited or the distance of these barriers from the instrument should be at least twice their height.

Table 2.8 Kinetic energy flux E

Class	$V_l = 14.2D^{1/2}$ (mm/s)	22/2/2013 7:48:00 p.m.	22/2/2013 7:50:00 p.m.	22/2/2013 7:52:00 p.m.	22/2/2013 7:54:00 p.m.	22/2/2013 7:56:00 p.m.	22/2/2013 7:58:00 p.m.	22/2/2013 8:00:00 p.m.	22/2/2013 8:02:00 p.m.
1	2,691	0.00	0.00	0.00	0.00	0.00	0.00	0.00	0.00
2	3,031	0.01	0.01	0.01	0.01	0.00	0.01	0.04	0.12
3	3,335	0.14	0.06	0.10	0.08	0.06	0.07	0.14	0.49
4	3,637	0.44	0.23	0.39	0.26	0.21	0.28	0.56	1.17
5	3,943	0.96	0.53	0.93	0.77	0.45	0.78	1.17	1.48
6	4,292	2.29	1.85	3.90	2.95	2.20	3.31	4.06	5.51
7	4,744	3.93	6.24	9.77	5.89	5.89	11.05	9.33	10.46
8	5,181	5.17	7.75	7.85	8.35	10.23	22.65	14.90	6.76
9	5,511	3.58	8.47	9.45	7.17	7.98	23.78	10.91	2.28
10	5,793	6.56	9.48	7.05	8.26	11.42	21.14	6.80	0.24
11	6,209	15.23	21.15	13.12	16.08	20.31	32.15	6.77	0.42
12	6,749	20.61	11.54	11.54	14.84	23.91	18.14	1.65	0.00
13	7,218	14.11	2.82	5.65	14.11	12.70	9.88	1.41	0.00
14	7,606	4.29	2.14	6.43	8.58	4.29	6.43	0.00	0.00
15	8,030	0.00	3.31	0.00	16.56	0.00	0.00	0.00	0.00
16	8,455	0.00	5.00	5.00	5.00	0.00	0.00	0.00	0.00
17	8,886	0.00	0.00	0.00	0.00	0.00	0.00	0.00	0.00
18	9,366	0.00	0.00	0.00	0.00	0.00	0.00	0.00	0.00
19	9,876	0.00	0.00	0.00	0.00	0.00	0.00	0.00	0.00
20	10,409	0.00	0.00	0.00	0.00	0.00	0.00	0.00	0.00
Kinetic energy flux E (J/m^2 h)		77.33	80.58	81.18	108.91	99.66	149.67	57.75	28.92

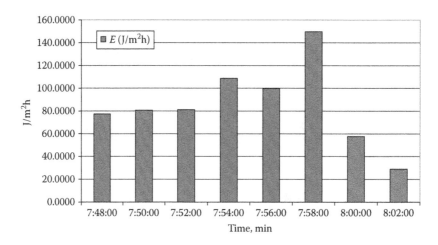

Figure 2.9 Kinetic energy flux *E*.

2.5.7 Space measurements

2.5.7.1 Satellites

Satellites are classified depending on their orbits around the Earth. There are a variety of different possible orbits: sun-synchronous, mid-inclination and low-inclination orbits. Each one of them has advantages and disadvantages.

The sun-synchronous orbit has an approximate inclination of 98.6° and has the advantage of providing samples at the same local time each day. The constant angle between the sun, the satellite and the observed spot on the Earth allows for a simple solar array and thermal design. However, the retrograde orbit (an orbit of a satellite orbiting the Earth in which the projection of the satellite's position on the Earth's equatorial plane revolves in the direction opposite to that of the rotation of the Earth) requires more launcher capability and therefore is more expensive. Also, in the case of launching multiple satellites, it is difficult to configure the launching patterns so that the satellites will be distributed through different ascending nodes at the same altitude.

Mid-inclination orbits (35°–70° inclination) provide short revisit intervals around the inclination latitude. However, due to the 70° constraint, the polar regions are not covered at all by these satellites. It is easier to distribute multiple launched satellites to the desired orbits, but they require more complex solar array and thermal design and may require periodic manoeuvres to maintain the desired orbit.

Finally, for low-inclination orbits (up to 25°–30° inclination), the revisit intervals around the tropics (and nowhere else) are very satisfactory and the limited range of sun angles simplifies solar array and thermal design. The problem with using low-inclination orbits is that satellites with mid-inclination or sun-synchronous orbits must also complement in order to have coverage beyond the tropics. Since the mid-inclination and sun-synchronous orbited satellites will also cover the tropics, the result will be perfect coverage up to 30° latitude but will have many gaps from 30° latitude to 90° latitude.

The optimization process, which involves the orbits' architecture, is basically a trade-off between coverage and resolution (see Figures 2.13 and 2.14). The greatest coverage is achieved with high-altitude satellites (i.e. 833 km) since the swath width will be large. However, this implies poor resolution since the resolution is the swath width divided by the number of

Table 2.9 Radar reflectivity Z

Class	$U_1 = 14.2D^{1/2}$ (mm/s)	22/2/2013 7:48:00 p.m.	22/2/2013 7:50:00 p.m.	22/2/2013 7:52:00 p.m.	22/2/2013 7:54:00 p.m.	22/2/2013 7:56:00 p.m.	22/2/2013 7:58:00 p.m.	22/2/2013 8:00:00 p.m.	22/2/2013 8:02:00 p.m.
1	2,691	0.00	0.00	0.00	0.00	0.00	0.00	0.01	0.00
2	3,031	0.04	0.02	0.05	0.02	0.01	0.04	0.13	0.43
3	3,335	0.67	0.27	0.46	0.38	0.30	0.32	0.69	2.33
4	3,637	2.74	1.46	2.45	1.64	1.31	1.72	3.47	7.27
5	3,943	7.64	4.17	7.37	6.13	3.55	6.22	9.32	11.72
6	4,292	23.47	18.96	39.94	30.24	22.57	33.85	41.52	56.42
7	4,744	54.30	86.20	135.07	81.45	81.45	152.72	128.96	144.57
8	5,181	93.01	139.52	141.31	150.25	184.24	407.83	268.31	121.63
9	5,511	77.63	183.49	204.66	155.26	172.90	515.17	236.41	49.40
10	5,793	165.19	238.61	177.43	208.02	287.55	532.28	171.31	6.12
11	6,209	472.11	655.71	406.54	498.34	629.48	996.68	209.83	13.11
12	6,749	820.43	459.44	459.44	590.71	951.69	721.98	65.63	0.00
13	7,218	687.34	137.47	274.93	687.34	618.60	481.14	68.73	0.00
14	7,606	244.40	122.20	366.61	488.81	244.40	366.61	0.00	0.00
15	8,030	0.00	222.02	0.00	1110.10	0.00	0.00	0.00	0.00
16	8,455	0.00	391.25	391.25	391.25	0.00	0.00	0.00	0.00
17	8,886	0.00	0.00	0.00	0.00	0.00	0.00	0.00	0.00
18	9,366	0.00	0.00	0.00	0.00	0.00	0.00	0.00	0.00
19	9,876	0.00	0.00	0.00	0.00	0.00	0.00	0.00	0.00
20	10,409	0.00	0.00	0.00	0.00	0.00	0.00	0.00	0.00
Radar reflectivity Z (mm^6/m^3)		2649.0	2660.8	2607.5	4399.9	3198.1	4216.5	1204.3	413.0
Radar reflectivity ζ (dBZ)		34.23	34.25	34.16	36.43	35.05	36.25	30.81	26.16

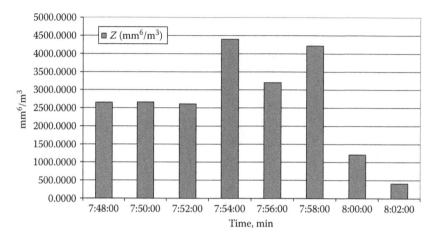

Figure 2.10 Radar reflectivity Z in mm^6/m^3.

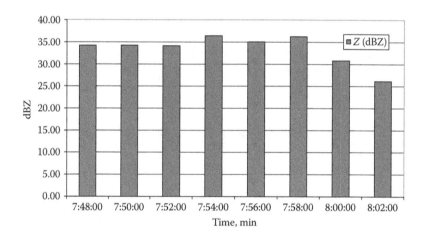

Figure 2.11 Radar reflectivity ζ in dBZ.

Table 2.10 N_0 and Λ values

Time		N_0 (1/m^3 mm)	Lambda (1/mm)
22/2/2013	7:48:00 p.m.	2,948.697	2.599
22/2/2013	7:50:00 p.m.	3,735.054	2.687
22/2/2013	7:52:00 p.m.	4,819.560	2.794
22/2/2013	7:54:00 p.m.	2,715.052	2.389
22/2/2013	7:56:00 p.m.	4,398.741	2.679
22/2/2013	7:58:00 p.m.	10,300.557	2.908
22/2/2013	8:00:00 p.m.	11,668.312	3.541
22/2/2013	8:02:00 p.m.	21,994.359	4.517

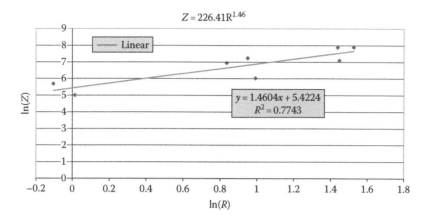

Figure 2.12 Linear regression of the Z–R relationship.

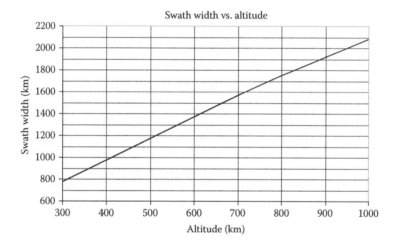

Figure 2.13 Swath width vs. altitude (at 140° sector).

beams that the instruments are using. Since the number of successive beams per scan is constant for a particular instrument, the resolution is higher when the swath width is lower.

The satellite Tropical Rainfall Measuring Mission (TRMM) was launched into orbit around the Earth on 27 November 1997 (Figure 2.15). It is the result of a joint effort between the United States and Japan, with the sole aim of measuring precipitation from space. Due to the uniqueness of the project and the wide acceptance of the products, which are freely available to the scientific community, the operational lifetime of the satellite was extended from 3 years and 2 months, as it was originally designed, to 6 years and 2 months (Enright, 2004). This increase was made possible by minimizing the energy consumption. NASA further extended the duration of the mission until the year 2012, with a potential risk of uncontrolled re-entry of the satellite into the atmosphere, as the fuel would have been completely exhausted by then (National Aeronautics and Space Administration, 2007).

The satellite has a total weight of 3620 kg. The satellite rotates asynchronously to the sun at an altitude of 350 km above the ground and an angle of 35°. The small angle has

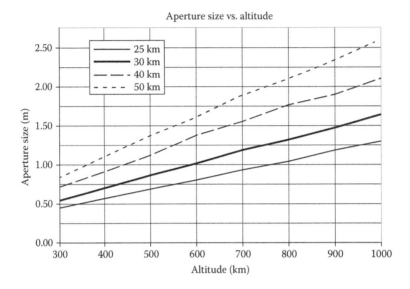

Figure 2.14 Aperture for footprint vs. altitude.

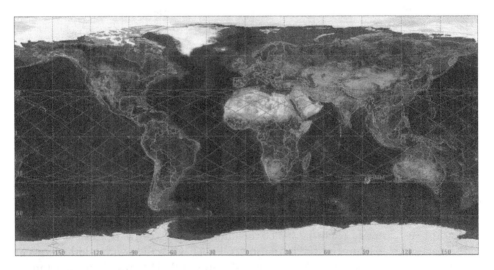

Figure 2.15 The orbit of Tropical Rainfall Measuring Mission in a day period.

the advantage of very dense measurements around the tropics as shown in Figures 2.16 and 2.17. Because of the small angles of the sun's rays, the design of solar collectors is greatly simplified (Fotopoulos, 2002). These collectors provide most of the energy consumed by its instruments.

2.5.7.2 Installed instruments

Five different instruments are installed in the TRMM satellite. Three of them aim to assess the precipitation and the other two are used to monitor other weather parameters. The instruments used to record the precipitation are the precipitation radar (PR), the microwave imager (TMI) and the visible and infrared scanner (VIRS). The other two instruments are

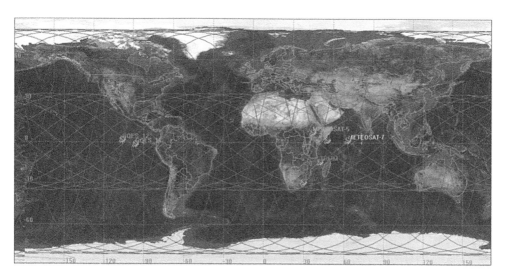

Figure 2.16 The orbits of all satellites in a day period.

Figure 2.17 The position of Tropical Rainfall Measuring Mission on January 1, 2007, 13:40:52.

the clouds and the Earth's radiant energy system (CERES) and the lightning imaging sensor (LIS) (Everett, 2001). A schematic view of the scan geometries of primary rainfall sensors is given in Figure 2.18.

2.5.7.3 Advanced precipitation radar

The PR was the first rain radar in space (Table 2.11). Its purpose is to: provide a 3D structure of the rainfall, particularly of its vertical distribution; obtain quantitative rainfall measurements over land and over ocean and improve the overall precipitation

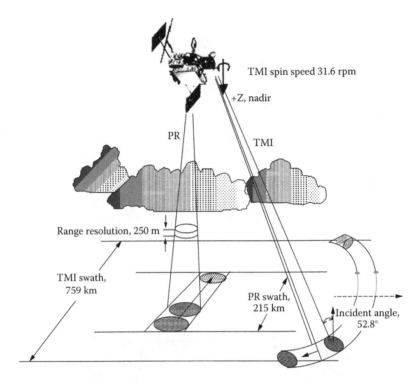

Figure 2.18 Schematic view of the scan geometries of primary rainfall sensors.

Table 2.11 Major parameters of the precipitation radar

Item	Specification
Frequency	13.796, 13.802 GHz
Sensitivity	≤ ~0.7 mm/h (S/N/pulse ≈ 0 dB)
Swath width	215 km
Observable range	Surface to 15 km altitude
Horizontal resolution	4.3 km (nadir)
Vertical resolution	0.25 km (nadir)
Antenna	
Type	128-element WG planar array
Beam width	0.71° × 0.71°
Aperture	2.0 × 2.0 m
Scan angle	±17° (cross track scan)
Transmitter/receiver	
Type	SSPA and LNA (128 channels)
Peak power	≥500 W (at antenna input)
Pulse width	1.6 μs × 2 channels (transmitted pulse)
PRF	2776 Hz
Dynamic range	≥70 dB
Number of independent samples	64
Data rate	93.2 kbps

retrieval accuracy. The National Space Development Agency of Japan (NASDA) has developed the PR in cooperation with the Communications Research Laboratory, Ministry of Posts and Telecommunications (NASDA, 2001). The measurements taken by this instrument, combined with those of TMI, provide the vertical profile of rainfall. This profile can be used to estimate the emission of the Earth's latent heat. The width of the swath of the unit on the ground does not exceed 215 km, while the vertical resolution reaches up to 250 m starting from the ground and reaching a height of about 20 km. The horizontal resolution of the recordings ranges from 4.18 to 4.42 km, depending on the position of the satellite.

The PR has only a Ku band, which is sufficient to measure the rainfall in the tropics. For mid-latitudes, where weak rain and snow occurs, another band has to be added to the instrument. The Ka band will operate simultaneously with the Ku band. The Ka band will have improved accuracy and will be used to measure weak rainfall and snowfall and to separate snow from rain. Since this is the first time that this band will operate, its specifications will be based on preliminary requirements as shown in Table 2.12, which are subject to change. To achieve these preliminary operational requirements of the Ka band (separation of snow and ice from rain, accurate estimation of the rain rate and the DSD and computation of the effect of non-uniformity of raindrop distribution), information from both bands is needed at a given location at a given time. The radar has to use special scan patterns to combine both bands (Ka and Ku) within a single sweep.

It would be ideal to use the same scan patterns for both bands, since the target would be the same in both cases and no shifting algorithms would be necessary. However, this is not possible because the two bands operate at different wavelengths and they have different resolutions and swath widths, so the scanning patterns cannot be identical within a sweep. Also, an exact match of the two beams is technically impossible. The currently proposed scan pattern is given in Figure 2.19. The arrow in Figure 2.19 indicates the direction of the satellite's movement. The instrument scans from left to right and the satellite moves from

Table 2.12 Original operational requirements of the Ka band

Item	Specification
Frequency	35.5 GHz
Sensitivity	11 dBZ or better
Swath width	20–40 km
Observable range	Surface to 15 km altitude
Horizontal resolution	4.0 km (nadir)
Vertical resolution	0.25 km (nadir)
Measurable rain	
Minimum	0.3 mm/h
Maximum	10 mm/h (near surface)

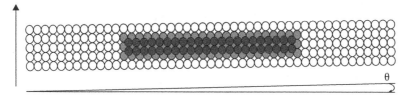

Figure 2.19 Schematic view of APR conical scan.

the bottom to the top of the figure. The result is a rotated scan at an angle Θ with respect to the plane perpendicular to the flight path. There are 49 horizontal Ku beams (transparent circles), yielding a swath of 245 km. There are 25 Ka beams (light coloured circles), yielding a swath of 125 km. The number of Ka beams is subject to change. Finally, the dark coloured circles are Ka interlaced beams and should be one less in number than Ka footprints in every scan. This scan pattern should not be considered as final.

2.5.7.4 Microwave imager

The TMI is a nine-channel (frequencies) passive microwave radiometer based upon the special sensor microwave/imager, which has been flying aboard the U.S. Defense Meteorological Satellite Program satellites since 1987 (Table 2.13). Its purpose is to provide data related to the intensity of rainfall over the oceans. The accuracy decreases when measurements are made over land, since the heterogeneous emission from the land surface complicates the interpretation of the measurements. The swath of the instrument on the ground is conical shaped, with a width of 760 km and a horizontal resolution of 6–50 km, depending on the position of the satellite.

The TMI antenna is an offset parabola, with an aperture size of 61 cm (projected along the propagation direction) and a focal length of 50.8 cm. The antenna beam views the Earth's surface at a *nadir* angle of 49°, which results in an incident angle of 52.8° at the Earth's surface. The TMI antenna rotates around a nadir axis at a constant speed of 31.6 rpm. The rotation draws a *circle* on the Earth's surface. Only 130° of the forward sector of the complete circle is used for measurements. The rest is used for calibrations and other instrument maintenance purposes.

2.5.7.5 Visible and infrared radiation scanner

This instrument measures the emitted radiation in five spectral bands, operating in the area of visible and infrared radiation. The combined use of the measurements of VIRS with those of TMI provides a more accurate assessment of precipitation, when compared to separate

Table 2.13 TMI instrument specifications

Channel No.	1	2	3	4	5	6	7	8	9
Centre frequency (GHz)	10.65	10.65	19.35	19.35	21.3	37.0	37.0	85.5	85.5
Polarization	V	H	V	H	V	V	H	V	H
Bandwidth (MHz)	100	100	500	500	200	2000	2000	3000	3000
Stability (MHz)	10	10	20	20	20	50	50	100	100
Beam width (degree)	3.68	3.75	1.90	1.88	1.70	1.00	1.00	0.42	0.43
IFOV-CT (km)	35.7	36.4	18.4	18.2	16.5	9.7	9.7	4.1	4.2
EFOV-DT (km)	63.2	63.2	30.4	30.4	22.6	16.0	16.0	7.2	7.2
Time (ms)/sample	6.60	6.60	6.60	6.60	6.60	6.60	6.60	3.30	3.30
# EFOVs/scan	104	104	104	104	104	104	104	208	208
Samples/beam	4	4	2	2	2	1	1	1	1
Beam EFOV (km²)	63 × 37	63 × 37	30 × 18	30 × 18	23 × 18	16 × 9	16 × 9	7 × 5	7 × 5
# EFOVs/scan	26	26	52	52	52	104	104	208	208

measurements of all the parameters (NASDA, 2001). That is because the distribution of clouds is estimated by VIRS. The scanning angle of the instrument is ±45° which is translated to a footprint with a width of 720 km on the ground. The horizontal analysis does not exceed 2 km.

2.5.7.6 Clouds and the Earth's radiant energy system

This instrument is designed to reduce uncertainty in the prediction of long-term climate change. The instrument measures the Earth's radiant energy, which is separated from the clouds' radiation (NASA, 2008). It is based on the deviation of radiant energy used in physical models of climate prediction and the determination of the balance of surface emissivity, which plays an important role in atmospheric processes and the transport of energy from the air to the sea and vice versa. Although the natural processes have not been simulated successfully, the measurements of CERES are considered necessary for the success of the research in this scientific field.

2.5.7.7 Lightning imaging sensor

An LIS is an optical telescope combined with a filtered imaging system, which records not only the lightnings which occur inside the clouds but also those which occur from clouds to the ground. In conjunction with the measurements of PR, TMI and VIRS, important steps have been taken to link lightning with rainfall and other properties of storms (Petersen et al., 2005; Pessi and Businger, 2009).

2.6 INSTALLATION OF NETWORK OF POINT MEASUREMENT DEVICES

Installing a point measuring device, even a simple one such as a rain gauge, means actually installing a metering station.

The meteorological instruments of the stations, their density and their location of installation are a subject of a special study and depend mainly on geomorphological and climatic factors as well as on the use of rainfall data. For this reason, there are no general rules for the density of the stations and the types of the instruments at each station.

Generally, the denser the network, the more representative the estimated surface rainfall (Manning, 1997). The heterogeneities of the geomorphology require a dense network. Moreover, the cost of installation, maintenance and ease of accessibility of the observer are the factors which are taken into account.

In general, the measurement errors of rainfall increase as the surface rainfall increases and decrease with increasing density of the network, the rainfall duration and the size of the area. Larger deviations occur at the scale of rainfall event, than at month, season or year and during the summer season. In summer, the network should be two to three times denser than in winter (Singh, 1992).

The adequacy of a network of rain gauges is determined statistically. The optimum number of rain gauges corresponding to a specified percentage of error in the estimation of surface rainfall is

$$N = \left(\frac{C_u}{\varepsilon} \right)^2 \tag{2.16}$$

where
   N is the optimum number of rain gauges
   C_u is the variation coefficient of the rainfall at the measuring devices
   ε is the allowed percentage of error (%)

The standard value of ε is 10%. If this value should decrease, then more rain gauges are required.

If there are m rain gauges in a basin and P_1, P_2, P_3,..., P_m are the rainfall depths for a specific time interval, then the coefficient C_u is

$$C_u = 100 \, S/P \tag{2.17}$$

where
   P is the mean value of the rainfall recorded in the rain gauges

$$P = \frac{1}{m} \sum_{i=1}^{m} P_i \tag{2.18}$$

S is the standard deviation

$$S = \left[\sum_{i=1}^{m} \frac{(x_i - \bar{x})^2}{m-1} \right]^{0.5} \tag{2.19}$$

The World Meteorological Organization (WMO) has proposed the following regarding the density of the network in relation to the general hydro-meteorological conditions:

- One rain gauge per 600–900 km² in flat areas for mild Mediterranean and tropical zones
- One rain gauge per 100–250 km² in mountainous areas for mild Mediterranean and tropical zones
- One rain gauge per 25 km² in semi-mountainous areas with intense variation in rainfall
- One rain gauge per 1,500–10,000 km² in dry and polar zones

If the purpose is the estimation of the hydrograph or the peak discharge, then the density of the network may be different. The densities of rain gauges as a function of the area of agricultural basins, according to the WMO, are given in Table 2.14. The number and type of devices to be placed in a basin depend on economic, climatic and topographic (accessibility)[139] as well as on the methodology of data analysis. The density of a network must be determined primarily based on the degree of irregularity of precipitation and the purpose to be served.

Table 2.14 Density of rain gauges depending on the area of agricultural basins

Basin area (× 1000 m²)	Ratio (km²/station)	Minimum number of stations
0–120	0.13	1
120–140	0.20	2
400–800	0.25	3
800–2,000	0.40	1 per 0.4 km²
2,000–10,000	1.00	1 per 1 km²
10,000–20,000	2.50	1 per 2.5 km²
>20,000	7.50	1 per 7.5 km²

Table 2.15 Annual rainfall per rain gauge

Rain gauge	1	2	3	4	5	6
Annual rainfall (cm)	48	75	81	63	104	89

Example 2.2

A basin consists of a network of six rain gauges. The annual rainfall recorded from these rain gauges is listed in Table 2.15

Calculate the optimum number of rain gauges for this basin, with 10% error in the calculation of the mean surface rainfall.

Solution

The average of the annual rainfall from the six rain gauges is

$$P = \frac{1}{m} \sum_{i=1}^{m} P_i = 76.67 \text{ cm}$$

The standard deviation is

$$S = \left[\sum_{i=1}^{m} \frac{(x_i - \bar{x})^2}{m-1} \right]^{0.5} = 19.64 \text{ cm}$$

The coefficient C_u is

$$C_u = \frac{100 \times 19.64}{76.67} = 25.62\%$$

The optimum number of rain gauges to the basin under study is

$$N = \left(\frac{25.62}{10} \right)^2 = 6.56 \cong 7 \text{ rain gauges}$$

Thus, one more rain gauge should be installed.

2.7 TEST OF DATA HOMOGENEITY AND ANALYSIS OF DOUBLE CUMULATIVE CURVES

The probability of a storm occurring in a particular region is the same if all the meteorological conditions remain stable. Such an area is considered meteorologically homogeneous. Additionally, a region is homogeneous if it has the same annual rainfall and roughly the same range of rain. The factors which determine the annual rainfall and the meteorological homogeneity are the distance from the coast, the wind direction which determines the direction of weather systems, the mean annual temperature, the altitude and the topography. If one area is meteorologically homogeneous, then the probability for a storm of the same duration and depth to occur is the same throughout the region.

Before analyzing the rainfall data of a station, the quality and completeness of data should be checked. The homogeneity test of the data includes, apart from the *logical* testing, the test of the data quality. What needs to be checked is whether the measurements have been obtained under the same conditions. Changing the position of the instrument, and replacement of the instrument and the observer usually lead to non-homogeneous data.

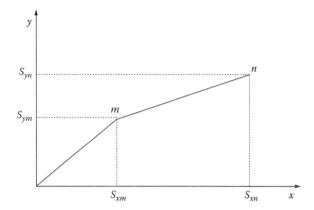

Figure 2.20 Point of slope change in a double cumulative curve.

The testing of the homogeneity of a station data is feasible through methods such as the double cumulative curve, the cumulative deviations and the von Neumann ratio. The most common way to test the homogeneity of rainfall time series is by plotting the double cumulative curve (Dingman, 1994; Shaw, 1994). The test of homogeneity is applied only on the annual time series of rainfall.

The double cumulative curve is derived as follows: consider two neighbouring stations X and Y, where the annual rainfall time series is denoted by x and y, respectively. The cumulative time series is calculated as follows:

$$sx_j = \sum_{i=1}^{j} x_i \quad \text{and} \quad sy_j = \sum_{i=1}^{j} y_i \tag{2.20}$$

Each value is equal to the sum of the rainfall of all the preceding years. The sum is commonly calculated beginning from the more recent year. The pairs of the values (sy_j, sx_j) are plotted on a diagram, leading to the double cumulative curve.

The standard form of a double cumulative curve with a change of slope point m is shown in Figure 2.20. In this case, the points of the double cumulative curve are not well described by a single straight line, but by two lines, forming an angle. This heterogeneity is corrected by performing a procedure, in which the values of the heterogeneous time series are multiplied with a coefficient, such that the double cumulative curve becomes a straight line. Specifically, if the time series x and y have n values and the double cumulative curve presents a break point m, then the correcting coefficient of the time series y is calculated as follows:

$$\lambda = \frac{sy_n - sy_m}{sx_n - sx_m} \cdot \frac{sx_m}{sy_m} \tag{2.21}$$

Then, the values from y_{m+1} to y_n are multiplied with λ, and the time series y is corrected.

Example 2.3

The annual rainfall of station X and the average rainfall of 10 neighbouring stations are given in Table 2.16. Determine the homogeneity of the measurements of Station X. In which year did the change occur? Calculate the average annual rainfall at station X for a period of 30 years with and without the homogeneity of the measurements.

Table 2.16 Annual rainfall of station X and average values of the 10 neighbouring stations

Hydrological year	Station X (mm)	Average value of 10 stations (mm)	Hydrological year	Station X (mm)	Average value of 10 stations (mm)
1950–1951	470	290	1965–1966	360	340
1951–1952	240	210	1966–1967	350	280
1952–1953	420	360	1967–1968	280	230
1953–1954	270	260	1968–1969	290	330
1954–1955	250	230	1969–1970	320	330
1955–1956	350	300	1970–1971	390	350
1956–1957	290	260	1971–1972	250	260
1957–1958	360	260	1972–1973	300	290
1958–1959	370	260	1973–1974	230	280
1959–1960	350	280	1974–1975	370	340
1960–1961	580	400	1975–1976	340	330
1961–1962	410	260	1976–1977	300	350
1962–1963	340	240	1977–1978	280	260
1963–1964	200	220	1978–1979	270	250
1964–1965	260	250	1979–1980	340	350

Solution

It is assumed that the most recently recorded values are more accurate than older values, which will be corrected. Therefore, the values are classified in reverse chronological order, as shown in Table 2.17 (columns 4, 5 and 6).

Then, the cumulative values of the two stations are calculated in Table 2.18 (columns 7 and 8), and a double cumulative curve is plotted, as shown in Figure 2.21. A change in the slope of the double cumulative curve is observed in 1963–1964. Therefore, the values of station X will be corrected from the year 1950–1951 up to the year 1962–1963.

Applying linear regression between the data of columns 7 and 8, the slope of the curve for the years 1979–1980 ~ 1963–1964 is $\lambda_1 = 1.0192$ and for the years 1963–1964 ~ 1950–1951, it is $\lambda_2 = 1.2669$. The ratio of the two slopes is $m = \lambda_1/\lambda_2 = 0.8045$. The values in column 5 are multiplied with this value from the years 1950–1951 to the years 1962–1963, in order to get the corrected values for station X, which are given in Table 2.18 (column 9). The new cumulative values of Station X are presented in column 10.

The new double cumulative curve (based on data of columns 10 and 11 of Table 2.19) has a uniform slope, as shown in Figure 2.22.

The average annual rainfalls at station X before and after the homogeneity of the measurements are 328 and 297 mm, respectively.

2.8 COMPLETION OF RAINFALL MEASUREMENTS: ADAPTATION TO DIFFERENT ALTITUDES

The main sources of error in measurements of rainfall are

1. Instrumentation errors
2. Improper positioning of the rain gauge
3. Human errors

Table 2.17 Rank of values in reverse chronological order

(1)	(2)	(3)	(4)	(5)	(6)
	Initial data			*Reverse chronological order*	
Year	*Station X*	*Average value of 10 stations*	*Year*	*Station X*	*Average value of 10 stations*
1950–1951	470	290	1979–1980	340	350
1951–1952	240	210	1978–1979	270	250
1952–1953	420	360	1977–1978	280	260
1953–1954	270	260	1976–1977	300	350
1954–1955	250	230	1975–1976	340	330
1955–1956	350	300	1974–1975	370	340
1956–1957	290	260	1973–1974	230	280
1957–1958	360	260	1972–1973	300	290
1958–1959	370	260	1971–1972	250	260
1959–1960	350	280	1970–1971	390	350
1960–1961	580	400	1969–1970	320	330
1961–1962	410	260	1968–1969	290	330
1962–1963	340	240	1967–1968	280	230
1963–1964	200	220	1966–1967	350	280
1964–1965	260	250	1965–1966	360	340
1965–1966	360	340	1964–1965	260	250
1966–1967	350	280	1963–1964	200	220
1967–1968	280	230	1962–1963	340	240
1968–1969	290	330	1961–1962	410	260
1969–1970	320	330	1960–1961	580	400
1970–1971	390	350	1959–1960	350	280
1971–1972	250	260	1958–1959	370	260
1972–1973	300	290	1957–1958	360	260
1973–1974	230	280	1956–1957	290	260
1974–1975	370	340	1955–1956	350	300
1975–1976	340	330	1954–1955	250	230
1976–1977	300	350	1953–1954	270	260
1977–1978	280	260	1952–1953	420	360
1978–1979	270	250	1951–1952	240	210
1979–1980	340	350	1950–1951	470	290

The rainfall data are usually incomplete. The lack of data may involve a few days to several years. Popular methods of data completion are the method of the arithmetic average, the method of the normal ratios, the method of inverse distances, the contour method, the Lagrange method, interpolation and the Kriging family of methods.

Data completion is essential to any hydrological study and is based on data from adjacent rainfall stations. Besides, the total operating period of a rain gauge station may be short, while other stations with a longer operating period may be in the neighbourhood of the station. In this case, an expansion of the station's data could be achieved, using the data of the neighbouring stations (Maidment, 1993). The general methodology for completion and expansion is the same; however, the case of expansion normally concerns longer periods than completion and needs more attention.

Table 2.18 Cumulative values

	(7)	(8)
		Cumulative
Year	Station X	Average value of 10 stations
1979–1980	340	350
1978–1979	610	600
1977–1978	890	860
1976–1977	1190	1210
1975–1976	1530	1540
1974–1975	190	1880
1973–1974	2130	2160
1972–1973	2430	2450
1971–1972	2680	2710
1970–1971	3070	3060
1969–1970	3390	3390
1968–1969	3680	3720
1967–1968	3960	3950
1966–1967	4310	4230
1965–1966	4670	4570
1964–1965	4930	4820
1963–1964	5130	5040
1962–1963	5470	5280
1961–1962	5880	5540
1960–1961	6460	5940
1959–1960	6810	6220
1958–1959	7180	6480
1957–1958	7540	6740
1956–1957	7830	7000
1955–1956	8180	7300
1954–1955	8430	7530
1953–1954	8700	7790
1952–1953	9120	8150
1951–1952	9360	8360
1950–1951	9830	8650

This set of values constitutes the point of slope change.

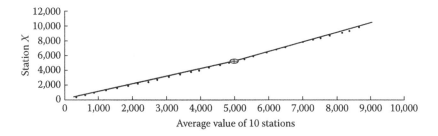

Figure 2.21 Double cumulative curve.

Table 2.19 New cumulative values

	(9)	(10)	(11)
	New values	Cumulative	
Year	Station X	New cumulative values (station X)	Average value of 10 stations
1979–1980	340	340	350
1978–1979	270	610	600
1977–1978	280	890	860
1976–1977	300	1190	1210
1975–1976	340	1530	1540
1974–1975	370	1900	1880
1973–1974	230	2130	2160
1972–1973	300	2430	2450
1971–1972	250	2680	2710
1970–1971	390	3070	3060
1969–1970	320	3390	3390
1968–1969	290	3680	3720
1967–1968	280	3960	3950
1966–1967	350	4310	4230
1965–1966	360	4670	4570
1964–1965	260	4930	4820
1963–1964	200	5130	5040
1962–1963	274	5404	5280
1961–1962	330	5733	5540
1960–1961	467	6200	5940
1959–1960	282	6482	6220
1958–1959	298	6779	6480
1957–1958	290	7069	6740
1956–1957	233	7302	7000
1955–1956	282	7584	7300
1954–1955	201	7785	7530
1953–1954	217	8002	7790
1952–1953	338	8340	8150
1951–1952	193	8533	8360
1950–1951	378	8911	8650

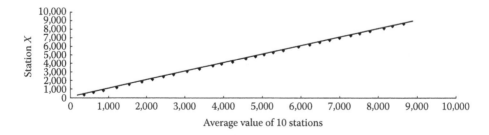

Figure 2.22 New double cumulative curve.

The methods which will be studied in the context of this course are the arithmetic mean, the normal ratios, the inverse distance and the linear regression.

2.8.1 Method of arithmetic mean

This method uses the average value of the measurements of the stations, which surrounds the station with missing data. This method is used when the annual measurements of the station with missing data diverge less than 10% from the average annual rainfall depth:

$$P = \sum_{i=1}^{N} \alpha_i P_i \tag{2.22}$$

where
α_i is $1/N$
N is the number of the stations
P_i is the rainfall recorded on station i
P is the average rainfall of the stations

2.8.2 Method of normal ratios

This method is very simple and widely used, and uses a weighting coefficient for each station. Three stations surrounding the station whose distance from the station with missing data is almost the same are usually selected. The general relationship is the following:

$$P_x = \frac{1}{m}\left(\frac{N_X}{N_1} P_1 + \frac{N_X}{N_2} P_2 + \frac{N_X}{N_3} P_3 + \cdots \frac{N_X}{N_m} P_m \right) = \sum_{i=1}^{m} \left(\frac{N_X}{mN_i} \right) P_i \tag{2.23}$$

where
P is the rainfall depth
N is the annual rainfall depth
X is the index which denotes the station with missing data
The indices $i = 1,2,3, ..., m$ denote the surrounding stations, m in total.

2.8.3 Inverse distance method

This method involves the surrounding stations in relation to the distance from the station with missing data. These distances are calculated assuming that the origin of the axes is the station with missing data. The rainfall is calculated using the following equations:

$$P_y = \sum_{i=1}^{k} w_i P_i \tag{2.24}$$

and

$$w_i = \frac{d_i^{-b}}{\sum_{j=1}^{k} d_j^{-b}} \tag{2.25}$$

where

    k is the number of the stations

    P_y is the estimated rainfall depth at the station with missing data

    P_i is the rainfall depth at each station

    d_i is the distance of each station from the station with missing data

    b is the exponent with a typical value of 2

    w_i is the weighting coefficient of each station i

2.8.4 Correlation and regression

Additionally, the completion of rainfall measurements may be achieved based on the linear correlation of the measurements of the station with missing data with other base stations if the degree of linear correlation between the two stations is high.

Correlation implies the determination of the relationship between two or more random variables. For example, if X and Y are the random variables which represent the precipitation in two adjacent stations, then the variables are associated with the relationship:

$$Y = f(X) \tag{2.26}$$

The correlation of the variables consists of determining the relationship f.

Regression is the correlation which is based on the least squares method, which practically, is identical with the correlation since the least squares method is used in almost all cases. The most common one is the linear regression, in which the function f is linear.

All time series, independent and dependent, should have a common measurement period without missing data, in order to carry out the regression.

The simple linear regression will be described here. According to this method, the value for completion $y = P_y$ is estimated from the corresponding value $x = P_x$ of the neighbouring station X (for the period with lack of data at station Y) based on the linear relationship:

$$y = \alpha + bx \tag{2.27}$$

where a and b parameters which are estimated so as to minimize the sum of square errors of the estimate. If x_i and y_i are simultaneous measurements at stations X and Y, respectively, at the time period I (usually year or month), then the following holds:

$$b = \frac{n\sum_{i=1}^{n} x_i y_i - \sum_{i=1}^{n} x_i \sum_{i=1}^{n} y_i}{n\sum_{i=1}^{n} x_i^2 - \left(\sum_{i=1}^{n} x_i\right)^2} = \frac{\sum_{i=1}^{n}(x_i - \bar{x})(y_i - \bar{y})}{\sum_{i=1}^{n}(x_i - \bar{x})} \tag{2.28}$$

$$\alpha = \bar{y} - b\bar{x} \tag{2.29}$$

where $\bar{x}$ and $\bar{y}$ are the average values of x_i and y_i, respectively, i.e.

$$\bar{x} = \frac{1}{n}\sum_{i=1}^{n} x_i; \quad \bar{y} = \frac{1}{n}\sum_{i=1}^{n} y_i \tag{2.30}$$

and n is the (common for x_i and y_i) sample time period (Koutsoyiannis and Xanthopoulos, 1997).

The degree of suitability of the method for the specific data is calculated by the estimate of the correlation coefficient r:

$$r = \frac{n\sum_{i=1}^{n} x_i y_i - \sum_{i=1}^{n} x_i \sum_{i=1}^{n} y_i}{\sqrt{\left[n\sum_{i=1}^{n} x_i^2 - \left(\sum_{i=1}^{n} x_i\right)^2\right]\left[n\sum_{i=1}^{n} y_i^2 - \left(\sum_{i=1}^{n} y_i\right)^2\right]}} = \frac{\sum_{i=1}^{n}(x_i - \bar{x})(y_i - \bar{y})}{\sqrt{\sum_{i=1}^{n}(x_i - \bar{x})^2 \sum_{i=1}^{n}(y_i - \bar{y})^2}}$$

(2.31)

The closer the correlation coefficient to the unit, the more suitable the method is. Usually, the following is required for the application of the linear regression method:

$$r \geq \frac{2}{\sqrt{n}}$$

(2.32)

In simpler terms, it is required that $r \geq 0.7$.

Example 2.4

The rain gauge station X was out of operation during a storm event. The measurements of rainfall, which were recorded at three adjacent stations (meteorologically similar) A, B and C, were 117, 99 and 124 mm. The annual rainfall at the stations X, A, B and C are 958, 1130, 985 and 1220 mm, respectively. Calculate the rainfall at station X.

Solution

The data of the exercise are summarized in Table 2.20.

The rainfall at station X is determined using the normal ratio method as follows:

$$P_x = \frac{1}{m}\sum_{i=1}^{m} \frac{N_x}{N_i} P_i = \frac{1}{3}\left(\frac{N_x}{N_A} P_A + \frac{N_x}{N_B} P_B + \frac{N_x}{N_C} P_C\right) \Rightarrow$$

$$P_x = \frac{1}{3}\left(\frac{958}{1130}117 + \frac{958}{985}99 + \frac{958}{1220}124\right)\text{mm} \Rightarrow$$

$$P_x = 97.62 \text{ mm} \cong 98 \text{ mm}$$

Table 2.20 Annual rainfall data

Station	Rainfall depth (mm)	Station	Annual rainfall (mm)
A	117	X	958
B	99	A	1130
C	124	B	985
		C	1220

Example 2.5

Calculate the rainfall at station A in Figure 2.23 using the method of inverse distances. The distances of five adjacent stations from station A and the corresponding recorded rainfall depths are given in Table 2.21.

Solution

According to the method of inverse distances, the rainfall at station A is given by the relationship

$$P_a = \sum (P_i \cdot w_i)$$

where P_i is the rainfall depth at each neighbouring station and w_i is the coefficient at each station which is a function of the distance d_i from station A, where

$$w_i = \frac{1/d_i^2}{\sum_{j=1}^{k} 1/d_j^2}$$

$k = 5$. The product $P_i \cdot w_i$ is calculated as given in Table 2.22.

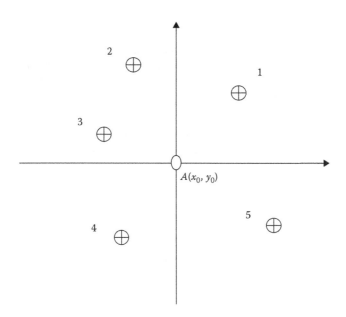

Figure 2.23 Rain gauge stations.

Table 2.21 Distances of the neighbouring stations from station A

Station	$X_i - X_0$ (km)	$Y_i - Y_0$ (km)	Rainfall (mm)
1	1.3	0.8	25
2	0.6	1.1	35
3	0.7	0.4	14
4	0.6	1.3	23
5	1.2	0.9	19

Table 2.22 Procedure of the inverse distance method

n	P_i (mm)	$x_i - x_0$ (km)	$y_i - y_0$ (km)	d_i^2 (km²)	d_i^{-2} (km⁻²)	w_i	$P_i \cdot w_i$ (mm)
1	25	1.3	0.8	2.33	0.43	0.12	3.03
2	35	0.6	1.1	1.57	0.64	0.18	6.30
3	14	0.7	0.4	0.65	1.54	0.43	6.09
4	23	0.6	1.3	2.05	0.49	0.14	3.17
5	19	1.2	0.9	2.25	0.44	0.13	2.39

The requested rainfall depth is

$$P_a = \sum (P_i \times w_i) = 21 \text{ mm}$$

Example 2.6

The annual rainfall in millimetres at two stations for the period 1970–1971 to 1989–1990 is given in Table 2.23. It has gaps in some years due to problems in one of the two stations. Calculate the missing data for station 2 by applying the statistical method of simple linear regression.

Solution

The linear regression between the rainfall depths of the two stations (without taking into account the years for which there are no data at station 2) gives the straight line:

$$y = 0.695x + 591.12$$

with a coefficient of linear correlation $r = 0.752$.

Table 2.23 Annual rainfall depth of the stations

Hydrological year	Station 1	Station 2
1970–1971	1145.8	1354.7
1971–1972	1311.5	1466.2
1972–1973	1078.5	
1973–1974	1278.5	1455.8
1974–1975	1262.8	1422.7
1975–1976	1088.6	1469.2
1976–1977	1175.6	1366.7
1977–1978	963.4	1368.9
1978–1979	1066.8	
1979–1980	985.2	1128.6
1980–1981	1066.9	1208.6
1981–1982	1306.4	
1982–1983	1285.6	1502.3
1983–1984	1139.6	1402.5
1984–1985	1297.5	1544.3
1985–1986	958.3	
1986–1987	1144.2	1499.8
1987–1988	978.3	1275.4
1988–1989	1145.2	1399.8
1989–1990	955.3	1255.9

Table 2.24 Completion of values using linear regression

Hydrological year	Station I (x)	Station 2	Station 2 (y)
1970–1971	1145.8	1354.7	
1971–1972	1311.5	1466.2	
1972–1973	1078.5	—	1340.55
1973–1974	1278.5	1455.8	
1974–1975	1262.8	1422.7	
1975–1976	1088.6	1469.2	
1976–1977	1175.6	1366.7	
1977–1978	963.4	1368.9	
1978–1979	1066.8	—	1332.42
1979–1980	985.2	1128.6	
1980–1981	1066.9	1208.6	
1981–1982	1306.4	—	1498.91
1982–1983	1285.6	1502.3	
1983–1984	1139.6	1402.5	
1984–1985	1297.5	1544.3	
1985–1986	958.3	—	1257.02
1986–1987	1144.2	1499.8	
1987–1988	978.3	1275.4	
1988–1989	1145.2	1399.8	
1989–1990	955.3	1255.9	

The completion of the requested values is done by applying the last relationship for the respective values of x. The results are presented in Table 2.24.

2.8.5 Correction of rainfall with altitude

Useful for the supplement of the rainfall data of a station is the fact that point rainfall increases with increasing altitude. Actually, this can be generally applied to any rainfall data processing for stations at different altitudes.

The average increase in the annual rainfall at a station, per 100 m increase of altitude, is called rainfall gradient. The rainfall gradient is usually obtained for each region from the graph of the average annual rainfall heights of the stations in relation to the altitude of the stations.

Oftentimes, the linear regression line which is plotted for the determination of the rainfall gradient has a low degree of linear correlation. In this case, the potential exception of some stations may be considered or the region may be split into subregions, for which there is a straight regression line with satisfactory coefficient of linear correlation.

2.9 SURFACE INTEGRATION OF AREAL RAINFALL FROM POINT MEASUREMENTS

The rainfall measurements taken by rain gauges are point measurements, and thus, represent the point where precipitation was measured. In most cases, however, such as in the assessment of the water balance, the surface rainfall is of particular importance, representing the entire watershed under consideration. For this reason, a network of rain gauges is installed at the watershed, whose positions should be such as to describe as best as possible the spatial variability of rainfall. Then, the point measurements are used for the calculation of surface rainfall, using surface integration methods.

There are numerous methods which have been developed and used to estimate the average surface rainfall. These can be divided into direct integration methods and methods of areal adjustment (Koutsoyiannis and Xanthopoulos, 1997). The direct integration methods calculate the areal rainfall directly from the values of the point rainfall. The best-known methods which belong to this category are the averaging method, the Thiessen method, the two-axis method of Bethlahmy and the optimal integration method (Kriging). On the contrary, the surface adaptation methods first estimate the geographic variability of rainfall in the region under study and based on this, they calculate the surface rainfall. Methods of this category are the method of isohyetal curves, the method of linear interpolation, the method of inverse distance, the method of multisquare interpolation, the least squares method with polynomials, the Lagrange polynomial method, the adaptation method of splines and the optimal interpolation method (Kriging). The methods of areal adaptation are further divided into interpolation and smoothing methods. The averaging method, the Thiessen method and the isohyetal method are presented next.

Regardless of which method is used, the reliability of the final result depends primarily on the density of the point information: the integration is more successful when the density of the network of the rainfall stations is dense. Unfortunately, networks are usually not quite dense and in some mountainous inaccessible areas, there is a scarcity of stations.

2.9.1 Averaging method

This is the simplest method, according to which the weights of all stations are taken to be equal, i.e. $w_i = 1/k$. The method may be used for the first rough estimates of rainfall due to its simplicity. The accuracy is tolerable only when the area is relatively flat, the stations are evenly distributed and the rainfall depth does not vary greatly from one station to another. The surface precipitation is given by the following equation:

$$P_s = \sum_{i=1}^{k} w_i P_i \tag{2.33}$$

2.9.2 Thiessen method

According to this classical method, the total area A is divided into geometric zones A_i, one for each station, so that

$$\sum_{i=1}^{k} A_i = A \tag{2.34}$$

The weighting coefficient depends on the area of the station's zone, i.e.

$$w_i = \frac{A_i}{A} \tag{2.35}$$

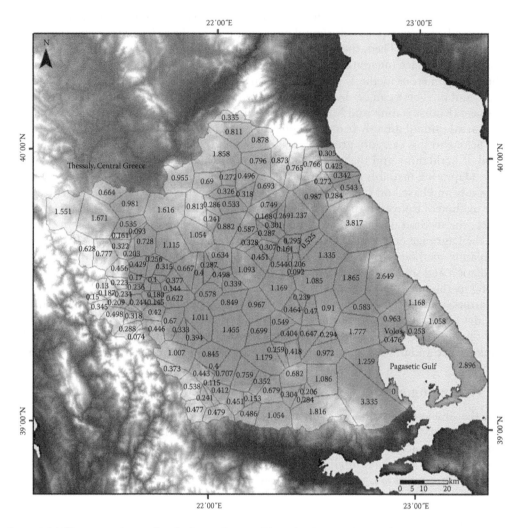

Figure 2.24 Thiessen polygons for the Pinios Basin in Thessaly, Greece. (From Mimikou, M., National Bank of Hydrology and Meteorology Data (NDBHMI), Final Report, NTUA, Athens, Greece, 2000.)

The geometric zones are determined in a way that the distance of each point of the zone of station i to station j is lower than it is from any other station in the area. This principle leads directly to a simple geometric construction of zones based on the segments perpendicular to the line segments connecting the two stations. This creates the Thiessen polygons (Figure 2.24). Despite its prolonged use, the method remains widespread today because of ease of implementation and reliable estimates. The estimates of the method are better when the network of rainfall stations is dense and the timescale of the study is long (e.g. the estimates are more accurate at hyperannual scale than the estimates at storm event scale). The altitude adjustment of the areal rainfall of the method is conducted through an altitude correction coefficient, based on the linear relation between the rainfall and the elevation.

2.9.3 Isohyetal curve method

A rainfall isohyetal curve is defined by the group of points where the rainfall depth has a specific value. Depending on the range of rainfall, isohyetal curves are plotted with an interval DP. The exact plot of an isohyetal curve depends on the available data and the experience of the hydrologist. After the plotting of the curves, the areas between successive curves corresponding to rainfall depths P_i and P_{i-1} are calculated. The areal mean rainfall of the region can be computed as follows:

$$P_S = \sum_r \frac{P_i + P_{i-1}}{2} \frac{A_i}{A}$$

(2.36)

The rainfall isohyetal curves for the region of Thessaly are presented in Figure 2.25, which are based on the hyperannual rainfall values (from 1/10/1955) and interval of 100 mm (Mimikou, 2000).

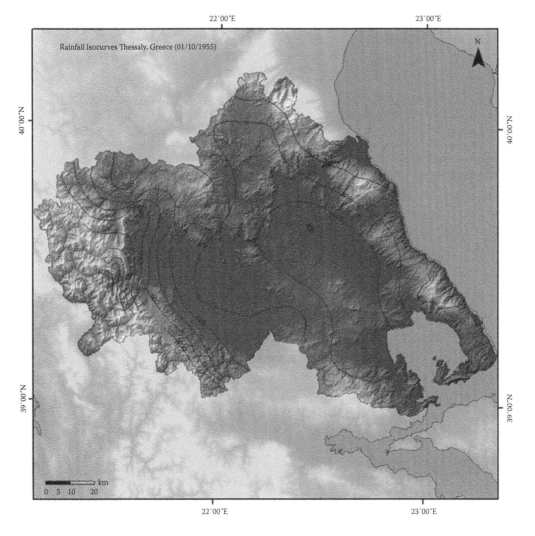

Figure 2.25 Rainfall isohyetal curves in the region of Thessaly. (From Mimikou, M., National Bank of Hydrology and Meteorology Data (NDBHMI), Final Report, NTUA, Athens, Greece, 2000.)

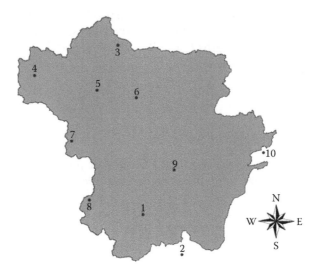

Figure 2.26 Stations' locations.

Example 2.7

The monthly rainfall data and the respective altitudes for 10 rain gauges in a sub-basin of Pinios River in Thessaly (Figure 2.26) are shown in Table 2.25. Calculate the average surface rainfall using the methods of arithmetic mean, Thiessen and isohyetal curves. Moreover, an adjustment of the average surface rainfall calculated by the Thiessen method with respect to the average elevation of the basin should be made. The area of the basin is equal to 2940 km², and the average elevation of the basin is equal to 532 m.

Solution

1. Method of arithmetic mean

   The surface rainfall is equal to the average of the point measurements of the 10 stations:

$$P = \frac{(74 + 73 + 92 + 101 + 83 + 87 + 110 + 103 + 82 + 70)}{10} = 88\,mm$$

Table 2.25 Altitude and monthly rainfall at stations

α/α	Station	Altitude (m)	Rainfall (mm)
1	MOUZAKI	226	74
2	TAVROPOS	220	73
3	AGIOFILLO	581	92
4	MALAKASIO	842	101
5	MEGALI KERASIA	500	83
6	METEORA	596	87
7	PALAIOCHORI	1050	110
8	STOURNAREIKA	860	103
9	TRIKALA	116	82
10	FARKADONA	87	70

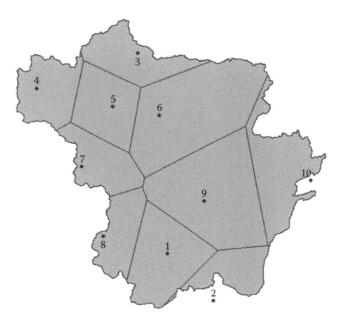

Figure 2.27 Drawing the Thiessen polygons.

2. Thiessen method

The design of Thiessen polygons is made by drawing lines perpendicular and at midpoint to the ones connecting the stations one by one, as shown in Figure 2.27. These polygons are defined uniquely for a particular network of rain gauges.

The procedure for the calculation of the surface rainfall is summarized in Table 2.26. The percentage w is equal to the ratio of the Thiessen polygon of the station to the total area of the basin.

The surface rainfall is estimated as 86 mm.

3. Isohyetal curve method

The rainfall isohyetal curves have been drawn at 5 mm interval in Figure 2.28. Ten intervals are formed.

The drawing of the rainfall isohyetal curves is not unique, as in the Thiessen method, but depends on the interpolation method. The simplest way is the graphic drawing of the isohyetal curves, which involves the discretion and experience of the

Table 2.26 Thiessen method

Station	Area (km²)	Percentage w	Rainfall P (mm)	w × P
1	316	0.107	74	7.955
2	153	0.052	73	3.800
3	216	0.073	92	6.760
4	212	0.072	101	7.270
5	238	0.081	83	6.720
6	582	0.198	87	17.225
7	194	0.066	110	7.260
8	177	0.060	103	6.202
9	550	0.187	82	15.342
10	302	0.103	70	7.191
Total	2940	1		86

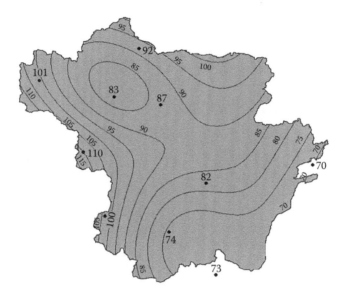

Figure 2.28 Drawing of rainfall isohyetal curves.

Table 2.27 Procedure of the isohyetal curve method

Interval (mm)	Area (km²)	Percentage w	Rainfall P (mm)	w × P
65–70	8	0.003	67.5	0.180
70–75	444	0.151	72.5	10.960
75–80	295	0.100	77.5	7.783
80–85	405	0.138	82.5	11.370
85–90	729	0.248	87.5	21.688
90–95	506	0.172	92.5	15.911
95–100	327	0.111	97.5	10.833
100–105	149	0.051	102.1	5.179
105–110	55	0.019	107.5	2.010
110–115	22	0.007	112.5	0.841
Total	2940	1		87

designer, while in geographical information systems, the commonly used methods are the inverse distance or the nearest neighbour. The procedure for the calculation of the surface rainfall is given in Table 2.27.

The surface rainfall is estimated at 87 mm.

4. Rainfall adaptation to the mean altitude of the basin

The surface rainfall calculated by the method of Thiessen ignores the actual mean altitude of the basin. It only takes into account the altitude of the stations. Therefore, in a basin where most stations have been installed in the lowlands, the Thiessen method would underestimate the true surface rainfall and vice versa. Therefore, a correction of the surface rainfall resulting from the Thiessen method would be required based on the actual mean altitude of the basin, which, in this example, is equal to 532 m. The correction involves assessing the rainfall gradient and the difference between the weighted mean altitudes of the stations and the actual mean altitude of the basin.

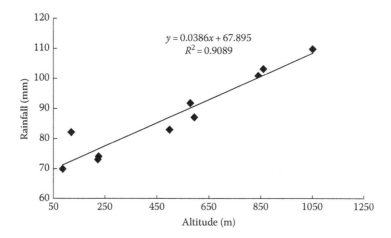

Figure 2.29 Linear regression between the rainfall and the altitude.

The rainfall of the stations and their altitude are linearly correlated to estimate the monthly rainfall gradient, as shown in Figure 2.29. The slope of the line formed ($\lambda = 0.039$) is the monthly rainfall gradient and denotes the increase in the monthly amount of rainfall with altitude. The rainfall gradient is usually expressed in millimetres of rainfall per 100 m of altitude change, i.e. here it is equal to 3.86 mm/100 m.

The weighted mean altitude of the stations may be derived using the Thiessen coefficients which were calculated in the previous step using the procedure presented in Table 2.28.

The adjusted surface rainfall is as follows:

$$\overline{P}_t = P_t + \lambda \cdot \Delta h$$

where
$P_t = 86$ mm, the rainfall which resulted from the Thiessen method
$\lambda = 0.039$ (mm/m), the rainfall gradient
$\Delta h = H_{KB\ surface} - H_{weighted} = 532 - 449 = 83$ m

Table 2.28 Calculation of the weighted altitude of the stations

Station	Thiessen percentage w	Altitude H (m)	w × H
1	0.107	226	24.29
2	0.052	220	11.45
3	0.073	581	42.69
4	0.072	842	60.61
5	0.081	500	40.48
6	0.198	596	118.00
7	0.066	1050	69.30
8	0.060	860	51.78
9	0.187	116	21.70
10	0.103	87	8.94
Total	1		449

The adjusted surface rainfall is

$$\overline{P_t} = 86 + 0.039 \cdot 83 = 89.24 \text{ mm} \cong 89 \text{ mm}$$

The difference in the estimated rainfall is 3 mm in relation to the respective amount, which resulted from the Thiessen polygon method.

2.9.4 Optimum interpolation method (Kriging)

Interpolation is a method of constructing new data points within the range of a discrete set of known data points by minimizing the expected error fluctuations. Kriging is a geostatistical estimator which derives the value of a random field at an unobserved location from point measurements (e.g. rainfall). It is used in the spatial estimation of rainfall in areas without rain gauge measurements. Kriging actually comprises a group of methods named after the mining engineer D. G. Krige who pioneered in their development. Kriging is based on the principle that the value at an unknown point should be the average of the known values of its neighbours, which is weighted by the neighbours' distance to the unknown point. A characteristic of these methods is that an important role plays a function called the variogram (or sometimes the semivariogram) which quantifies the extent to which the difference between values tends to become significant as the time between observations becomes longer, assuming that this does in fact occur. In the case of rain gauges, pairs of gauges are selected from a wider area, taking into account that gauges are positioned in places with similar climate characteristics.

Plotting consists of values of distances between the two stations in the horizontal axis and differences of measurements $D_{ij} = 0.5(y_i - y_j)^2$ in the vertical axis.

It appears to be a tendency as measurements become far apart and the difference between them grows. A plot of this type is called a variogram. The variogram itself is a curve through the data that gives the mean value of D_{ij} as a function of the distance. An experimental or empirical variogram is obtained by smoothing the data to some extent to highlight the trend. A model variogram is obtained by fitting a suitable mathematical function to the data, with a number of standard functions being used for this purpose.

Typically, a model variogram looks like the one in Figure 2.30. Even two points which are close together may tend to have different values, so there is a *nugget effect*; the expected value of $0.5(y_i - y_j)^2$ is greater than zero even with $h = 0$. The maximum height of the curve is called the 'sill'. This is the maximum value of $0.5(y_i - y_j)^2$ which applies for two points which are far apart in the study area. Finally, the *range of influence* is the distance at which two points have independent values. This is defined as the point in which the curve's value is equal to 95% of the difference between the nugget and the sill. There is a number of standard mathematical models for variograms. One is the Gaussian model with the following equation:

$$C(h) = c + (S - c)\left[1 - \exp\left(\frac{-3h^2}{a^2}\right)\right] \tag{2.37}$$

where
   c is the nugget effect
   S is the sill
   a is the range of influence

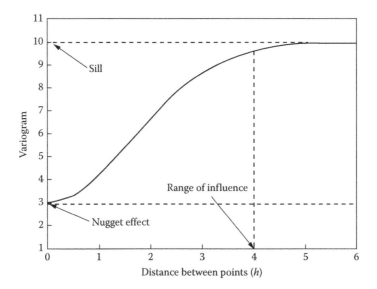

Figure 2.30 A typical variogram showing the sill, the nugget effect and the range of influence.

Here are other models which are often considered:

1. The spherical model

$$Y(h) = \begin{cases} c + (S-c) \ [1.5(h/a) - 0.5(h/a)^3], & h \le a \\ c, & \text{otherwise} \end{cases}$$ (2.38)

2. The exponential model

$$Y(h) = c + (S-c)\left[1 - \exp\left(\frac{-3h}{a}\right)\right]$$ (2.39)

3. The power model

$$Y(h) = c + Ah^w$$ (2.40)

For all of these models, c is the nugget effect. The spherical and the exponential models also have a sill at S, but the power model increases without limit as h increases. In applied hydrology, the models which are usually preferred are the Gaussian and the exponential model, based on X^2 test. Suppose that in the study area, sample values y_1, y_2,..., y_n are known at n locations, and it is desired to estimate the value y_0 at another location. For simplicity, assume that there are no underlying trends in the values of Y. Then, Kriging estimates y_0 using a linear combination of the known values:

$$Y_0 = \sum a_i y_i$$ (2.41)

where the weights a_1, a_2,..., a_n for these known values are selected so that the estimator for y_0 is unbiased, with the minimum possible variance for prediction errors. The equations for determining the weights to be used in this equation are somewhat complicated. They are a function of the assumed model of the variogram.

Some variations of the method are as follows:

Ordinary simple Kriging: This is the most common form and is based on the following assumptions: (1) The variable is normally distributed, (2) the estimate is unbiased, and (3) there is a continuity of the second degree, with (d_1) when the local mean is known (simple) and (d_2) when the local mean is unknown (ordinary). The use of ordinary Kriging for rainfall estimation is limited due to particular assumptions. Rainfall is not normally distributed due to intermittence (large numbers of zero rainfall); the estimated weights and variances are also independent of the data values. These assumptions may cause Kriging to overestimate in the no-rain and low rainfall situations and to underestimate in high rainfall situations.

Indicator Kriging: The indicator approach is one overcoming both these limitations and therefore of obtaining better estimates of rain areas. An indicator function is a binary variable representing zero and non-zero rainfall amounts. So, the departure from normality in rainfall should be greatly reduced when rain and no-rain intervals are separated. Also, the problem of data independence of the Kriging variance is reduced since indicator Kriging depends on the data values.

Neighbourhood Kriging: If the local mean and variance are constant throughout the region (cases of permanency and isotropic field), in most applications, the data contain variations on a local scale. For this reason, the nearest points or those contained in the surrounding area are involved in the estimation of the value in an unknown area.

Universal Kriging: This is applied when the data contain trends.

Cokriging: The estimation with normal Kriging is improved significantly when the variable being tested is connected with another variable for which there are measurements.

Examples of the use of the Kriging method are shown in Figures 2.31 and 2.32. Figure 2.31 depicts the piezometric map produced through the Kriging method in ArcView, while Figure 2.32 depicts the mean areal rainfall for western Greece.

2.9.5 Time distributions of rainfall

For design purposes, a matter of great importance is the definition of spatial and time distributions of the design storm, whose importance in hydrologic design will be discussed thoroughly in Chapter 6. The categorization of the spatial distribution of rain is done according to scales, discussed next.

2.9.5.1 Limited-scale phenomena

In this limited scale, mostly convective rains are dominant in cell type (cells of compacted vapours, which are moving with the help of the wind). These cells have an average length from 1 to a few kilometres and mean rain duration from 30 min to 1 h. For a point observer, this type of rain has a short duration of few minutes. These clouds follow a random movement which tends to be independent of the wind's average speed.

2.9.5.2 Medium-scale phenomena

The aforementioned small cells can be gathered in groups and separated from the medium cells with average characteristic dimensions from 8 to 50 km.

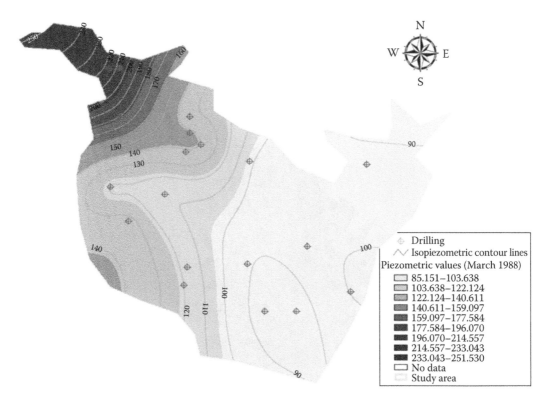

Figure 2.31 The piezometric map for Central Greece by using the ordinary Kriging method.

2.9.5.3 Synoptic-scale phenomena

Storm systems which come together with weather fronts and intense barometric low pressures can cover usually an area of hundreds of kilometres and are big enough to be presented in continental weather maps. These systems are called synoptic-scale phenomena.

Taking for granted the climatologic conditions of an area, it is very likely for phenomena of the same scale to show similar distributions if they are normalized for their duration and size. Based on this fact, a time distribution of rains was conducted and expressed by the percentage cumulative curve. This curve shows the cumulative depth of rain divided into sections, where the percentage of rain duration is plotted on the x-axis and the percentage of total rainfall depth is plotted on the y-axis. The rain is categorized into four standard types with their respective group of curves and according to their respective quartiles (first, second, third or fourth), when the rain height reaches its peak, with an extra parameter which is the level of trust of occurrence for each curve inside a certain quartile. These distributions have been derived after the experimental analysis of many storms (Huff, 1970), and the respective levels of trust show the percentage of rain events which correspond to each curve. In general, convective and frontal rains tend to reach their peak in the beginning of the rain (a first quartile distribution), while stratiform rains usually reach their peak during the third quartile (Laurenson, 1960; Figure 2.33).

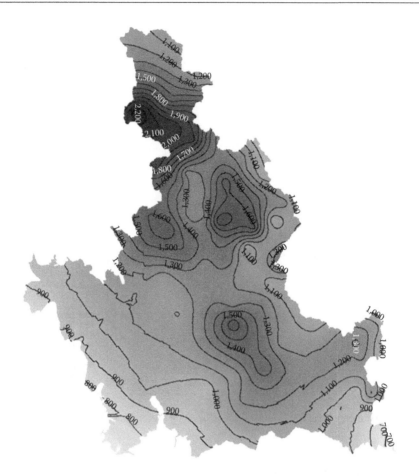

Figure 2.32 Mean annual rainfall for western Greece using the ordinary Kriging method.

2.10 HYDROLOGICAL LOSSES

2.10.1 General

Water from precipitation is delayed or retained in many different ways during its course to a stream. First, many obstacles get in its way, like trees and plants. This is called interception. The excess water fills the ground cavities (detention), while a film of water is also retained in the surface. The latter is known as surface detention. Part of the water returns to the atmosphere by evaporation and plant transpiration, while the rest enters the ground by infiltration. Figure 2.34 shows a qualitative view of the distribution of quantities of water held by interception, retention, infiltration and surface runoff.

Infiltration is an important process in the hydrological cycle because it determines the percentage of water reaching a stream and also the percentage of water entering the ground recharging the aquifers. Infiltration rate has time and spatial variations, depending not only on the physical properties of the soil but also on vegetation and rainfall intensity and time distribution. In the following paragraphs, infiltration analysis and the respective empirical relationships are described.

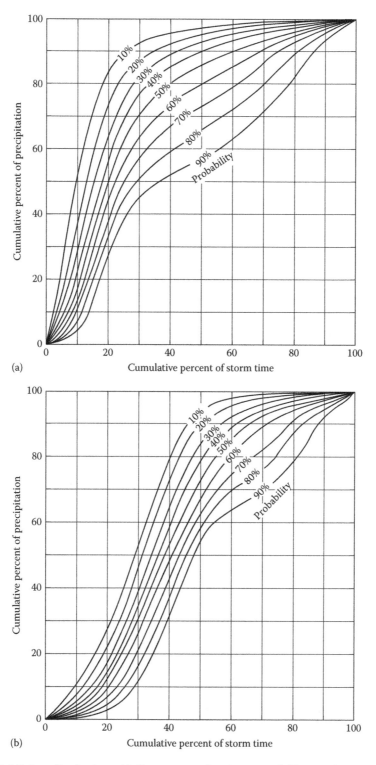

Figure 2.33 Rainfall time distributions. (a) First quarter distribution and (b) second quarter distribution.

(*Continued*)

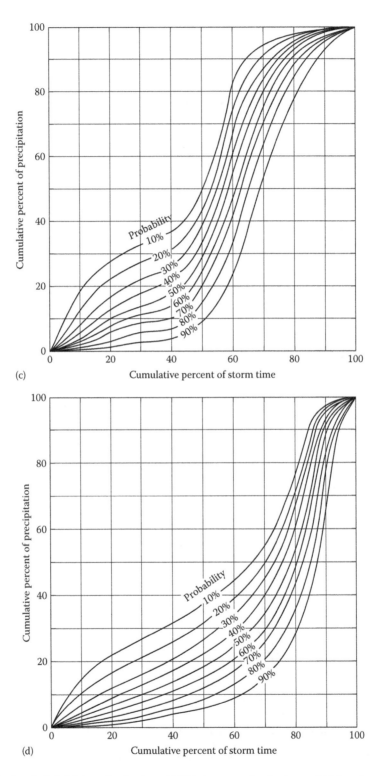

Figure 2.33 (Continued) Rainfall time distributions. (c) Third quarter distribution and (d) fourth quarter distribution. (From Huff, F.A., Water Resour. Res., 6, 447, 1970.)

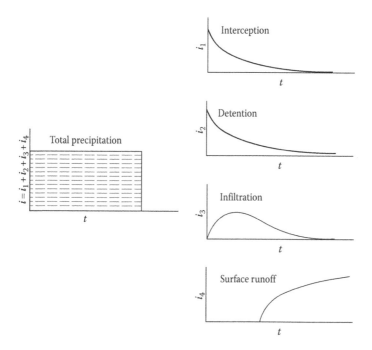

Figure 2.34 The distribution of interception, retention, infiltration and surface runoff.

2.11 EVAPORATION

Evaporation is the process of water transportation from the Earth and water surfaces to the atmosphere in the form of water vapour. This process is mainly dependent on solar radiation, temperature, water vapour pressure, wind speed and type of the evaporation surface.

Solar radiation is a major factor affecting evaporation, because it supplies the energy required for the change in the water phase. Moreover, for constant water temperature, evaporation depends on (i) the wind speed and (ii) the difference between the saturation pressure of water vapour at the water temperature and the partial pressure of water vapour of the overlaying air. All these factors are interdependent, and thus, the effect of each cannot be estimated accurately.

Evaporation, as it depends on solar radiation, varies with latitude, season, altitude, time of day and cloud cover. The evaporation of water from a surface also depends on the available amount of water. Therefore, the potential for evaporation is unlimited from water surfaces, while it varies at soil surfaces, where it ranges from unrestricted when the ground is saturated to nearly zero for dry soil.

Evaporation is usually measured as the mass of water per unit area and time. Alternatively, it is measured as the equivalent water depth in millimetres at a certain point of time.

2.11.1 Water balance methods

The water balance for the estimation of evaporation in a watershed can be written as follows (Sakkas, 1985):

$$E = I + P - O - Os + \Delta s \qquad (2.42)$$

where

E is the evaporation
I is the input
P is the precipitation
O is the outflow
Os is the infiltration
Δs is the change in storage

The infiltration Os can be measured or calculated directly, and the degree of accuracy of this quantity affects the accuracy of measurement of the actual amount of evaporation. The inflow, the outflow, the precipitation and the change in storage can be measured with sufficient accuracy. This method, used for assessing long-term evaporation, can be used as a reference for comparing the efficiency of other methods.

2.11.1.1 Thornthwaite's water balance method

Thornthwaite developed a water balance method for the estimation of available water content in soil, which is described by the following equations:

$$A_n = \min[(A_{n-1} + P_n - AET_n), AWC]$$

$$\text{If} \quad P_n \geq PET_n \quad E_n = PET_n \quad \text{and}$$

$$RO_n = \max(A_{n-1} + P_n - PET_n - AWC, 0) \tag{2.43}$$

$$\text{If} \quad P_n \leq PET_n \quad E_n = P_n + [A_{n-1}(1 - e^{(P_n - PET_n)/AWC})]$$

$$\text{and} \quad RO_n = 0$$

where

A_n and A_{n-1} are the available water contents in the soil for the time intervals n and $n-1$, respectively
P_n is the precipitation
AET_n is the actual evapotranspiration
PET_n is the potential evapotranspiration
AWC is the water capacity of the soil
RO_n is the runoff

This method's prerequisite is the start of calculation at the beginning of the wet season and that an iterative approach be used to correct for the first cycle. The disadvantages of the method are the use of average values in monthly intervals, the assumption that all rainfall is evaporated and the soil moisture drying curve being approximated by a simplistic exponential relation.

2.11.2 Energy balance methods

The total net radiant energy at the Earth's surface is

$$R_n = S_n - L_n \tag{2.44}$$

This quantity of energy is converted into latent heat Λ, i.e. energy spent for the evaporation of water, and sensible heat, H, i.e. the quantity of energy that is induced by the water body to the atmosphere. Neglecting the smaller energy losses (e.g. conduction to the ground, biochemical processes and temporary storage), the equation (Koutsoyiannis and Xanthopoulos, 1997) can be approximately stated as

$$R_n = H + \Lambda \tag{2.45}$$

The most essential parameter in energy processes is the latent heat Λ, since it provides the energy required for evaporation. The Bowen ratio connects latent heat with sensible heat according to the following equation:

$$B = \frac{H}{\Lambda} \tag{2.46}$$

Combining these equations, the evaporation can be calculated as the ratio of the latent heat Λ to the latent heat for water vaporization λ:

$$E_t = -\frac{dh}{dt} = \frac{\Lambda}{\lambda} = \frac{R_n}{\lambda(1+B)} \ [\text{kg/(m}^2 \ \text{d)}] \tag{2.47}$$

where
   $E_t = dh/dt$ is the evaporation in kg of water per day and m² of area
   λ is the latent heat for water vaporization (kJ/kg)
   R_n is the net radiation to the mass of water (kJ/m² d)
   B is the Bowen ratio [dimensionless parameter]

2.11.3 Mass transfer methods

These methods consider evaporation as mass transfer or diffusion of water vapour from higher to lower concentrations. As a diffusion process, it can be described by the Fick diffusion law, which in the case of evaporation takes the following form:

$$E_a = -M\frac{de}{dz} \tag{2.48}$$

where
   E_a is the evaporation rate
   e is the partial pressure of water vapour
   M is the transfer coefficient
   z is the altitude

Based on this equation, complex equations have been developed, which calculate the evaporation rate, using the logarithmic profile of wind speed near the surface, which depends on the roughness of the ground. In practice, the empirically determined relationship of Dalton has been used since 1802. According to this relationship, the evaporation rate is given by the following equation:

$$E_a = (e_s - e)F(u) = D \cdot F(u) \ [\text{kg/(m}^2 \ \text{d)}] \tag{2.49}$$

where

$D = e_s-e$ (hPa) is the deficit of water vapour
$F(u)$ is the wind speed function

The wind speed function $F(u)$ contains the wind speed and is given by the following linear equation (Penman):

$$F(u) = 0.26(1+0.54u)\,[\text{hPa kg/(m}^2\text{ d)}]$$ (2.50)

where u is the wind speed in (m/s) measured at an altitude of 2 m above the ground. The latter relationship is one of the many approaches formulated for the estimation of the linear relationship of $F(u)$ with the wind speed u.

2.11.4 Combination methods: Penman method

The combination methods for assessing evaporation integrate energy balance with mass transfer methods. The first combinational method for calculating the evaporation from a water surface was given by Penman in 1948. Later, Penman equation was modified by Monteith (1965), in order to compute evapotranspiration from soil surfaces. The meteorological data used by the Penman method are measurements of temperature, relative humidity, relative sunshine and wind speed at a height of 2 m above the ground.

According to Penman, evaporation from a water surface is given by the following equation:

$$E = \frac{\Delta}{\Delta+\gamma}\frac{R_n}{\lambda} + \frac{\gamma}{\Delta+\gamma}F(u)D \quad (\text{kg/m}^2\text{ d})$$ (2.51)

where

- γ is the psychrometric coefficient (hPa/°C)
- Δ is the slope of the saturation water vapour curve (hPa/°C) given by the relationship
  $\Delta = \dfrac{4098e_s}{(T+237.3)^2}$ (hPa/°C), where T is the mean air temperature (°C) and e_s is the saturated vapour pressure (hPa) estimated from $e_s = 6.11\exp\left(\dfrac{17.27T}{T+237.3}\right)$ (hPa)
- R_n is the net total radiant energy (kJ/m² day)
- $R_n = S_n-L_n$ (kJ/m²/day), where S_n is the short-wave net radiation and L_n is the long-wave net radiation. The short-wave net radiation can be described by
  $S_n = (1-a)S_0\left(0.29\cos\varphi+0.55\dfrac{n}{N}\right)$ (kJ/m²/day), where S_0 is the solar radiation reaching the atmosphere ($S_0 = 40731.0$ kJ/m²/day), φ the latitude of the site, n the mean daily sunshine (h), n = total monthly sunshine (h/days of the month) and N the mean day duration (h)
- L_n, the long-wave net radiation, is given by the expression
  $L_n = \sigma\cdot T_k^4\left(0.9\dfrac{n}{N}+0.1\right)(0.56-0.09\sqrt{e_sU/100})$ (W/m²), where σ is Stefan–Boltzmann constant ($\sigma = 4.9\times10^{-6}$ [kJ/m²/K⁴/d]), T_K is the mean temperature in K ($T_K = T + 273$) and U is the mean relative humidity (%)
- λ is the latent heat of evaporation (kJ/kg), where $\lambda = 2501 - 2.361\cdot T$ (kJ/kg) and T in °C
- $F(u)$ is the function of wind speed (kg/hPa m² d) where $F(u)$ is given by Equation 2.50
- D is the water vapour saturation deficit (hPa), $D = e_s - e_sU/100$ (hPa)

The ratios $\Delta/(\Delta + \gamma)$ and $\gamma/(\Delta + \gamma)$ function as weighing coefficients for the participation of the energy term and the mass transfer term, respectively, in the evaporation from water surface and aggregate one to the unit.

Using the units reported in the earlier equations, evaporation is given in kg of water per square meter and day. Dividing by the water density (kg/m^3), the evaporation is given in m/day. Usually, the final results are given in mm/day or mm/month.

2.12 EVAPOTRANSPIRATION

Evapotranspiration includes both transpiration from vegetation and evaporation from water surfaces, soil, snow, ice and vegetation. There are several methods for assessing evapotranspiration, some of which are accurate and reliable and others provide simple approaches. The selection of the method depends mainly on the type of the surface and the purpose of the study. Consequently, the selection of the method determines the scales of time and space and the precision requirements. Other equally important factors are the cost and technique of each method.

2.12.1 Water balance methods

These methods are related to the water balance equation and contribute directly or indirectly to the determination of the evapotranspiration. In most studies based on the water balance equation, evapotranspiration is calculated as the residual term of the equation while the other components are either measured or calculated. If there is no irrigation, evapotranspiration is given by the equation

$$ET = P + \Delta SW \pm RO - D \tag{2.52}$$

where

P is the precipitation
ΔSW is the change of the water content of the soil
RO is the surface runoff
D is the deep infiltration

This equation can be applied at any scale.

The disadvantages of this method are the low-level accuracy of the measurements and the difficulties in assessing the ET during periods of rain. Even with the use of careful measurements, it is difficult to detect changes in soil water with precision greater than 2 mm. When this method is applied over large areas, the main problem is the lack of good average spatial values of its components, due to the variability of rainfall over large areas and the diversity of topography and the soils beneath the area. The errors associated with this method make its use insufficient on a daily basis.

2.12.2 Methods for the determination of potential evapotranspiration from climatic data

2.12.2.1 Penman–Monteith method

The Penman method assumes that water vapour is at saturation at the water surface. Therefore, it cannot be used in the case of transpiration from plant surfaces. To address this weakness, Monteith introduced the surface resistance of foliage in evaporation.

The Penman–Monteith method is suitable for the estimation of evapotranspiration from soil surfaces and is given by the following equation:

$$E' = \frac{\Delta}{\Delta + Y'} \frac{R_n}{\lambda} + \frac{Y'}{\Delta + Y'} F(u)D \quad [\text{kg}/(\text{m}^2 \text{ d})]$$
(2.53)

where

$$Y' = (1 + 0.33u_2)\gamma \quad [\text{hPa}/^\circ\text{C}]$$

This is the modified term of psychrometric coefficient which takes into account the resistance of foliage. Wind speed u is given in (m/s):

$$F(u) = \frac{90}{T + 275} u_2 \quad (\text{kg}/[\text{hPa}/\text{m}^2/\text{d}])$$

This is the modified wind speed equation with T in (°C) and u in (m/s). The rest of the variables are calculated just as in the Penman method, except of the reflectance coefficient a (albedo), which is modified, taking greater values, thus changing the net radiation R_n.

2.12.2.2 Thornthwaite's method

Thornthwaite (1948), except of the estimation of available water content in soil described earlier in this chapter, created an equation which can be used for the estimation of monthly evapotranspiration on limited water availability based on the average monthly temperature, which has the following form:

$$E_p = 16\left(\frac{10t_i}{J}\right)^a \frac{\mu N}{360}$$
(2.54)

where
E_p is the potential evapotranspiration in mm/month
t_i is the average monthly temperature in °C
μ is the number of days
N is the mean astronomical duration of the day
J is the annual temperature indicator
a is an empirical parameter which depends on J ($a = 0.016 \cdot J + 0.5$)

The temperature indicator, J, is given by the following equation (Koutsoyiannis and Xanthopoulos, 1997):

$$J = \sum_{i=1}^{12} j_i$$
(2.55)

The monthly temperature indicator j_i is a function of the average monthly temperature, according to the following equation:

$$j_i = 0.09t_i^{3/2}$$
(2.56)

The monthly percentages (%) of day hours with respect to the day hours of the year are given in Table 2.29.

Table 2.29 Monthly percentages of day hours to the day hours of the year

N. latitude	January	February	March	April	May	June	July	August	September	October	November	December
							Month					
24	7.58	7.17	8.40	8.60	9.30	9.20	9.41	9.05	8.31	8.09	7.43	7.46
26	7.49	7.12	8.40	8.64	9.38	9.49	9.10	8.31	8.06	9.30	7.36	7.35
28	7.40	7.07	8.39	8.68	9.46	9.38	9.58	9.16	8.32	8.02	7.27	7.27
30	7.30	7.03	8.38	8.72	9.53	9.49	9.67	9.22	8.34	7.99	7.19	7.14
32	7.20	6.97	8.37	8.75	9.63	9.60	9.77	9.28	8.34	7.93	9.11	7.05
34	7.10	6.91	8.36	8.80	9.72	9.70	9.88	9.33	8.36	7.90	7.02	6.92
36	6.99	6.86	8.35	8.85	9.81	9.83	9.99	9.40	8.36	7.85	6.92	6.79
38	6.87	6.79	8.34	8.90	9.92	9.95	10.10	9.47	8.38	7.90	6.82	6.66
40	6.76	6.73	8.33	8.95	10.02	10.08	10.22	9.54	8.38	7.75	6.72	6.52
42	6.62	6.65	8.31	9.00	10.14	10.21	10.35	9.62	8.40	7.70	6.62	6.38
44	6.40	6.58	8.30	9.05	10.26	10.38	10.49	9.70	8.41	7.63	6.49	6.22
46	6.33	6.50	8.29	9.12	10.39	10.54	10.64	9.79	8.42	7.58	6.36	6.04
48	6.17	6.42	8.27	9.18	10.53	10.71	10.80	9.89	8.44	7.51	6.22	5.86
50	5.98	6.32	8.25	9.25	10.69	10.93	10.99	10.00	8.44	7.43	6.07	5.65

2.12.2.3 Blaney–Criddle method

Blaney and Criddle (1962) developed an empirical formula which connects evapotranspiration with mean air temperature and mean percentage of daylight hours per day. Evapotranspiration directly depends on the sum of products of the average monthly temperatures and the percentage of hours of daylight in a month, in an actively growing crop with sufficient soil moisture, according to the following relationship:

$$ET = kF = k\frac{(1.82 + 32)p}{3.94} \tag{2.57}$$

where
  ET is the monthly potential evapotranspiration in mm
  k is an empirical factor, referring to a particular crop (crop factor)
  T is the average monthly air temperature in °C
  p is the percentage of daylight hours per month

p is calculated as follows:

$$p = 100\frac{N \cdot \mu}{365 \cdot 12} \tag{2.58}$$

where
  N is the average astronomical duration of the day in hours
  μ is the number of days of every month

Values for k are given in Table 2.30 for a variety of crops, as estimated by Blaney and Criddle. This method was initially designed for the estimation of the seasonal water needs during the growing period of each crop. In Table 2.31, the values of monthly k are given for different crops. From the comparison of Tables 2.30 and 2.31, one can conclude that there are considerable differences between seasonal and monthly crop factors k. This is due to the difference in the development of the root system and the aboveground part of the plant,

Table 2.30 Values of seasonal crop factor k (Blaney–Criddle method)

Crop	Growing duration in months	k
Clover	Between frost	0.80–0.85
Corn	4	0.75–0.85
Cotton	7	0.60–0.70
Cereals	3	0.75–0.85
Citrus	12	0.45–0.55
Deciduous fruit trees	Between frost	0.60–0.70
Grass meadow	Between frost	0.75–0.85
Potato	3–5	0.65–0.75
Rise	3–5	1.00–1.10
Sugar beet	6	0.65–0.75
Tomato	4	0.65–0.70
Vegetables	2–4	0.60–0.70

Table 2.31 Monthly crop factors of irrigated crops (Blaney–Criddle method)

Crop	Area	J	F	M	A	M	J	J	A	S	O	N	D
Clover	Warm, lowland	0.35	0.45	0.60	0.70	0.85	0.95	1.00	1.00	0.95	0.80	0.55	0.30
//	Coastal highland	—	—	—	0.37	0.56	0.75	0.92	1.00	1.03	0.98	0.82	—
//	Inland lowland	—	—	0.57	0.78	0.93	1.02	1.01	0.95	0.84	0.63	0.42	—
Avocado	Inland lowland	0.15	0.25	0.40	0.52	0.63	0.73	0.75	0.69	0.60	0.48	0.32	0.19
Corn	Inland lowland	—	—	—	—	0.12	0.40	0.60	0.62	0.45	—	—	—
Cotton	Inland lowland	—	—	—	—	0.30	0.45	0.90	1.00	1.00	—	—	—
Grass	Coastal	0.24	0.38	0.55	0.70	0.88	0.92	0.94	0.92	080	0.72	0.54	0.35
Melons	//	—	—	—	—	—	0.45	0.70	0.74	0.64	—	—	—
Fruit trees	//	—	—	0.23	0.45	0.70	0.85	0.88	0.85	0.47	0.20	—	—
Deciduous lemon	//	—	—	0.40	0.40	0.50	0.55	0.60	0.60	0.60	0.50	0.40	—
Orange	//	0.27	0.34	0.40	0.46	0.50	0.53	0.54	0.54	0.52	0.48	0.43	0.30
//	Mid	0.33	0.39	0.45	0.50	0.54	0.56	0.57	0.57	0.56	0.53	0.47	0.38
//	Inland	0.37	0.44	0.49	0.54	0.57	0.60	0.62	0.62	0.60	0.57	0.51	0.43
Nuts	Inland	—	—	0.13	0.30	0.55	0.84	0.98	0.88	0.60	0.37	0.20	—
Meadow	Inland	—	—	0.10	0.27	0.42	0.52	0.57	0.55	0.35	0.15	—	—
//	Inland, highland	—	—	0.16	0.45	0.65	0.75	0.78	0.74	0.55	0.20	—	—
Potato	Inland	—	—	—	0.45	0.80	0.95	0.90	—	—	—	—	—
Millet	Inland, dry	—	—	—	—	—	0.40	1.00	0.85	0.70	—	—	—
Sugar beet	Inland	—	—	—	0.31	0.69	0.96	1.01	0.83	—	—	—	—
//	Mid	—	—	—	—	0.40	0.67	0.76	0.70	0.50	0.29	—	—
//	Coastal	—	—	—	0.37	0.42	0.43	0.44	0.43	0.38	—	—	—
Barley	Inland, dry	0.32	0.60	0.98	1.08	0.45	—	—	—	—	—	—	0.15
Wheat	Inland, dry	0.20	0.40	0.80	1.10	0.60	—	—	—	—	—	—	—
Tomato	Inland	—	—	—	—	0.41	0.74	0.93	0.98	0.89	—	—	—
Vegetables	Coastal	—	—	—	0.23	0.49	0.67	0.78	0.78	0.64	0.40	—	—

Note: The symbol "//" means the same as above.

depending on the growing phase. For this reason, for the estimation of the monthly water needs, the use of the monthly crop factors is suggested.

Remark: Smaller values of k are applied in coastal areas and larger values are applied in dry climates.

2.12.2.4 Jensen–Haise method

By analyzing 3000 measurements of evapotranspiration collected by a certain procedure for a period of 35 years, Jensen and Haise (1963) developed the following equation:

$$ET = CT(T - Tx)R_s \tag{2.59}$$

where
   R_s is the solar radiation in ly/day
   CT is a temperature constant equal to 0.025
   Tx is the x ordinate of the temperature curve in zero value, equal to 3 if temperature is given in °C

These coefficients are constant for a certain area. CT is given by the following equation:

$$C_T = \frac{1}{C_1 + C_2 C_H}, \quad C_H = \frac{50mb}{e_2 - e_1} \tag{2.60}$$

In these relationships, e_2 and e_1 are, respectively, the saturation point of vapour in the maximum and minimum temperature during the warmest month of the year, with $C_2 = 7.6$°C. C_1 is defined as follows:

$$C_1 = 38 - (2°C \times \text{altitude in m}/305) \tag{2.61}$$

and

$$T_x = -2.5 - 0.14 \, (e_2 - e_1)°C/mb - (\text{altitude in m}/550)$$

2.12.2.5 Makkink's method

Makkink's method (1957) is based on the theory that much of the energy consumed for evapotranspiration almost entirely comes from two sources: radiation energy and energy of air which is warmer than the surface. These two energy sources are actually transformed into solar energy. So, evapotranspiration is correlated with solar radiation and moreover is dependent intensely on short-wave radiation in a linear manner. The dependence of evapotranspiration on radiation is not constant throughout the year as the climate and surface conditions change. It must be noted that in dry areas, the energy transfer on the horizontal level and its downward transfer form an important percentage of the respective evapotranspiration in subtropical and semiarid areas. In these cases, even though the transferred heat comes from the sun, it leads to a non-linear correlation of ET_p and short-wave radiation R_s. Many of the methods based on solar radiation have embedded a term dependent on temperature. For this reason, Makkink (1957) has suggested the following relationship for the estimation of ETp (mm/day) from solar radiation measurements:

$$ET_p = R_S\left(\frac{\Delta}{\Delta+\gamma}\right)+0.12 \tag{2.62}$$

where R_s is converted into equal units of vaporized water. Makkink's relationship has offered satisfying results in dry areas. According to the modified form for grass, the equation of ET_0 in mm/day is

$$ET_0 = C(W - R_s) \tag{2.63}$$

where
   R_s is the solar radiation in equivalent evaporation in mm/day
   W is a coefficient which depends on temperature and geographical latitude of the area
   C is a factor which depends on average humidity and wind conditions only during daytime

2.12.2.6 Hargreaves method

Hargreaves (1974) developed a method for ET_p estimation which is simple and needs minimum climate data. This method can be expressed in the form of an equation:

$$ET_p = MF(18Ta + 32)CH \tag{2.64}$$

where
   ET_p is estimated in mm/month
   MF is a monthly factor dependent on geographical latitude and given in tables
   Ta is the average monthly temperature (in °C)
   CH is a correction factor for the relative humidity RH which applies only when the
      average relevant humidity exceeds 64%

CH is estimated by the following formula:

$$CH = 0.166(100 - RH)^{1/2} \tag{2.65}$$

For mean relative humidity less than 64%, CH is equal to unit. Hargreaves compared his method with data collected by lysimeters in various areas around the world and developed equations (from the various linear regressions) connecting actual ET with ET_p for different climatic conditions.

2.12.2.7 Priestley–Taylor method

The Priestley–Taylor model (Priestley and Taylor, 1972) is a modification of Penman's theoretical equation (University of Waterloo, 2001). Used in areas of low moisture stress, the two equations have produced estimates which differ approximately by 5% (Shuttleworth and Calder, 1979). An empirical approximation of the Penman combination equation is made by the Priestley–Taylor model to eliminate the need for input data other than radiation. The adequacy of the assumptions made in the Priestley–Taylor equation has been validated by a review of 30 water balance studies in which it was commonly found that in vegetated areas with no water deficit or very small deficits, approximately 95% of the annual evaporative demand was supplied by radiation (Stagnitti et al., 1989).

The reason is that under ideal conditions, evapotranspiration would eventually reach a rate of equilibrium for an air mass moving across a vegetation layer with an abundant supply of water. The air mass would become saturated and the actual rate of evapotranspiration (AET) would be equal to the Penman rate of potential evapotranspiration. Under these conditions, evapotranspiration is referred to as equilibrium potential evapotranspiration (PETeq). The mass transfer term in the Penman combination equation approaches zero and the radiation terms dominate. Priestley and Taylor (1972) found that the AET from well-watered vegetation was generally higher than the equilibrium potential rate and could be estimated by multiplying the PETeq by a factor (α) equal to 1.26:

$$PET = \alpha \frac{s(T_a)}{s(T_a) + \gamma}(K_n + L_n)\frac{1}{\rho_w \lambda_v}$$ (2.66)

where
   K_n is the short-wave radiation
   L_n is the long-wave radiation
   $s(T_a)$ is the slope of the saturation vapour pressure versus temperature curve
   γ is the psychrometric constant
   ρ_w is the mass density of water
   λ_v is the latent heat of vaporization

Although the value of α may vary throughout the day (Munro, 1979), there is a general agreement that a daily average value of 1.26 is applicable for humid climates (Stewart and Rouse, 1976; De Bruin and Keijman, 1979; Shuttleworth and Calder, 1979) and temperate hardwood swamps (Munro, 1979). Morton (1983) notes that the value of 1.26, estimated by Priestley and Taylor, was estimated using data from both moist vegetation and water surfaces. Morton has recommended that the value should be increased slightly to 1.32 for estimates from vegetated areas as a result of the increase in surface roughness (Brutsaert and Stricker, 1979; Morton, 1983). Generally, the coefficient α for an expansive saturated surface is usually greater than 1.0. This means that the true equilibrium potential evapotranspiration rarely occurs; there is always some component of advection energy which increases the actual evapotranspiration. Higher values of α, ranging up to 1.74, have been recommended for estimating potential evapotranspiration in more arid regions (ASCE, 1990).

The α coefficient may also have a seasonal variation (De Bruin and Keijman, 1979), depending on the climate being modelled. The study by DeBruin and Keijman (1979) indicated a variation in α with minimum values occurring during the mid-summer when radiation inputs were at their peak and maximum during the spring and autumn (winter values were not determined), but in comparison to advective effects, radiation inputs were large. The equation was successfully applied not only for open water bodies but also for vegetated regions. The satisfactory performance of the equation is probably due to the fact that the incoming solar radiation has some influence on both the physiological and the meteorological controls of evapotranspiration. A value of 1.26 has been used for α throughout. Temporal variations in α as suggested by researchers are emulated by the conversion factors used in the calculation of AET from the PET which is described in the following.

Estimates of PET using the Priestley–Taylor equation have been adjusted as a function of the difference in albedo at the site where measurements of radiation have been made (albe) and the land classes with differing albedo (alb). In the adjustment, it is assumed that the ground heat flux (which should be included in the net total radiation data if it is available) contributes 5% of the overall energy. The remaining 95% of the potential evapotranspiration estimate is scaled as a function of the difference in albedo:

$$PET = 0.05PET + 0.95PET \frac{1 - alb}{1 - albe}$$ (2.67)

Example 2.8

In a close distance to a natural lake, a meteorological station is installed (the station is sited in latitude 38°N). Based on the measurements of a year, the monthly values of four variables are estimated according to Figure 2.35.

The following are asked:

1. Estimate the evaporation of the lake according to the Penman method for July. The albedo of the wet surface is estimated at $a = 0.05$. Also given are $\gamma = 0.67$ hPa/°C, psychrometric constant; $\sigma = 4.9 \times 10^{-6}$ kJ/m^2/K^4/d, Stefan–Boltzmann constant; and $\rho = 1000$ kg/m^3, the density of water. July has 14.4 h mean day duration for latitude 38°N, and the solar radiation which reaches the atmosphere from the outer space has the value $S_0 = 40,731.0$ kJ/m^2/d.
2. Estimate the potential evapotranspiration on the soil using the Penman–Monteith method for July. Soil albedo is estimated at $a = 0.25$.
3. In the particular area, an irrigation network is about to be developed with a water consumption coefficient of $K_c = 0.80$. Estimate the potential evapotranspiration of this crop using the Blaney–Criddle method.
4. Comment on the differences between the results of the various methods.

Solution

1. The latitude of $\varphi = 38°$ has $\cos\varphi = 0.79$. The table in Figure 2.36 is filled using Equation 2.51 and the subsequent analytical expressions of each term. The light coloured cells are filled with given values, while the dark ones are estimated

Month	Temperature T (°C)	Relative humidity U (%)	Wind speed u_2 at 2 m (m/s)	Monthly sunshine duration (h)
October	17.6	60.4	1.9	169.1
November	11.7	66.4	1.9	125.2
December	8.4	67.9	2.3	117.7
January	6.3	66.6	2.5	119.8
February	7.6	64.4	2.6	111.2
March	11.4	62.2	2.6	153.1
April	15.3	57.9	2.8	208.4
May	20.9	54.2	2.4	262.3
June	26.1	48.3	2.7	315.2
July	28.9	44.8	2.5	338.2
August	27.9	47.2	2.6	305.2
September	23.5	51.4	2.4	240.2

Figure 2.35 Meteorological data.

Month	July
Days	31
Temperature (°C)	28.9
Relative humidity U (%)	44.8
Wind speed u_2 (m/s)	2.5
Monthly sunshine (h)	338.2
Y (hPa/°C)	0.67
T_K (K)	301.9
e_s (hPa)	39.84
Δ (hPa/°C)	2.30
N (h)	14.4
n (h)	10.91
S_0 (kJ/m²/d)	40731.0
σ (kJ/m²/K⁴/d)	4.9×10^{-6}
Albedo a	0.05
ρ (kg/m³)	1000
$\cos \phi$	0.79
S_n (kJ/m²/d)	24966.13
L_n (kJ/m²/d)	5721.57
R_n (kJ/m²/d)	19244.56
$F(u)$ (kg/hPa/m²/d)	0.61
D (hPa)	21.99

Figure 2.36 Penman method.

(in a logical order from top to bottom). Evaporation in *mm/d* units is derived by multiplication with the density of water, while in *mm/month*, it is obtained by multiplication with the days of the month.

2. The potential evapotranspiration on the soil is estimated by the modified Equation 2.53 and the following expressions of γ' and $F'(u)$. The results are presented in Figure 2.37, where the values which are different from section (1) of the example are given in dark gray.

3. In this section, Equations 2.57 and 2.58 of the Blaney–Criddle method are used, and the results are presented in Figure 2.38.

4. Although temperature-based methods like Blaney–Criddle are useful when data of other meteorological parameters are unavailable, the estimates produced are generally less reliable than those which take other climatic factors into account like the Penman and Penman–Monteith methods.

2.12.3 Methods for the determination of actual evapotranspiration

For the determination of actual evapotranspiration for a relatively small duration, the soil–plant–atmosphere system is taken into account. For long time periods, the average actual evapotranspiration can be evaluated with sufficient accuracy based on the mean values of rainfall and air temperature. The average actual evapotranspiration of an area depends on many factors like total rainfall, rainfall distribution, plant cover, geological features and meteorological conditions affecting evaporation and transpiration. In an area which is geologically and climatically homogeneous, the annual actual evapotranspiration can be expressed as a function of annual rainfall and annual temperature. The Turc and Coutagne methods were developed on this principle (Sakkas, 1985).

Month	July
Days	31
Temperature (°C)	28.9
Relative humidity U (%)	44.8
Wind speed u_2 (m/s)	2.5
Monthly sunshine (h)	338.2
Y' (hPa/°C)	**1.22**
T_K (K)	301.9
e_s (hPa)	39.84
Δ (hPa/°C)	2.30
N (h)	14.4
n (h)	10.91
S_0 (kJ/m²/d)	4,0731.0
σ (kJ/m²/K⁴/d)	4.9×10^{-6}
Albedo a	0.05
ρ (kg/m³)	1,000
$\cos \phi$	0.79
S_n (kJ/m²/d)	**1,9710.10**
L_n (kJ/m²/d)	5,721.57
R_n (kJ/m²/d)	**13,987.35**
$F(u)$ (kg/hPa/m²/d)	**0.74**
D (hPa)	21.99
λ (kJ/kg)	2,432.77
E (kg/m²/d)	**9.40**
E (mm/d)	**9.40**
E (mm/month)	**291.42**

Figure 2.37 Penman–Monteith method.

Month	July
Days	31
Temperature (°C)	28.9
K_c	0.8
N (h)	14.4
P (hours percentage/days on the month/days of the year)	10.19
E (mm/month)	173.87

Figure 2.38 Blaney–Criddle method.

2.12.3.1 Turc method

Turc (1961) analyzed data collected from 254 watersheds, sited worldwide; he correlated evaporation with rainfall and temperature, according to the following relationship:

$$E = \frac{P}{\left[0.9 + (P/I_T)^2 \right]^{0.5}} \tag{2.68}$$

where
   E is the annual evaporation or evapotranspiration in mm
   P is the annual rainfall in mm, I_T or $L = 300 + 25T + 0.05T_2$
   T is the average annual temperature in °C

Turc has also formulated another equation which embodies the effect of soil humidity variance as follows:

$$E = \frac{P + E_{10} + K}{\left[I + \left(\frac{P + E}{I_T} + \frac{K}{2I_T} \right) \right]^{0.5}} \tag{2.69}$$

where
   E is the evaporation in mm for a period of 10 days
   E_{10} is the estimated evaporation (for a period of 10 days) from plain soil, assuming that there is no precipitation or that it is less than 10 mm
   K is a crop factor given by

$$K = 25 \ (cM/G) \ 0.5 \tag{2.70}$$

where 100 M is the final performance of dried mass in kg/ha, 10 G is the value of growing period in days and c is a factor for the crop. IT is the ability of air evaporation:

$$IT = (T + 2) \ Rs \ 0.5/16 \tag{2.71}$$

In this equation, T is the average air temperature for the 10-day period in °C and R_s is the incident solar radiation in cal/cm²·day.

   For the climatic conditions of western Europe, Turc estimated the evapotranspiration in mm/day for 10-day periods as follows:

$$ET = 0.013 \frac{T}{T + 15} (R_s + 50) \tag{2.72}$$

for relative humidity (RH) > 50% and

$$ET = 0.013 \frac{T}{T + 15} (R_s + 50) \left(1 + \frac{50 - RH}{70} \right) \tag{2.73}$$

for relative humidity (RH) < 50%, with T the average temperature in °C and R_s the solar radiation in ly/day.

2.12.3.2 Coutagne method

The Coutagne method is based on the same assumptions of the Turc method for the estimation of the average annual value of the actual evapotranspiration (Sakkas, 1985). The empirical formula given by Coutagne is

$$E = P \left(1 - \frac{P}{I} \right) \tag{2.74}$$

where E and P have been defined before and I is a temperature function:

$$I = 800 + 140T \text{ (mm)} \tag{2.75}$$

where T stands for the average annual temperature in °C. These equations are valid only for the depth of precipitation within the limits:

$$\frac{I}{8} \leq P \leq \frac{I}{2} \tag{2.76}$$

If the rainfall depth is lower than $I/8$, then the lack of flow is equal to the rainfall, and there occurs no flow. Therefore,

$$E \equiv P \quad \text{for} \left(P \leq \frac{I}{8} \right) \tag{2.77}$$

If the rainfall depth is greater than $I/2$, then the lack of flow is independent of the rain and it can be determined by the following equation:

$$E = \frac{I}{4} = 200 + 35T \tag{2.78}$$

The results of the equations using both the Turc and Coutagne methods are given by preconstructed charts to demonstrate the simplicity of the procedure.

2.13 INFILTRATION RATE ESTIMATION

2.13.1 Infiltration

Infiltration is a process of water movement beneath the surface of the ground. Although it is different from percolation, very often, these two processes coincide.

In the first stage, water accumulates on the ground and then moves downwards into the soil, mostly by gravity, and also due to molecular forces and capillary potential.

Capillary potential is caused by the presence of infinite pipelines in the ground in different directions. In an unsaturated ground, these pipelines have little or no quantity of water, and the capillary potential is great. On the contrary, in moist ground, the capillary potential is low. For this reason, after rain, the infiltration capacity is lower.

At the beginning of a rainfall, the downward direction of the capillary potential and gravity is the same, which helps the water to recharge the aquifers from a few metres to tens of metres below the surface, as described in Chapter 5. Upon ground saturation, the capillary potential diminishes. The methods of infiltration estimation (Horton, etc.) are discussed in this section, and the known φ-index method is given in Chapter 3.

By the end of the rainfall, the upper ground moisture starts to decrease due to evaporation and transpiration, and the capillary potential reverses and contributes to the movement of water upwards. This ground zone varies from a few centimetres to a few metres above the aquifer, and the hydrostatic pressure in this zone is lower than the atmospheric pressure.

The importance of capillary potential for plant survival is easily understood in intense drought periods. In different cases, water could only exist at great depths in the ground.

The role of infiltration is important in a storm event. When intense rain exceeds the infiltration capacity of the ground, part of the water quantity reaching the ground becomes surface runoff. Also, water sufficiency in the upper ground layers defines evapotranspiration, while water quantity in the aquifer supplies streams during the dry season. Infiltration also plays an important role in seasonal distribution of water supply. If an impermeable layer exists in the ground, infiltration is minimum, but on the other hand, water can stay longer in the ground and sustain base flow in a stream. So, it is critical to determine the infiltration rate. The factors which influence it are the extent and type of plant cover, the saturation and type of ground, the temperature, the rain intensity and water quality. The measurement of infiltration rate is possible through infiltrometers and respective hydrograph analysis.

The infiltration rate estimation varies in complexity: from simple observations of different categories of vegetation to the use of differential equations. The infiltration process is complicated, so many empirical relationships based on saturated flow equations have been proposed.

2.13.2 Horton's model (1930)

Horton's equation is as follows:

$$f_p = f_c + (f_0 - f_c)e^{-kt} \tag{2.79}$$

where
f_p is the infiltration rate (mm or cm/h) in time t
k is the constant depending mostly on soil and vegetation
f_c is the final constant rate of infiltration capacity
f_0 is the initial rate of infiltration capacity

Horton's equation is shown in Figure 2.39 in conjunction with a rain episode. The area under the curve represents the water infiltration capacity of the ground at any time t.

Using this graph, f_0 and k can be evaluated by choosing two pairs of values and solving a 2×2 system of equations with successional approaches.

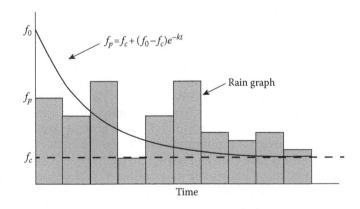

Figure 2.39 Horton's infiltration curve and the rain graph.

2.13.3 Green–Ampt model (1911)

This was proposed originally in 1911 and after various modifications has been found to have immense applications in simulation models like the stormwater management model.

In Figure 2.40, the assumptions of the model are shown: the water moves downwards in a linear front, and the ground is covered with a thin layer of water.

K_s is the factor of infiltration capacity in the saturated zone and under Darcy's law:

$$f_p = \frac{K_s(L + S)}{L} \tag{2.80}$$

where
  L is the distance from the surface
  S is the capillary potential in the waterfront

If F is the accumulated infiltration depth and IMD is the initial lack of moisture, using the equation (2.80),

$$f_p = K_s \left(1 + \frac{S \cdot IMD}{F} \right) \tag{2.81}$$

Considering $f_p = dF/dt$

$$\frac{dF}{dt} = K_s \left(1 + \frac{S \cdot IMD}{F} \right) \tag{2.82}$$

and then by integrating

$$F - S \cdot IMD \cdot \ln \left(\frac{F + S \cdot IMD}{S \cdot IMD} \right) = K_s t \tag{2.83}$$

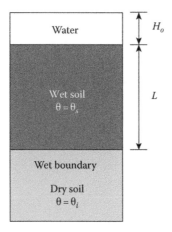

Figure 2.40 Green–Ampt model.

This form of Green–Ampt model is easier in simulation modelling due to the connection between aggregated infiltration and time. The values of θ_s ($\theta_s - \theta_i = IMD$, where θ_s is the porous coefficient and θ_i is the initial content of water), S and K_s can be taken by USDA, depending on the soil characteristics (Maidment, 1993).

2.13.4 Huggins–Monke model (1966)

Many researchers omitted the variable of time. An example is the following equation by Huggins and Monke:

$$f = f_c + A\left(\frac{S-F}{T_p}\right)^P \tag{2.84}$$

where
A and P are the coefficients
S is the storage capacity over the impermeable ground layer (T_p minus moisture)
F is the total infiltration water volume
T_p is the total porosity of the soil over the impermeable ground layer

The coefficients are evaluated by infiltrometer experiments. F must be calculated for every time step, while if $t = 0$, $F = 0$.

2.13.5 Holtan model (1961)

This model is given by the following equation:

$$f = \alpha F_p^n + f_c \tag{2.85}$$

where
f is the infiltration capacity (in./h)
α is the infiltration capacity [(in./h)/in.$^{1.4}$] of the available storage volume (coefficient relevant to porous surface)
F_p is the available storage capacity in the surface layer (first 6 in. of layer, in inches)
f_c is the ultimate infiltration rate (in./h)
n is the coefficient usually equal to 1.4

2.13.6 Kostiakov model (1932)

Kostiakov has developed the following relationship:

$$f = (ab)\, t^{b-1} \tag{2.86}$$

The total infiltration is obtained by

$$F = at^b \tag{2.87}$$

where
f is the infiltration capacity (in./h)
F is the accumulated infiltration (in.)
a and b are the coefficients

It can also be expressed in the logarithmic form:

$$\log F = \log a + b \log t \qquad (2.88)$$

Parameters a and b are estimated using a double logarithmic paper with linear regression among accumulated infiltration and time.

2.13.7 Philip model (1954)

Philip has proposed the following equation:

$$F = 0.5st^{-0.5} + A \qquad (2.89)$$

Therefore, the accumulated infiltration is

$$F = st^{0.5} + At \qquad (2.90)$$

where
  f is the infiltration capacity (in./h)
  F is the accumulated infiltration (in.)
  s is the soil storage capacity
  A is the constant relevant to hydraulic characteristics of the ground

2.13.8 Soil Conservation Service method (1972)

Soil Conservation Service (SCS) uses an empirical process, involving CN curve number, not directly estimating infiltration but with an indirect evaluation of it (see also Chapter 7).

This method estimates the active precipitation depth, that is, the precipitation which gives direct runoff according to the following relationship:

$$h_e = \begin{cases} 0 & h \le 0.2S \\ \dfrac{(h - 0.2S)^2}{h + 0.8S} & h > 0.2S \end{cases} \qquad (2.91)$$

where
  h_e is the active precipitation depth
  h is the total precipitation depth

S is a parameter connected to CN curve number according to the relationship:

$$S \text{ (mm)} = 254 \left(\frac{100}{CN} - 1 \right) \qquad (2.92)$$

The SCS method uses empirical matrices of direct runoff in conjunction with soil types. A CN curve number (1–100) is computed. The categorization of soil types is as follows:

  A. Strong infiltration (sand or gravel)
  B. Medium infiltration (sandy clay)
  C. Low infiltration (clay)
  D. Very low infiltration (clay over impermeable formation)

Table 2.32 Curve numbers for the hydrological status II

Land use	Hydrological status	Soil class			
		A	B	C	D
Grass land	Poor	68	79	86	89
	Medium	49	69	79	84
	Good	39	61	74	80
Forest	Poor	45	66	77	83
	Medium	36	60	73	79
	Good	25	55	70	77
Soil	—	72	82	87	89
Land (hard surface)	—	74	84	90	92

A weighted CN curve number can be estimated by different percentages of soil types and land uses.

CN curve numbers are divided according to soil moisture (antecedent moisture condition, AMC) in different soil statuses:

AMC I: Dry soils
AMC II: Average moisture of a wet season
AMC III: Intense precipitation in the last 5 days; soil in saturation

Values of CN for AMC II are given in Table 2.32. Those for AMC I or III are derived from Table 2.32 using the AMC II values.

REFERENCES

American Society of Civil Engineers, Committee on Irrigation Water Requirements of the Irrigation and Drainage Division of the ASCE, 1990, *Evapotranspiration and Irrigation Water Requirements: A Manual*, ASCE, New York, 332pp.

Anderson, M.G. and Burt, T.P., 1985, *Hydrological Forecasting*, Vol. 372, Chichester, Wiley.

Brutsaert, W. and Stricker, H., 1979, An advection-aridity approach to estimate actual regional evapotranspiration, *Water Resources Research*, 15(2), 443–450.

Collier, C.G., 1989, *Applications of Weather Radar Systems*, Ellis Horwood, Wiley.

De Bruin, H.A.R. and Keijman, J.Q., 1979, The Priestley-Taylor evaporation model applied to a large, shallow lake in the Netherlands, *Journal of Applied Meteorology*, 18, 898–903.

Dingman, S.L., 1994, *Physical Hydrology*, Prentice Hall, Inc., New Jersey.

Distromet Ltd., 2011, Basel, Switzerland. http://www.distromet.com/, last accessed January 5, 2016.

Eagleson, P.S., 1970, *Dynamic Hydrology*, McGraw-Hill, New York.

Enright, L., 2004, Low-cost re-architecturing of NASA's TRMM mission control center, *Proceedings of Ground System Architectures Workshops*, Manhattan Beach, CA.

Everett, D., May 2001, GPM satellites, orbits and coverage, Goddard Space Flight Center, Greenbelt, MD.

Fotopoulos, F., May 2002, Simulation of the sampling properties of the global precipitation mission, MSc thesis, Massachusetts Institute of Technology, Cambridge, MA.

Green, W.H., and Ampt, G.A., 1911, Studies on Soil Phyics. The Journal of Agricultural Science, 4(01), 1–24.

Holtan, H.N., 1961, Concept for infiltration estimates in watershed engineering. U.S. Dept. Agr., Agr. Res. Serv., ARS 41–51, 25pp.

Huff, F.A., 1970, Time distribution characteristics of rainfall rates, *Water Resources Research*, 6, 447–454.

Huggins, L.F., and Monke, E.J., 1966, The mathematical simulation of the hydrology of small watersheds. Tech. Rep. No. 1, Purdue Water Resources Reasearch Centre, Lafayette.

Kostiakov, A.N., 1932, On the dynamics of the coefficient of water-percolation in soils and on the necessity for studying it from a dynamic point of view for purposes of amelioration. Trans, 6, 17–21.

Koutsoyiannis, D. and Xanthopoulos, T., 1997, *Engineering Hydrology*, NTUA, Athens, Greece (in Greek).

Laurenson, E.M., 1960, Temporal pattern of Sydney storms, Seminal rain, Paper 7/4, Sydney, Australia.

Maidment, D.R., 1993, *Handbook of Hydrology*, McGraw-Hill, New York.

Manning, J.C., 1997, *Applied Principles of Hydrology*, Vol. 276, Upper Saddle River, NJ: Prentice Hall.

Mimikou, M., 2000, National Data Bank of Hydrological and Meteorological Information (NDBHMI), Final report, NTUA, Athens, Greece.

Mimikou, M. and Baltas, E., 2012, *Engineering Hydrology*, Papasotiriou, Athens, Greece (in Greek).

Morton, F.I., 1983, Operational estimates of areal evapotranspiration and their significance to the science and practice of hydrology, *Journal of Hydrology*, 66, 1–76.

Munro, D.S., 1979, Daytime energy exchange and evaporation from a wooded swamp, *Water Resources Research*, 15(5), 1259–1265.

National Aeronautics and Space Administration, 2007, TRMM turns Ten, http://www.eorc.jaxa.jp/TRMM/documents/data_use/text/handbook_e.pdf, last accessed January 5, 2016.

National Space Development Agency of Japan, February 2001, *TRMM Data Users Handbook*, Earth Observation Center, Saitama, Japan.

Pessi, A. and Businger, S., 2009, Relationships between lightning, precipitation, and hydrometeor characteristics over the North Pacific Ocean, *Journal of Applied Meteorology and Climatology*, 48, 833–848.

Petersen, W.A., Christian, H.J., and Rutledge, S.A., 2005, TRMM observations of the global relationship between ice water content and lightning, *Geophys. Res. Lett*, 32(1–4), L14819, 4pages.

Philip, J.R., 1954, An infiltration equation with physical significance. Soil Science, 77(2), 153–158.

Sakkas, J., 1985, *Engineering Hydrology-Surface Water Hydrology*, DUTH, Hanthi Greece (in Greek).

Shaw, E.M., 1994, *Hydrology in Practice*, 3rd Edition, Chapman and Hall, London, U.K., 569pp.

Shuttleworth, W.J. and Calder, I.R., 1979, Has the Priestley-Taylor equation any relevance to the forest evaporation?, *Journal of Applied Meteorology*, 18, 639–646.

Singh, V.P., 1992, *Elementary Hydrology*, Pearson College Division.

Soil Conservation Service, 1972, SCS national engineering handbook, section 4: hydrology. The Service. United States.

Stagnitti, F., Parlange, J.Y., and Rose, C.W., 1989, Hydrology of a small wet catchment, *Hydrological Processes*, 3, 137–150.

Stewart, R.B. and Rouse, W.R., 1976, A simple method for determining the evaporation from shallow lakes and ponds, *Water Resources Research*, 12(4), 623–662.

University of Waterloo, 2001, Watflood model. http://www.civil.uwaterloo.ca/watflood/manual/manualstart.htm, last accessed January 5, 2016.

Viessman W., Jr and Lewis, G.L., 1996, *Introduction to Hydrology*, Fourth edition, Harper Collins College Publishers, New York.

Chapter 3

Runoff

3.1 GENERAL

Part of the amount of water that reaches the ground as precipitation is held by the canopy of plants covering the ground. The amount of water intercepted in this way is not constant but is dependent on the plant type and the percentage of cover in the area and the characteristics of the precipitation. Some amount of water is retained by soil irregularities; part of this quantity goes back into the atmosphere through the process of evaporation and transpiration. The remaining moves on the ground or percolates into the soil. One part of the infiltrated water moves laterally, immediately beneath the surface, and reappears downstream on the surface or stream banks (interflow). The remaining moves into deeper layers and enriches the aquifers, becoming part of groundwater; the groundwater, again moving sideways, can reach the bed of a stream or even move outside the boundaries of the basin. Above-ground runoff starts almost immediately when precipitation intensity exceeds infiltration and surface detention capacities or is delayed and starts in accordance with the upper soil layer saturation level provided the precipitation continues unabated. The water moving above and below the ground surface following any of these paths forms the runoff. In particular, the part of the water moving on the surface makes the surface runoff, which combines with interflow to form the direct runoff. The water moving underground as groundwater also flows to stream beds, forming the main runoff or baseflow.

3.2 RIVER BASIN

All points covering the ground surface receiving rainfall and through which the water runoffs supplying the stream at a specific point are considered as the drainage basin of this stream or river at that point. In nature, the boundaries of the area which contributes groundwater to a stream may not be identical to those of the region which contributes surface runoff. For small basins, groundwater can move from one catchment to the neighbouring one or even further away. This causes some uncertainties in defining the overall boundaries of basin. To overcome these uncertainties, the area contributing water as direct runoff into a stream has been defined as the stream or river basin.

The boundary which separates a river basin from an adjacent one is called the ridge. The basin boundary follows the ridge around the basin and crosses the stream only at the outlet. Often, it is necessary for practical reasons to divide a large basin into smaller ones,

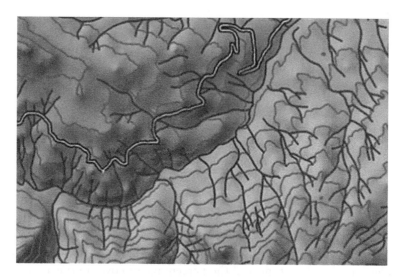

Figure 3.1 Way of plotting the drainage network on a topographic map.

which are called sub-basins and are defined by internal basins (Bedient and Huber, 1992). Figure 3.1 shows the way of plotting the drainage network on a topographic map.

Basins have certain characteristics that modulate their hydrological behaviour to a large extent. These features are the following:

1. *The drainage surface of the basin*: This is the plan view of the area bounded by the watershed boundary; it is defined by the area calculated from topographic maps, and it is expressed usually in square kilometres or hectares.
2. *Stream order*: The order of streams is a characteristic which reflects their degree of branching in the basin. Horton (1945) describes that streams which are small and have no branches as the first-order streams, streams which have branches as the second-order streams and streams which have branches of the second order as third-order streams. Thus, the order of the main stream shows the extent of branching streams within a catchment. Other classifications of streams are known by Shreve and Strahler. The classification by Strahler is similar to that by Horton, whereas in the classification by Shreve, the streams are summed each time and, as a result, the final main stream has a great order. Figure 3.2 shows the classification of streams by Shreve and Strahler.

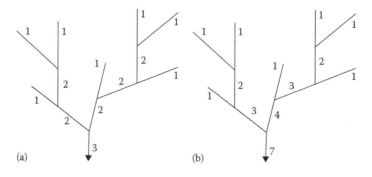

Figure 3.2 Stream order by (a) Horton (1945) and Strahler and (b) Shreve.

3. *The drainage network density*: This is a parameter of the basin, which is calculated, if the total length of streams is divided by the area of the basin. It is expressed in the following form:

$$D_d = \frac{\Sigma L}{A_d} \tag{3.1}$$

where L is usually expressed in kilometres and A_d in square kilometres. The drainage network density shows the length to be travelled on the ground surface by the water up to a certain stream. Generally, a river basin is considered to have a low coefficient of hydrographic network, which is proportional to the drainage of the basin, if $D_d \leq 0.5$, while it is considered to have a high coefficient of hydrographic network, implying a very good drainage if $D_d \geq 3$.

4. *The hypsometric curve*: Hypsometric curve is the distribution of surface elevation of a basin. It refers to the relationship between an altitude and the percentage of the surface of the basin which has a value above or below it. It is usually presented in charts derived in geographic information systems (GISs) (Heywood and Carver, 1998), as shown in Figure 3.3. These charts are useful because they give a picture of the average ground slope of the basin and they can set the basis for comparisons between catchments. Also, the hypsometric curve is an important guide in the design and installation of a network of hydrometeorological stations.

The calculation of the average elevation of the basin is based on the hypsometric curve using the following equation:

$$Z_s = \sum_r \frac{Z_r + Z_{r+1}}{2} \Delta L_r \tag{3.2}$$

where the range of values of the hypsometric curve has been divided into intervals of length ΔL_r, and the Z values corresponding to the edges of the interval ΔL_r are Z_r and Z_{r+1}.

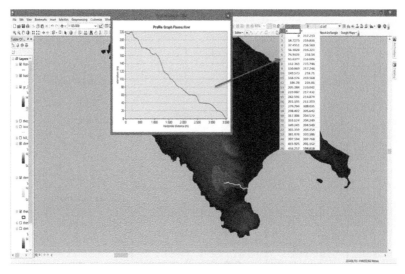

Figure 3.3 Hypsometric curve using GIS methods. (From National Data Bank of Hydrological and Meteorological Information (NDBHMI), Athens, Greece, http://ndbhmi.chi.civil.ntua.gr/.)

5. *The alignment and longitudinal profile of the stream*: The alignment of the stream is the path that the stream follows on a map. The longitudinal section of the main stream of the basin is a graph of elevation along the length (alignment) of this stream. Based on this, the slope of the main stream may be defined in many ways. One of them is to divide the difference of the elevation between the highest and lowest points of the stream by the total length. This slope, in general, is not very representative because of the significant slope variation along the stream. More realistic and useful are nearly uniform slope values obtained for various reaches along the stream, based on its longitudinal section.

6. *Average slope of the basin*: This parameter is calculated with a procedure which begins by the construction of a grid over the catchment. Then, the number of contour lines which intersect each horizontal grid line is counted in one direction, and this number is divided by the total length of the lines in this direction (Figure 3.4). If h is the contour interval, N is the number of intersections of the contour lines with the grid lines in one direction and L is the length of all lines in this direction, the slope S is given by

$$S = \frac{hN}{L} \tag{3.3}$$

7. *The surface of the streams of the river basin*: It is obvious that such a calculation is very complex, and therefore, some simplifications are needed. If we assume that the width of the main stream at the outlet of the basin or the width of any stream at the point of its confluence with higher-order stream is b and the width at the beginning of the streams is practically zero, then the wet surface A_W of each stream may be estimated by a relation of the following form:

$$A_w = \frac{bL}{2} \tag{3.4}$$

where L is the length of the stream. It has been observed that in most river basins, the total surface of the streams does not exceed a percentage of 5% of the basin area.

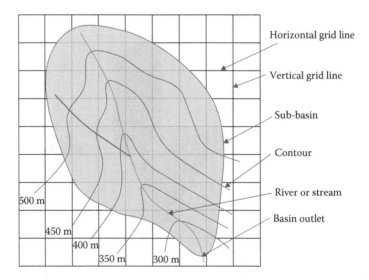

Figure 3.4 Grid over the catchment area to determine the slope of the basin.

3.3 HYDROGRAPHS

The discharge of a river basin at a specific location is defined as the change in the volume of water per unit time, measured in m³/s. The variation of the discharge of a stream (which is a result of the rainfall–runoff process in the basin) with respect to time is called 'hydrograph'. The basic components (parts) of a hydrograph are the surface runoff, the intermediate runoff (or interflow) and the baseflow. Figure 3.5 shows these three components of a typical hydrograph. The surface runoff (component) includes the water flowing over the ground surface. The intermediate runoff (component) includes the water which moves laterally beneath the ground surface in the unsaturated zone and returns after some travelled distance back to the ground surface or directly to the stream banks. The base flow (component) includes the groundwater drainage from the saturated zone. The surface and intermediate runoff components constitute the direct runoff which is related to a rainfall event. The simple sketch given in Figure 3.6 shows the factors which modulate the discharge of a stream.

3.3.1 Characteristics of the hydrograph

The shape of a hydrograph, caused by a relatively short duration rainfall event which covers the whole catchment area, typically follows a general pattern. This pattern is bell shaped, as shown in Figure 3.5, at the beginning there is a period of rise, which means that during this time, the stream discharge increases to a maximum value (peak flow), and then, a

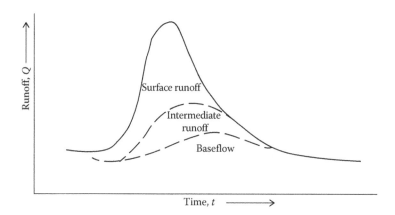

Figure 3.5 The three main runoff components which form a hydrograph.

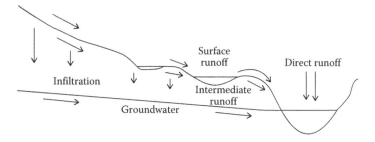

Figure 3.6 Diagram showing the components which modulate the stream discharge.

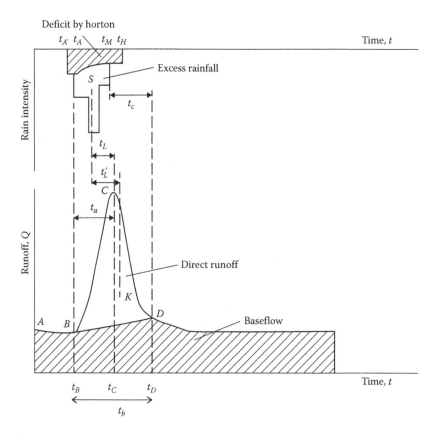

Figure 3.7 Typical form of hydrograph caused by an individual rainfall event.

period follows where the discharge continuously decreases, in which it can even reach zero, depending on the existence or not of baseflow. In small catchments, where the contribution of the intermediate and baseflow in stream discharge is usually limited or the groundwater table is low (e.g. arid areas), the hydrograph is formed essentially from surface runoff only.

Figure 3.7 shows a typical hydrograph (discharge of water from a basin with time) of a single flood event, in combination with the hyetograph, which caused the flood (rain intensity over time – which is presented at an inverted direction of ordinates, just as shown in the figure, but using the same timescale). The hyetograph shows that at time $t_{A'}$ an event of rain begins, and at a later time t_A, after an initial deficit, the excess rainfall begins, which is converted to direct runoff. The rain ends in time t_{NH}. The excess rainfall either ends at the same time t_H or in a previous time t_M, if the intensity at the end of the event is short.

At time $t_A \equiv t_B$ where the excess rainfall begins, the discharge of the river basin begins to rise at an increasing rate, until the discharge reaches its maximum value at time t_C. The branch BC of the hydrograph is called the rising limb, point C is called the peak of the hydrograph and the discharge at time t_C is called the peak flow discharge. The rising limb of a hydrograph begins with the surface runoff and ends at the point where the rate of the hydrograph is reduced. The shape of this curve depends on the characteristics of the basin, forming the travel time of water on the surface and streams, and on the duration, intensity and homogeneity of the rain. Its initial section is concave because only a part of the basin contributes and also at this stage, the infiltration, the retention of water in soil irregularities, the evaporation and the water retention from vegetation are proportionately higher than that in the next stages.

The (maximum) peak, caused by a rainfall with given intensity and duration, occurs when all parts of the basin contribute water, i.e. both the part of the basin near the outlet is still contributing water and the water from the furthest point of the basin has reached the outlet. In order for this to occur, the duration of the rainfall must be equal to or greater than the time of concentration of the basin.

Then, the decrease of the discharge follows in time, as shown in the falling limb CD. The descending limb includes the remaining part of the hydrograph, constituting the withdrawal of water which is temporarily stored at the surface of the basin after the actual completion of the rainfall event. The shape of this curve is independent of fluctuations in the intensity of the rainfall which caused the runoff and soil infiltration, depending almost exclusively on the characteristics of the stream bed. At time t_D, the direct runoff ceases, but the baseflow (presenting only contribution from groundwater) continues flowing in the stream, forming the DE branch. This branch, i.e. the baseflow, is not related to the specific rainfall event.

The direct runoff occurs at the time (t_B, t_D), while the baseflow is continuous. The duration $t_b = t_D - t_B$ is known as flood duration or hydrograph base time. Other typical durations of the hydrograph are the time of the rising limb, $t_a = t_c - t_B$, and the lag time (more precisely, the peak lag time), $t_L = t_c - t_S$, where t_S is the time corresponding to the centroid S of the active rainfall graph. More clearly, the lag time is defined as the time difference between the centroid of the direct runoff hydrograph and the excess rainfall graph, i.e. $t'_L = t_K - t_S$, where t_K is the time corresponding to the centroid K of the direct runoff hydrograph. For better understanding, time t'_L is also referred to as centroid lag time. For convenience, the factor t_L is used here instead of t'_L, without being regarded as the characteristic of the basin, as it also depends on the form of the rainfall graph.

Another approximately inalterable time period, which is also characteristic of the basin, is the time of concentration or confluence time. The time of concentration is defined as the time which is required from the water which contributes to direct runoff to reach from the hydrologically most distant point of the basin to the outlet of the basin. The time of concentration is shown in Figure 3.7 as the time distance from the end of active rain until the end of direct runoff, i.e. $t_c = t_D - t_M$.

In order to evaluate the aforementioned characteristic time durations, it is necessary to separate on the given hydrograph the two components of the runoff, base and direct flow. To do this, the following have to be determined: (1) the starting time of direct runoff, i.e. the point B; (2) the end time of direct runoff, i.e. the point D; and (3) the variation of the base flow during time.

The separation of excess rainfall and losses (mostly infiltration) in a given hyetograph follows the separation of direct and baseflow in the flood hydrograph. After separating the direct runoff from the baseflow, the volume $[L^3]$ of direct runoff can be calculated as follows:

$$V_d = \int_{t_B}^{t_D} Q_d(t)dt \tag{3.5}$$

where $Q_d(t)$ is the direct runoff at time t, which results from the difference of the total discharge $Q(t)$ minus the baseflow discharge $Q_b(t)$. As mentioned earlier, the direct runoff hydrograph constitutes a transformation of the active rainfall graph. The volume [L³] of water of the excess rainfall graph is

$$V_e = \int_{t_\Lambda}^{t_M} i_e(t)dt \; S = h_e A \tag{3.6}$$

where
$i_e(t)$ is the intensity of the active rain [L/T]
h_e is the cumulative vertical depth of excess rainfall [L]
A is the area of the basin [L²]

Mass conservation requires that

$$V_d = V_e \tag{3.7}$$

which implies that

$$h_e = \frac{V_d}{S} \tag{3.8}$$

Therefore, after determining the depth h_e, the rainfall excess graph can be subsequently established.

3.3.1.1 φ index

Infiltration indicators often assume that the infiltration is carried out at a fixed or average rate during the storm, something not true. Typically, these methods tend to underestimate the initial infiltration rates and overestimate the final ones. The best application is during large storms and moist (i.e. fully saturated) soils, i.e. storms where infiltration rates may be considered uniform.

The most well-known indicator is the φ *index*, according to which the total volume of losses during the storm is distributed evenly throughout the event (Figure 3.8). Therefore, the volume of precipitation over the line of the index is equal to the runoff. A variant of the φ index, which does not take into account the soil storage and retention, is the W *index*. Initial water quantities are often deducted from the first stages of the storm so as to exclude initial retention.

In order to determine the φ *index* for a given storm, the observed runoff is computed based on the measured direct runoff hydrograph using Equation 3.5 and then by subtracting this volume from the volume of the total measured precipitation. This resulting volume of losses (which include interception, depression storage and infiltration) is distributed evenly along the hyetograph, as shown in Figure 3.8.

The use of the φ *index* for calculating the volume of direct runoff from a given storm profile is the reverse process. The φ *index* is defined for a specific storm and is not generally applicable to other storms, and if it is not associated with the characteristic parameters of the basin, it has low value.

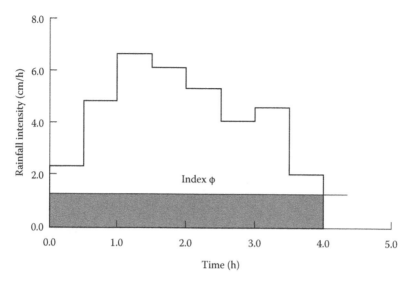

Figure 3.8 The φ index.

Example 3.1

Calculate the φ *index* for a basin with an area of 37,297 km². A rainfall event resulted in the runoff of 32 mm. The temporal evolution of the rainfall intensity is shown in Table 3.1.

Solution

The φ index will be calculated by a trial-and-error procedure. It is observed that the time step of the rainfall is not constant. The total rainfall is calculated for each time step by multiplying the volume with the corresponding period. Let us assume an initial value of φ = 3.0 mm/h. For this value, the losses are calculated at each time step and subtracted from the rainfall, resulting in the amount of rain flowing on the surface (net rainfall). One should be careful of the time steps when rainfall is less than the losses; the difference in this case should be set to 0.0, since there is no physical meaning for the surface runoff to have negative values. The procedure is presented in Table 3.2.

If the incremental runoff depths are summed, the total surface runoff results to 40.78 mm, which is greater than the amount of the given runoff (32 mm). Therefore, the assumed value of φ should be increased in the next trial. With successive trials, it was found that the value of φ is equal to 3.85 mm/h (Table 3.3), making surface runoff almost equal to the given one; therefore, the actual value of the φ index is 3.85 mm/h.

Table 3.1 Temporal evolution of the 15 December 1967 rainfall event

Time (min)	Intensity (mm/h)	Time (min)	Intensity (mm/h)
0–20	3	152–213	2
20–50	7	213–300	3.8
50–62	3.2	300–343	9
62–121	12	343–405	1
121–132	6	405–422	0.2
132–140	21	422–720	7
140–148	3	720–805	4
148–152	6		

Table 3.2 Runoff calculation with φ index: first trial

Time			Intensity	Depth	First trial		
					φ = 0.3		Surface runoff
T (min)	Δt (min)	Δt (h)	i (mm/h)	h (mm)	(mm/h)	Losses (mm)	(mm)
0.0	20.0	0.33	3.0	1.00	3.0	1.00	0.00
20.0	30.0	0.50	7.0	3.50	3.0	1.50	2.00
50.0	12.0	0.20	3.2	0.64	3.0	0.60	0.04
62.0	59.0	0.98	12.0	11.80	3.0	2.95	8.85
121.0	11.0	0.18	6.0	1.10	3.0	0.55	0.55
132.0	8.0	0.13	21.0	2.80	3.0	0.40	2.40
140.0	8.0	0.13	3.0	0.40	3.0	0.40	0.00
148.0	4.0	0.07	6.0	0.40	3.0	0.20	0.20
152.0	61.0	1.02	2.0	2.03	3.0	3.05	0.00
213.0	87.0	1.45	3.8	5.51	3.0	4.35	1.16
300.0	43.0	0.72	9.0	6.45	3.0	2.15	4.30
343.0	62.0	1.03	1.0	1.03	3.0	3.10	0.00
405.0	17.0	0.28	0.2	0.28	3.0	0.85	0.00
422.0	298	4.97	7.0	4.97	3.0	14.90	19.87
720.0	85	1.42	4.0	1.42	3.0	4.25	1.42
805.0			**Sum**	13.42			40.78

Table 3.3 Runoff calculation using the φ index: last trial

	Last trial	
φ = 0.59 (mm/h)	Losses (mm)	Surface runoff (mm)
3.85	1.50	0.00
3.85	2.25	1.58
3.85	0.90	0.00
3.85	4.43	8.01
3.85	0.83	0.39
3.85	0.60	2.29
3.85	0.60	0.00
3.85	0.30	0.14
3.85	4.58	0.00
3.85	6.53	0.00
3.85	3.23	3.69
3.85	4.65	0.00
3.85	1.28	0.00
3.85	22.35	15.65
3.85	6.38	0.21
	Sum	31.96

3.3.2 Hydrograph separation

During the process of hydrological analysis, it is sometimes necessary to separate a hydrograph from its components. This separation is not an easy task because it is difficult to identify and separate with accuracy the different components, especially in the case of surface runoff and intermediate flow. To separate the hydrograph, various methods have been devised, which, however, are to a degree arbitrary and subjective.

3.3.2.1 Methods of baseflow separation from the total hydrograph

With these methods, the hydrograph is divided into direct flow (surface runoff) and base-flow, and the separation process is based on identifying the falling limb of the hydrograph, the point where the direct runoff ends. The descending limb after this point only represents essentially the baseflow curve. Based on this, it is assumed that the baseflow curve has the same slope as a point just below the peak of the hydrograph. So, if this curve is extended backwards from the point where the direct runoff ends (point B) to a point below the peak (point C), this extension (straight line BC) can be assumed dividing the part of the hydrograph from its peak to point C into direct flow and baseflow underneath. Then point C is linked with a straight line with the point which starts the direct runoff at the beginning of the hydrograph (point A), so the separation is completed. This procedure is given schematically in Figure 3.9 and represented by the line ACB.

The difficulty in applying this method, similarly to the other two methods which follow, is to determine the point where the direct runoff ends, i.e. point B. One way to determine this is based on the examination of several hydrographs of the catchment area from which this point can be identified as the slope changing point of the descending curve. The definition of point B is essentially based on the fact that the base time of the direct flow hydrograph is limited and the amount of the baseflow is not comparatively large. The definition can also be aided by examining topographic and geological characteristics of the basin. In addition to these, several empirical relationships for the definition of point B have been occasionally proposed (Linsley et al., 1982). One of those given by Linsley et al. (1949) has the following form:

$$N = 0.8A^{0.2} \tag{3.9}$$

where
  N is the number of days after the peak flow the direct runoff stops
  A is the area of the basin in km^2

Another way to separate the hydrograph into two parts is just by connecting points A and B with a straight line, i.e. the points which indicate the beginning and the end of direct runoff. This process is shown in Figure 3.9 by the line AB. The third method of separation is to

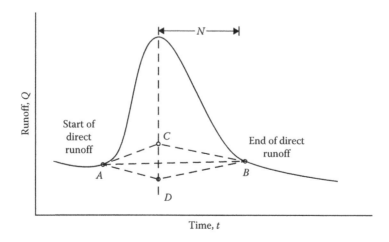

Figure 3.9 Hydrograph separation in direct runoff and base flow with three different graphical methods.

extend the curve at the beginning of runoff from point A until it crosses the vertical from the peak to a point D. Then, this point is connected to point B. This process is presented in Figure 3.9 by the line ADB.

It has to be noted that all these three methods are to a degree arbitrary, so it cannot be distinguished whether or not a certain one is advantageous. What should be clarified is that during a hydrological analysis, all hydrographs must be separated using the same method, regardless of which is used.

3.3.2.2 Hydrograph separation with the method of the logarithms

The descending limb, as previously mentioned, represents the recession of water which is stored on the basin surface after the rain stops. The shape of this curve is independent of fluctuations in the intensity of rainfall and resulting surface runoff and soil infiltration, but is dependent almost exclusively on the characteristics of the basin and the stream bed (i.e. slope, roughness, cross-sectional shape). Horner and Flynt (1936) and Barnes (1939) found that the descending limb can be described by the following equation (Papazafiriou, 1983):

$$Q_2 = Q_1 K^{-\Delta t} \tag{3.10}$$

where
$\quad Q_1$ and Q_2 are the discharges at times t_1 and t_2, respectively
$\quad K$ is a constant
$\quad \Delta t$ is the interval between the times t_1 and t_2

This equation, if drawn on semi-logarithmic paper for constant K, gives a straight line. However, it is observed that the value K is not constant for the entire length of the descent and this curve, if drawn on semi-logarithmic paper, does not represent a straight line, but a distinct point of slope break is seen. This slope break is attributed to the interruption of direct runoff contribution. It should be noted that many approaches have been tried and almost all lead to the conclusion that a constant K is not sufficient to describe the entire curve.

The procedure followed for the separation is, then, the following: First, the descending limb from the peak and below is plotted on semi-logarithmic paper as shown in Figure 3.10.

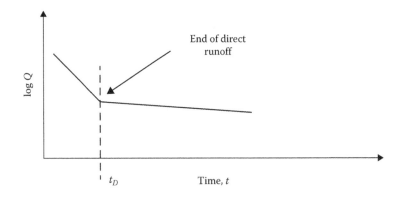

Figure 3.10 Descending limb of hydrograph plotted on semi-logarithmic paper.

Then, the two straight segments are separated. The break point is where the direct runoff ceases to contribute, defined in the graph by time t_D.

This hydrograph separation method, although strictly empirical, produces more consistent results and should be preferred to the previously mentioned methods, when a more detailed analysis is required.

Example 3.2

A rainfall event with intensity $I = 12.2$ mm/h and duration $t = 1$ h occurs in a catchment area of 50 km². The following values of discharges were measured at the exit of the basin (Table 3.4):

1. Plot the rainfall graph (hyetograph) and the flood hydrograph.
2. Separate the baseflow from the direct flood runoff.
3. Calculate the total flood volume at the basin outlet.
4. Determine the direct runoff at the basin outlet.

Solution

The hyetograph and the flood hydrograph are presented in Figures 3.11 through 3.13.

In order to separate the baseflow from the surface runoff, the logarithm of the discharge values with time is drawn and the point of slope shift is defined (in this example, the point is also obvious in the flood hydrograph). The slope changes at time $T = 6$ h, so it is considered that this time represents the end of the direct flood runoff.

It is considered that the baseflow varies linearly between the start point ($T = 0$ h) and the completion of flood runoff ($T = 6$ h), thus calculating the set of baseflow values Q_b. The flood runoff Q_f will be the difference of the baseflow from the total runoff, as shown in Table 3.5.

The total flood volume is derived by calculating the area enclosed by the curve of the total flow and the time axis. In this case, it can be calculated by the trapezoid method approximately equal to 871,200 m³. By removing the volume of baseflow, the flood volume is derived and it is equal to 588,960 m³ (Table 3.5).

Table 3.4 Discharges at the basin outlet

t (h)	0	1	2	3	4	5	6	7	8	9	10	11	12
Q (m³/s)	5.0	36.4	65.2	47.1	34.2	18.9	8.0	6.7	5.9	5.3	5.2	5.1	5.0

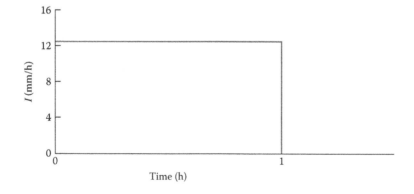

Figure 3.11 Hyetograph of the rainfall event.

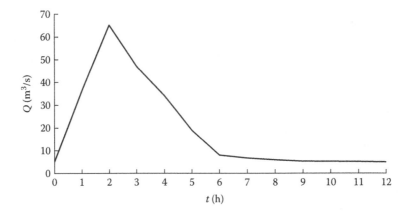

Figure 3.12 Flood hydrograph of the observed runoff.

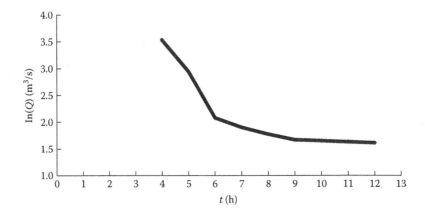

Figure 3.13 Separation of baseflow from the flood runoff.

Table 3.5 Calculation of the flood runoff

t (h)	Q (m³/s)	ln(Q)	Q_b (m³/s)	Q_f (m³/s)	V_f (m³)
0	5	1.6094	5	0	0
1	36.4	3.5946	5.5	30.9	55,620
2	65.2	4.1775	6	59.2	217,800
3	47.1	3.8523	6.5	40.6	397,440
4	34.2	3.5322	7	27.2	519,480
5	18.9	2.9392	7.5	11.4	588,960
6	8	2.0794	8	0	
7	6.7	1.9021	6.7	0	
8	5.9	1.7750	5.9	0	
9	5.3	1.6677	5.3	0	
10	5.2	1.6487	5.2	0	
11	5.1	1.6292	5.1	0	
12	5	1.6094	5	0	

3.3.3 Composite hydrograph separation

The methods presented earlier involve the separation of hydrographs which are caused by individual rains, i.e. hydrographs having a peak and the corresponding ascending and descending curves. However, during a rainfall event, it is possible to have some intervals with no rain. This results in a composite hydrograph, i.e. a hydrograph with two or more peaks and intermediate incomplete ascending and descending curves. Such a hydrograph, caused by two successive rainfall events, is shown in Figure 3.14.

The separation of a composite hydrograph into direct hydrograph and baseflow is more complex than in the simple case; first, the duration of the direct runoff after the peak should be calculated. This time period can be estimated by one of the methods mentioned, among which is Equation 3.9. A generalized separation process in such hydrographs is meaningless, because each case is unique. Instead of this, we present the separation of the hydrograph shown in Figure 3.14.

After the estimation of the time N from the peak until the end of the direct runoff, which is expressed by Equation 3.9, the descending limb AC after the first peak extends to the point D which is located N time from the peak A. This way, the composite hydrograph is analyzed in two simple ones which can be separated by any of the aforementioned methods, whereas there will be an overlap in time between the two peaks. For this period, the separation would be appropriate as best fits. In the case of Figure 3.14, the separation of the two simple hydrographs is made using the third method previously mentioned, and as a dividing line between the direct runoff and the baseflow of the composite hydrograph is the IHDKF line.

Instead of this method, any other method could be used from those we have mentioned with better fit; e.g. the one based on plotting the descending limb on semi-logarithmic paper. In this case, the descending limb BE is plotted on a logarithmic paper and the point F, which is highlighted, indicates the end of direct runoff (this is the second from the side of peak B point of slope shift of semi-logarithmic line). With this procedure, both the shape of the descending limb and the time N' in days that the direct runoff lasts (time between points B and F) are determined, without using Equation 3.9. Then, the descending limb is drawn after the first peak A, based on the form found by analyzing the BE segment of the curve, thereby resulting in two simple hydrographs. From here on, the procedure of the previous paragraph follows.

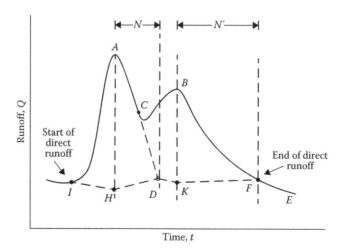

Figure 3.14 Composite hydrograph separation into direct flow and baseflow.

3.3.4 Factors influencing the hydrograph shape

The time distribution of runoff, which is expressed by the shape of the hydrograph, is affected by climatic factors as well as topographical and geological features of the basin (Papazafiriou, 1983). Generally, it can be said that the ascending limb of hydrograph is formed primarily by the characteristics of rain causing runoff, while the descending limb is almost independent of rain characteristics.

3.3.4.1 Climatic factors

The main climatic factors affecting the volume and distribution of runoff, i.e. the shape of the hydrograph, are

1. The intensity, duration and time distribution of rainfall
2. The rainfall distribution over the catchment
3. The direction which the rain moves
4. The form of the precipitation
5. The type of rain

3.3.4.1.1 Intensity, duration and time distribution of rainfall

These three characteristics of rain define the volume of the runoff, the duration of the surface runoff and the height of the hydrograph peak.

In the case of rain with steady intensity, the duration determines, to some extent, the magnitude of the peak and the duration of surface runoff. For a given rainfall event duration, an increase in the intensity would increase the peak and the volume of runoff, provided, of course, that the rainfall intensity exceeds the infiltration capacity of the soil. The effect of such an increase in the intensity does not affect the duration of the surface runoff. Finally, variations in the intensity of rain can affect the shape of the hydrograph of small catchment areas, while in larger basins, this effect is very limited.

Generally, a rain of high intensity and short duration causes high peak, ascending and descending hydrograph limbs with relatively steep slopes and limited duration of surface runoff (shorter time base of hydrograph). The rain of the same total depth but with low intensity and long duration causes low hydrograph peak, ascending and descending limbs with small slopes and longer duration of surface runoff. These cases are shown graphically in Figure 3.15.

3.3.4.1.2 Rain distribution in the catchment

If the catchment area is large enough, the distribution of a rain that falls on it may not be uniform. In other words, more rain may fall in one part of the basin when compared to another part. If this happens, there is an impact on the shape of the hydrograph.

Specifically, if most of the rainfall falls at the part of the basin which is near its exit, the hydrograph will normally have a strong peak and steep ascending and descending limbs, and if it falls at a distant part of the basin, then the peak of the hydrograph would be lower and more flattened. These cases are shown in Figure 3.16.

3.3.4.1.3 Rainfall direction

The direction in which the rain moves in relation to the main orientation of the network of streams of the hydrological basin can also affect the magnitude of the peak of the hydrograph

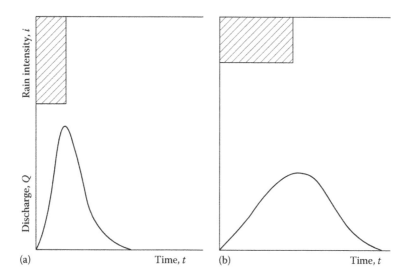

Figure 3.15 Hydrographs caused by rains of equal depths but of different intensities and durations. (a) Short duration, high intensity and (b) long duration, low intensity.

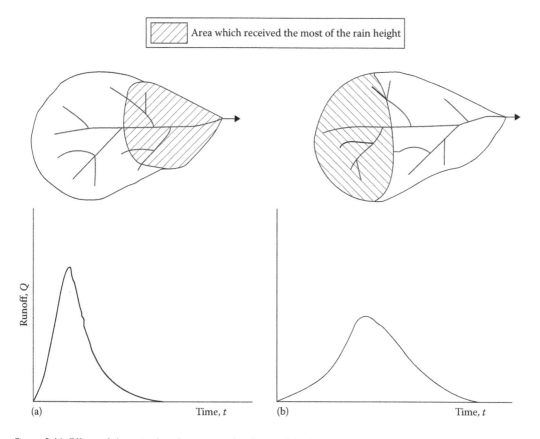

Figure 3.16 Effect of the rain distribution on the shape of the hydrograph. (a) Most of the rainfall has fallen near the outlet of the basin and (b) most of the rainfall has fallen away from the outlet of the basin.

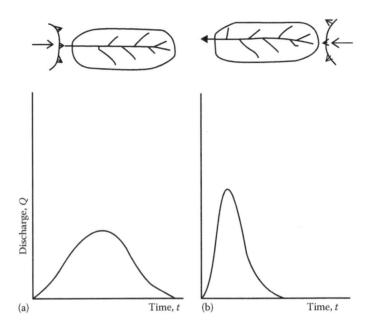

Figure 3.17 Effect of the rainfall direction on the shape of hydrograph. (a) Rainfall direction from the outlet to the upper part and (b) rainfall direction from the most distant end of the basin to the outlet.

and the duration of surface runoff. The effect of this factor is more intense in elongated basins. In such basins, if the direction of the rain is from the outlet to the upper part, the hydrograph will have a lower peak and a longer duration of surface runoff, compared to when the rain direction is from the most distant end of the basin to the outlet. These cases are qualitatively shown in Figure 3.17.

3.3.4.1.4 Forms of precipitation

Unlike the rainfall, which has a direct effect on the hydrograph shape, the effect of snowfall is not related to the amount and the duration but to the time and rate of snow melting layer covering the catchment. The runoff caused by snowmelt is a relatively slow process. This is due to the uneven distribution of the snow layer thickness over the basin and the daily temperature variations which result in the average rate of melting, which cannot, in most cases, exceed the soil infiltration rate. Thus, in these cases, most direct flow is derived from intermediate runoff, resulting in hydrographs having low peak but very long duration. These, of course, are relevant to basins which have formed a snow layer and the normal melting period is in spring. If the snow falls occasionally on a river basin and melts rather quickly, the shape of the hydrograph is similar to that caused by the rainfall. As for its final form, it will be a function of snow pack thickness and melting rate. Finally, in some cases, the snow melting is due to the fall of warm rain, causing high and fairly prolonged runoff. Such incidents sometimes cause very serious flooding.

3.3.4.1.5 Type of rainfall

The type of rainfall plays a major role in shaping the peak of the hydrograph in relation to the size of the catchment area. Rains which are caused by intense updrafts (spring and summer storms) produce the highest flow peaks in small basins, while their impact is not significant in large basins because the upward rainfalls are limited in extent. On the contrary, rains caused by widespread frontal systems combined with orographic factors give high and prolonged peaks.

3.3.4.2 Topographic factors

The direct runoff hydrograph of a catchment reflects the combined effect of all physical characteristics of the basin. Some of the main features which exert substantial influence on shaping the direct runoff of a river can be the following:

1. The size and shape of the basin
2. The distribution and density of streams
3. The slope of the banks of the basin
4. The slope of the main watercourses
5. The terrain
6. The percentage and type of vegetation cover

3.3.4.2.1 Size and shape of the basin

The main effect of the basin size on the shape of the direct runoff hydrograph is related to the duration of the hydrograph. So, regardless of other factors, a particular rain falling on a large basin will give hydrograph of longer duration, while the runoff per unit area at its peak will be smaller.

The shape of the basin modulates the rate at which water reaches the main streams, and it is an important factor which affects the ascending limb of the hydrograph and its peak. Thus, if the basin shape is such that the streams have a short length and converge towards the outlet, the ascending limb of the produced hydrograph will be steeper and the peak will be higher, compared to when the major part of the basin is relatively distant from the exit. In the second case, the basins do have an oblong shape. The effect of the basin shape on direct runoff is given qualitatively in Figure 3.18.

3.3.4.2.2 Distribution and density of streams

The denser and more uniform the network of the streams in a basin, the smaller the water paths over the basin surface and the shorter the time the water takes to reach the basin outlet (time of concentration). The opposite just holds in the case of a basin with sparse network of streams. Therefore, in the first case, the ascending limb of the hydrograph would be steeper, the peak higher and, generally, the direct runoff greater because the ponding time of water on the ground is shorter, and hence, less water infiltrates.

3.3.4.2.3 Slope of the banks of the catchment

The slope of the basin carries a substantial impact on surface runoff because it determines the contact time of the rainwater to the ground surface, thus affecting infiltration, and also it is the most important factor affecting the time of concentration since it regulates the water velocity over the surface. Accurate calculation of this velocity cannot be achieved because of the many factors which influence it. Generally, it can be said that the flow of water on

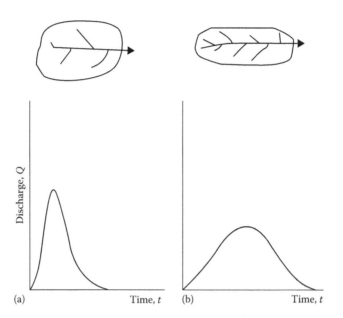

Figure 3.18 Hydrographs produced by the same rainfall in two basins with the same area but different shapes.

the surface can be expressed by an equation such as the one formulated by Butler (1957). This equation is the following (Papazafiriou, 1983):

$$q = aD^b S^c \tag{3.11}$$

where
  q is the flow (discharge) per unit width of area
  D is the flow depth
  a, b, and c are the constants which depend on the Reynolds number, the distribution of raindrops and the roughness of the soil surface
  S is the channel slope

This equation shows that the flow rate increases according to the slope. The greater the slope, the smaller is the time base of the hydrograph, the greater the peak, and the steeper the ascending and descending limbs of the hydrograph.

3.3.4.2.4 Slope of the main watercourse

When the water reaches the main stream of a basin, the time required to reach the outlet depends on the length and the slope of the stream. Generally, the flow velocity in open channels can be calculated by an equation of the following form:

$$V = CR^m S^n \tag{3.12}$$

where
  V is the flow velocity
  C is a parameter describing the roughness of the channel walls
  R is the hydraulic radius
  S is the channel slope
  m and n are the exponents of the R and S

When the Manning formula is used, m and n have the values 2/3 and 1/2, respectively. The time t required for water to travel a path of length L is

$$t = \frac{L}{V} \tag{3.13}$$

Thus, the travel time is inversely proportional to the slope of the stream, i.e. this time decreases with increasing slope. The effect of this factor on the hydrograph is associated with the time of the peak, the time base of hydrograph and the shape of the descending limb which becomes steeper as the slope increases.

3.3.4.2.5 Terrain of the catchment area

The rainfall which falls on the ground surface, before moving to the surface streams, must first fill any kind of territorial irregularities. It is, therefore, natural that the number and the form of soil cavities, together with the slope of the basin, will affect the shape of the hydrograph. In brief, it can be said that basins with little soil cavities and large bank slopes derive hydrographs with short time bases, steep ascending and descending limbs and high peaks. In contrast, basins with a large number of soil cavities give hydrographs with lower peaks and a long time bases.

3.3.4.2.6 Percentage and species of vegetation cover

The vegetation affects the runoff in two ways: one is direct and refers to the retention of rainwater from vegetation, which then evaporates back into the atmosphere, and the other is due to transpiration. Catchment areas, which have a large percentage of vegetation coverage, apart from trees with dense foliage, give very smooth hydrographs with long time bases and low peaks. In addition, the vegetation increases the basin and channel roughness, retards the flow and increases time of concentration and infiltration.

3.3.4.3 Geological factors

Geological factors affecting the shape of the hydrograph are those which regulate the flow of water to the streams, either in the form of intermediate runoff or in the form of groundwater flow (baseflow). These factors relate to the following characteristics:

1. Surface soil layers
2. Underlying geological formations

3.3.4.3.1 Surface soil layers

The ground surface condition, in combination with the texture and structure of the under-lying layers, determines the infiltration rate of water, directly affecting the surface runoff volume. Also, the layout of the soil layers affects the vertical and lateral rate of movement of water which has infiltrated into the ground. Thus, if the surface soil is sufficiently perme-able but it is underlain by an impermeable layer, a significant percentage of infiltrated water will be moved laterally, thus making the intermediate runoff an important factor in the final shape of the hydrograph. The hydrograph will have a low peak and a long duration.

If the surface layer of the basin consists of low permeability soils, most of the rainfall will become surface runoff collected in surface streams, resulting in a hydrograph with a high peak and a short time base. Finally, if all the soil layers are relatively permeable, most of the water will be deeply infiltrated, resulting in low peak and generally low volume of direct runoff.

3.3.4.3.2 Deeper geological formations

The type and the arrangement of geological formations beneath a river basin affect the base-flow. Depending on the layout of these formations, the area which contributes water to the streams of a basin can be greater or less than the surface area of the basin, as defined by the watershed. Moreover, this layout can control the location of the groundwater-level in such a way so as to provide the streams continuously with water, in which case the baseflow is continuous and significant; in other cases, the groundwater is permanently below the stream bed; in such cases, the stream enriches (recharges) the underground layers with water, and the baseflow is minimal or does not occur.

So far, the influence of the various factors on the hydrograph shape was examined separately. In nature, however, it is common for one factor to either cancel the effects of another factor, or factors to have a synergetic effect, i.e. the final hydrograph may be dependent on the cumulative effect of many factors as they act individually or in combination.

3.4 HYDROMETRY

The purpose of hydrometry is the measurement and the evaluation of flow parameters, i.e. water-level, velocity and river discharge. Hydrometry is a very complex and costly process, which requires skilled personnel for both the fieldwork and the storing and processing of collected data at the office. The fieldwork consists mainly of measuring the water-level and performing water measurements in order to estimate the flow discharge of the river. These data, after verification and required processing, are stored in the archives of the responsible regulatory or monitoring authorities (e.g. the Hydroelectric Power Corporation, the Ministry of the Water Resources and the Environment) which have established the monitoring locations. The data are finally processed in order to estimate the maximum or average discharge of the river at a specific time period.

3.4.1 Installation criteria for a hydrometric station

A fully equipped hydrometric station should be placed at an appropriate cross section of the watercourse. It includes instruments used for water-level and discharge measurement. One instrument needed in such station is the water-level meter. Often, the station has more than one water-level meters. Appropriate places to install these meters are the various structures that cross the stream, i.e. bridges. These should be checked frequently in order to observe any movement or deviation from the correct position. In places where we are interested in continuous monitoring of the level, the station includes a water-level recorder, which continuously monitors the variation of the level in time, thus allowing for a detailed evaluation of the temporal variation. In addition, the station may include hydrometric instruments to measure river flow velocity for the assessment of river discharge, including both regular and rare events, such as floods.

The following are the criteria of a site for the installation of a hydrometric station:

- The entire flow should be in a single branch.
- The river bed should have relatively stable and uniform geometry.
- The river reach should be relatively straight, which helps avoiding the influence of the flow from downstream obstacles.
- There should be reduced presence of channel erosion or deposition.
- The change in water-level should have a good relation to the discharge change.
- The station should be easily accessible, even during floods, in order to maintain it and mostly carry out the measurements.
- It should be favourable site for the construction of reservoirs, bridges, etc.

According to the recommendations of the World Meteorological Organization, the minimum density of hydrometric stations in the Mediterranean countries should be:

- In lowland/plain regions, 1 station per 1000–2500 km^2
- In mountainous regions, 1 station per 300–1000 km^2

The hydrological stations should be distributed in altitude zones with a height difference of about 500 m by zone.

3.4.2 Measurement of water-level

The most common measuring instruments of the water-level, as previously mentioned, are the level meters and recorders. The water-level meter is a single vertical ruler with an imprinted scale in centimetres. Its zero refers to a fixed reference altitude. In cases when the total flow of the river cannot be described by a single water-level meter, then more meters are installed in the same cross section (right, left, middle) or even upstream and downstream.

The water-level measurement is done manually by an observer every day at 8:00 a.m., while in the case of floods, the level measurement is usually done more frequently and during the flood event. The use of the water-level meter entails the risk that significant level changes which may occur in the time interval between two consecutive observations may not be recorded. Moreover, in many cases, it is necessary to make observations at positions which are not easily accessible by the observer. To overcome these difficulties, autorecording instruments, called water-level recorders, are used. A typical arrangement of a water-level recorder consists of the following components (Papazafiriou, 1983), as shown in Figure 3.19:

- A vertical stilling well of circular or rectangular cross section made of steel or reinforced concrete, whose bottom is set at least 50 cm below the lower point of the stream bed and which is located near the bank of the stream.
- Two or three tubes of relatively small diameter, horizontally levelled, connecting the stream with the well. These tubes have a small diameter in order to damp any variations and disturbances of the stream water-level due to surface waves and/or turbulence. A practical design rule is that when the length of the tubes is less than 5 m, the ratio of the tube diameter to the well diameter is 1:50, and for lengths exceeding 5 m, this ratio is 1:25.
- Since the pipes may be clogged by sediments, a purification system is usually necessary, which consists of a small water reservoir located above the ground surface, which is connected by piping to the horizontal tubes. The reservoir, from time to time, is flushing water into the tubes. This procedure is usually sufficient for cleaning.

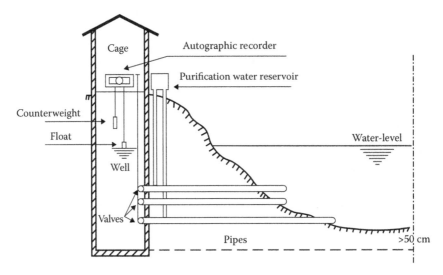

Figure 3.19 Typical water-level recorder in a stilling well.

- The water-level recording mechanism is placed in the well outside the maximum expected water and consists of a float in the well suspended through a cable and connected to a counterweight. This system, using a suitable structure, transfers level changes into the system logger. Another type of a water-level recorder has been used, instead of a rotating drum which has a perforating mechanism which transfers the level changes in a specific tape, for direct data transfer to a computer. However, these are old-type recorders; today the logger stores measurements in memory and transfers them to a personal computer.

A typical installation of a water-level recorder is shown in Figure 3.19.

3.4.3 Discharge measurement by the method of velocity field

To estimate the discharge of a stream, several methods are used, such as measurement by flow meter of the velocity field; the dilution method; floats; or estimates using empirical relationships. A very usual method is the method of velocity field, using a current meter. Note that the velocity profile in a river section is not uniform, but the greater velocity value appears at the maximum depth and the zero velocity at the cross-section solid boundary, as shown in Figure 3.20. The higher velocity is noted just below the water surface. The vertical velocity profile at the location of maximum depth is shown in Figure 3.21.

To calculate the discharge at a particular location of the river with the method of the field of velocities, it is absolutely essential to know the geometry of the wet section.

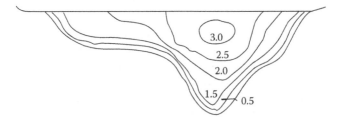

Figure 3.20 Velocities distribution profile in a river section.

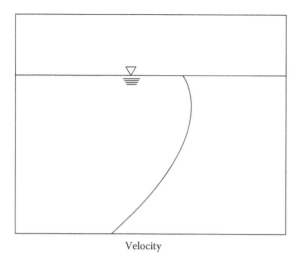

Velocity

Figure 3.21 Vertical velocity profile at the maximum depth of the cross section.

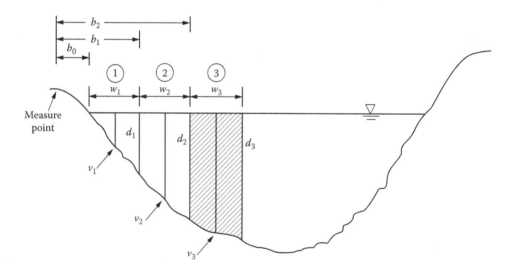

Figure 3.22 The cross section of a river.

This is determined by *dividing* the examined cross section into N subsections, usually one every 1–2 m, but depending on the width of the section and the desired accuracy, as shown in Figure 3.22. An empirical rule is applied to separate the river into subsections by dividing it into parts which allow no more than 10% of the total discharge in each of them.

To define the geometry of each subsection i, the height of its two vertical sides is measured (d_{i-1} and d_i), as well as the distance between them, which is the difference of the horizontal distance of the verticals ($b_i - b_{i-1}$), from a fixed point used to assess the measurements on one bank shore. With the help of a current meter, the mean velocity v_i is estimated for each subsection i, vertically in the middle of the interval ($b_i - b_{i-1}$). The current meter usually is equipped with a propeller, and when immersed in the cross section of the stream at that point, it rotates under the influence of the flow (Figure 3.23). Today, there are commonly

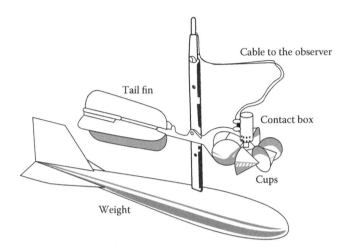

Figure 3.23 Typical configuration of a current meter.

used electromagnetic current meters which have no propeller, but their operation is based on the alteration by the flow of an electromagnetic field around the tip of the probe.

The access and the measurement of the velocity can be done in various ways depending on the flow conditions. In shallow streams of low flow rate, the process is done with wading and the lifting of the current meter is done manually, while in deep floating streams, a properly equipped boat can be used or the current meter can be suspended from a bridge. In cases where large flow velocities are developed, an overhead wiring with pulleys must be permanently installed. This allows the access to any point of the section from the bank of the stream. The necessary condition for the results to be valid is the correct orientation, horizontal and parallel to the flow direction, of the axis of rotation of the current meter. For this reason, current meters are equipped with fins to self-align to the flow direction, while a weight is suspended below the current meter to keep it in place, resisting the flow pressure force.

The velocity at each point of the vertical is a linear function of the rotation frequency of the propeller at that point (i.e. the angular velocity of the rotating propeller is transformed to linear velocity of the flow).

A question which arises at this point is on the proper depth to which the current meter should be immersed so that the measured point velocity, u_i, is representative of the velocity of the particular vertical and thus of the corresponding part and subsection, since it is known that the velocity varies within the section. According to the logarithmic law, it can be computed that the measured point velocity is more representative of the average, if measurement is made at a distance from the surface equal to 60% of the flow depth at the particular subsection. Therefore, if only one measurement is made in the subsection, the current meter should be submerged to a depth equal to 60% of the total depth from the surface. In practice, this is done when the subsection flow depth is less than 0.6 m. For larger depths, more than one velocity measurements are required, usually two (for depths up to 3 m). These are taken at distances from the surface equal to 20% and 80% of the depth, respectively. In this case, the average velocity in each vertical (and hence in each subsection) is estimated adequately by the average of these two values, as shown in the following equation:

$$v_i = \frac{u_{0.2} + u_{0.8}}{2} \tag{3.14}$$

When the water depth is more than 3 m, three velocity point measurements are performed to accurately approximate the mean subsection velocity. These are at distances from the surface equal to 20%, 60% and 80% of the depth, and the average velocity is given by the following equation:

$$v_i = \frac{u_{0.6}}{2} + \frac{u_{0.2} + u_{0.8}}{4} \tag{3.15}$$

Then, a measurement datasheet is filled on site by the observer with the data of water measurement and the total flow of the river, the date and time of measurement, as well as the corresponding level of the river which may vary during the measurement.

The average flow q_i of every segment results from the equation of continuity, as the product of the average speed i of the segment and the intersection A_i, namely,

$$q_i = v_i \cdot A_i \tag{3.16}$$

and the total flow at the particular cross section is given by

$$Q = \sum_{i=1}^{N} q_i \tag{3.17}$$

3.5 DISCHARGE ESTIMATION USING HYDROMETRIC DATA

To estimate the average flow of the river at a given period, water measurements at regular intervals are required (e.g. weekly, fortnightly). However, the frequency of measurements will never be the required one due to the special difficulties and the significant cost. The procedure for the estimation of the average flow rate for a shorter time interval (e.g. hourly, weekly) is to use a rating curve, i.e. a curve relating discharge to water-level, which can be prepared based on the following steps:

- Generation of a rating curve at this section of the river
- Estimation of the average level of the river at this point for the same time interval (e.g. hourly, daily)
- Extension of the rating curve
- Connection between the rating curves and the extension

It is obvious that in order to properly assess the average level (especially when the time step is less than one day), the station must necessarily have an automatic water-level recorder.

3.5.1 Preparation of a rating curve

The measurements of discharge and water-level at one station are plotted, forming the rating curve at that position. This curve has usually parabolic shape, and sometimes, it presents variations depending on the shape and variability of the section. In most streams, the computed rating curve may be changed over time, as a result of the change in the cross section and the slope, due to erosion or sedimentation of the bed. This fact shows that the curve discharge versus stage is not permanent, but it should be revised from time to time. Therefore, a

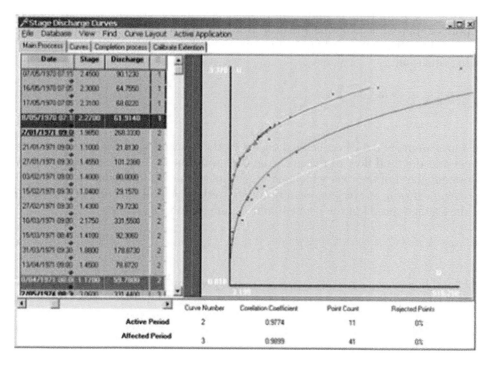

Figure 3.24 Rating curves in normal chart. (From National Data Bank of Hydrological and Meteorological Information (NDBHMI), Athens, Greece, http://ndbhmi.chi.civil.ntua.gr/.)

station has a set of curves, each one valid for a specific period of time, which varies and may be from several months to several years.

The first step for the preparation of the rating curve is to take and homogenize all the measurements collected at different time periods. The next step is to group the measurements which present the same behaviour and they could be represented by the same rating curve. This is a very tedious process that requires considerable experience. After grouping, the identification of curves follows (one for each subset). In each time period, the theoretical model describes a specific curve whose validity is defined by the start and the end of the time period. Typical rating curves for a particular river are presented in Figure 3.24.

For each group of points defined by the user, a different model for calculating each curve may be used. The rating curve can have various forms; one is the power equation:

$$Q = k(h - a)^b \tag{3.18}$$

where
  Q is the discharge
  h is the stage
  k and b are station-specific empirical coefficients
  a is the distance between zero altitude of water-level [meters] and altitude of zero flow of the section

Other equations which can describe the rating curve are polynomials:

$$Q = A_0 + A_1 H + A_2 H^2 \tag{3.19}$$

or second-degree polynomial logarithms

$$\ln(Q) = A_0 + A_1 \ln(H) + A_2 \ln(H^2) \tag{3.20}$$

where
   Q is the discharge
   H is the water-level
   A_0, A_1 and A_2 are the coefficients of the models

3.5.2 Extension of the rating curve

A major problem often encountered when using rating curves is that measurements, particularly at high flows (e.g. floods), are missing, so the rating curve is generated without using such data. So, the question is what to do to estimate the discharge during events of high flows. For this, the extension of the rating curve is required, which is usually done using the following Manning equation.

It must be pointed out that the Manning equation is valid only in the case of normal flow. For that reason, it only approximates real streamflow conditions and its use must be thoughtful.

$$Q = \frac{1}{n} A R^{2/3} J^{1/2} \tag{3.21}$$

where
   Q is the discharge
   A is the cross-sectional area
   R is the hydraulic radius
   J is the friction slope
   n is the Manning roughness coefficient

To define the constant term $J^{1/2}/n$, discharge data of the highest water measurements are used. This fixed term is calculated by fitting a regression equation between the discharge and the term ($AR^{2/3}$), which can be calculated from the geometry of the river cross section at the position of the rating curve; obviously, the cross section has to be surveyed. Alternatively, instead of the Manning equation, the Chézy formula can be used to extend the rating curve:

$$Q = CA(RJ)^{1/2} \tag{3.22}$$

where
   Q is the discharge
   C is the Chézy roughness coefficient

In this case, based on measuring the pairs (Q, $AR^{1/2}$), the term $CJ^{0.5}$ can be computed using linear regression. The way of linking between the rating curve and its extension concerns the way of fit. The fit of the curve is a straight line connecting the edge of each curve of the group with a point of the extension to always satisfy the relation $DQ/DH > 0$.

3.5.3 Remarks on the rating curves

Due to the variability of the curves, systematic discharge measurements are required throughout the period the station operates without any interruption. Otherwise, it will not be possible to identify changes in the rating curve. However, the assumption holds that the change in the curve usually takes place during a significant flood event (due to the erosive power of water); this contributes to the theory that the transition from one to the next curve concurs with the highest water-level ever recorded by the water-level meters at the station in the period between the two flow measurements from two consecutive subsets.

For a given cross section of a river and a certain time period, with no change in profile and characteristics of the channel, there is a unique match between stage and discharge. However, a more detailed study of the issue based on the principles of hydraulics shows that this is true only under the condition that the flow is steady. This condition is not valid during flood events, which are characterized by high flow variations over time.

In the case of unsteady flow, the rating curve has different rising and recession limbs; none of them concur with the curve of steady uniform flow, as typically shown in Figure 3.25. It is observed that at the initial stage of the flood event, the discharge increases significantly but the water-level increases at a lower rate compared to the uniform flow rating curve. The opposite is true during flood recession, i.e. the discharge declines at a faster rate than the corresponding rate of water-level reduction. It can, thus, be said that the level does not accurately monitor the change in the flow, but as shown in Figure 3.25, for the same value of water-level, the flow is greater in the rising phase of flood and smaller in the recession.

Usually, when calculating the time series of discharge from the corresponding stage time series, this phenomenon is ignored due to the complexity and lack of data. So, in the case of flood events, a unique rating curve applies, which creates a source of error on the estimation of flood hydrographs. The error is more significant when the slope of the basin is mild.

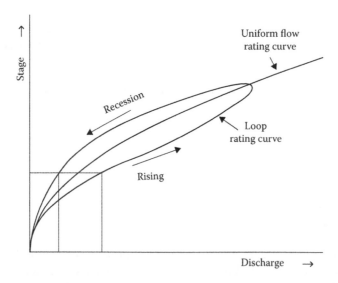

Figure 3.25 Loop rating curve in unsteady flow conditions.

3.5.4 Estimation of an average water-level for a specific time step

An important part of the processing of the recorded water-level information is related to the consolidation of the measurements from the water-level meter and the water-level recorder, if both instruments exist at the hydrometric station. It has been observed that simultaneous measurements from the water-level recorder and the water-level meter may be different. Generally, the water-level meter data comprise time series at a daily step, recorded on a suitable form by the observer, while water-level recorder data arise from decoding of the tape with a time step of half or 1 h.

Normally, simultaneous measurements of water-level gauges and water-level recorders should be identical. If this happened, then the data of the water-level gauge would be practically useless for the periods when both instruments operate in parallel, because the time series of water-level recorders would be more accurate and of greater time discretization. This is difficult to occur because the positions of the two instruments usually differ from each other and there are involved errors in the measurements. In the case of the water-level gauge, there is only an error in the reading of the level over the stage, while in the case of water-level recorders, the error sources are more (e.g. defects in the sensor, recording mechanism, clock mechanism, improper positioning of the film). For these reasons, the level of the water gauge is considered more accurate, while for the time series of water-level recorders, corrections are needed based on the measurements of water-level gauges. Another reason for the measurements of water-level gauges to be considered correct is that in the plotting of the rating curve, measurements of water-level gauges are only used.

The correction of water-level recorder measurements is a simple process. At times t_i where there are available simultaneous measurements of water-level gauges and recorders, the measurements of water-level gauges are taken into account, while the interim values available from water-level recorders are corrected by linear equation in function with time.

To estimate the average level for a specified period, the operation of water-level recorders is necessary. Initially, the reduction of the measurements of water-level recorders is based on the measurements of water-level meters, and then the average of the water-level measurements is obtained. For example, if the recording time step was 1/2 h, then the daily average level is derived as the average of 48 consecutive values, while if the recording time step was 1 h, then the daily average level is derived as the average of 24 values. To estimate the weekly or monthly average water-level, the average value of the corresponding daily values is taken into account.

3.5.5 Flow estimation from measurements of water-level meters/recorders

After having retrieved and processed all the data from the water-level gauges and the water-level recorders and having derived the rating curve groups, the calculation of discharges follows in time steps similar to the steps used in the estimate of the average water-level (half hour, hourly, daily, etc.). For this purpose, the appropriate rating curve is used. If $h_o(t)$ is the level recorded at time t from the water-level gauge or recorder, then the rating curve group that applies at time t is used for the conversion of level to a corresponding discharge. However, if at the same time t a flow measurement was carried out (with recorded level $h_p(t)$ and discharge $Q_p(t)$), then the application of the rating equation will not give the metered discharge $Q_p(t)$, but a different discharge, since the curve does not pass exactly through all the measured points (h_p, Q_p).

In order to remove this deficiency, the Stout correction can be used. This correction is done by applying the reverse process, i.e. for all time points with flow measurements, the estimated stage is calculated from the rating curve. Then, the difference between the estimated stage and the measured stage of the water-level meter is calculated for each of these time points.

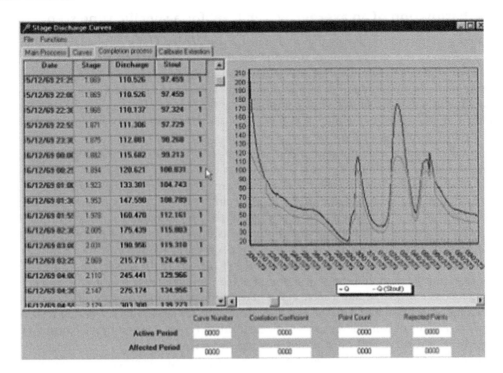

Figure 3.26 Estimation of discharge time series with a step of half an hour with and without correction of the level. (From National Data Bank of Hydrological and Meteorological Information (NDBHMI), Athens, Greece, http://ndbhmi.chi.civil.ntua.gr/.)

Therefore, a series of non-zero water-level differences Δ_h is developed, and it is assumed that the size Δ_h between times t_i and t_{i+1} varies linearly.

Having calculated now the level corrections at each time t, for which there is a stage measurement, the corrected level series h_d ($h_d = h_s - \Delta_h$) is formed, and thus, when applying the rating curve, the corrected series of discharge estimations is acquired. Figure 3.26 shows the discharge time series in half-hour time steps with and without correction of the level (Stout correction). In order to estimate the average daily discharge, the level is converted in high temporal resolution (e.g. every half hour) to corresponding discharges, and then again, these values are converted into daily, weekly, monthly, yearly, etc. The computing process of assessing the discharge with correction and without correction of the level is given in the following example.

Example 3.3

In a hydrometric station, systematic measurements of river stage and discharge have been made, as shown in Table 3.6, for a period without any significant changes in the geometry and the characteristics of the river section. Based on these measurements, plot the rating curve of the river for the relevant period. Using this rating curve and the data of measured flow and average daily levels for the period shown in Table 3.7, calculate the average daily discharge series of the station for the aforementioned period.

Solution

Setting up the rating curve
To plot the rating curve from the data of the water gauge, it is essential to initially calculate the logarithms of the values of stage and discharge, as shown in Table 3.8. This is done because the discharge is related to the level through an equation of the form of

Table 3.6 Measurements of water-level discharge

A/A	Date	Stage (m)	Discharge (m³/s)
1	20/3/1984	6.50	670.20
2	29/4/1984	5.70	510.30
3	27/5/1984	5.10	420.20
4	23/6/1984	4.60	370.70
5	20/7/1984	3.90	286.45
6	25/8/1984	3.70	278.20
7	20/9/1984	4.00	302.30
8	15/10/1984	4.42	319.79
9	13/11/1984	5.33	455.63
10	29/11/1984	6.10	582.98
11	23/12/1984	6.46	650.90
12	12/1/1985	7.19	820.70
13	30/1/1985	7.50	905.60
14	27/2/1985	7.74	984.84
15	4/3/1985	8.05	1098.04
16	10/3/1985	8.47	1296.04
17	28/4/1985	5.20	440.20

Table 3.7 Water-level measurements

A/A	Date	Stage (m)
1	27/2/1985	7.85
2	28/2/1985	7.89
3	1/3/1985	7.94
4	2/3/1985	7.99
5	3/3/1985	8.06
6	4/3/1985	8.09
7	5/3/1985	8.16
8	6/3/1985	8.23
9	7/3/1985	8.29
10	8/3/1985	8.32
11	9/3/1985	8.38
12	10/3/1985	8.41

Equation 3.18, i.e. $Q = A (h - h_o)^n$; thus, the logarithms of Q and h are linked through a linear equation.

Linear regression is used to estimate the coefficients between the logarithms of the discharge (y) and of stage (x), resulting in the following equation:

$$y = 3.100 + 1.840x \rightarrow Q = 22.20h^{1.840}$$

Estimation of average daily discharge

The estimation of daily discharges will be made with the use of the rating curve. If an effort is made to apply the equation of the previous question, it is observed that for 3 days, water gauging data are available (27/2, 4/3 and 10/3) and the resulting discharges differ from the corresponding measured ones. To avoid this phenomenon, the Stout correction

Table 3.8 Process of calculating the stage–discharge curve

A/A	Date	Stage H (m)	Discharge Q (m³/s)	ln(H)	ln(Q)
1	15/3/1988	6.5	670.20	1.871802	6.507576
2	19/4/1988	5.7	510.30	1.740466	6.234999
3	17/5/1988	5.1	420.20	1.629241	6.040731
4	22/6/1988	4.6	370.70	1.526056	5.915393
5	22/7/1988	3.9	286.45	1.360977	5.657564
6	24/8/1988	3.7	278.20	1.308333	5.62834
7	18/9/1988	4	302.30	1.386294	5.71142
8	17/10/1988	4.42	319.79	1.48614	5.767665
9	15/11/1988	5.33	455.63	1.673351	6.121681
10	27/11/1988	6.1	582.98	1.808289	6.368153
11	23/12/1988	6.46	650.90	1.865629	6.478356
12	08/1/1989	7.19	820.70	1.972691	6.710158
13	29/1/1989	7.5	905.60	2.014903	6.808598
14	27/2/1989	7.74	984.84	2.046402	6.892479
15	4/3/1989	8.05	1098.04	2.085672	7.001282
16	10/3/1989	8.47	1296.04	2.136531	7.167069
17	26/4/1989	5.2	440.20	1.648659	6.087229

Table 3.9 Stout correction

| A/A | Data | | Water metering | | | | | | |
	Date	Stage	Stage	Discharge	Q_{est}	H_{est}	Δh	H_{corr}	Q_{corr}
1	27/2/1989	7.85	7.74	984.84	983.81	7.855	−0.005	7.85	984.93
2	28/2/1989	7.89			993.06		−0.053	7.94	1005.26
3	1/3/1989	7.94			1004.67		−0.100	8.04	1028.13
4	2/3/1989	7.99			1016.34		−0.148	8.14	1051.23
5	3/3/1989	8.06			1032.78		−0.196	8.26	1079.38
6	4/3/1989	8.09	8.05	1098.04	1039.87	8.333	−0.243	8.33	1098.14
7	5/3/1989	8.16			1056.48		−0.321	8.48	1134.20
8	6/3/1989	8.23			1073.22		−0.399	8.63	1170.79
9	7/3/1989	8.29			1087.66		−0.476	8.77	1205.38
10	8/3/1989	8.32			1094.91		−0.554	8.87	1232.75
11	9/3/1989	8.38			1109.49		−0.631	9.01	1268.15
12	10/3/1989	8.41	8.47	1296.04	1116.81	9.119	−0.709	9.12	1296.16

is applied. With the use of this method, the measured levels will be corrected, and when using the rating curve, the calculated discharge for the days that measurements exist will be the same as that of the measured ones. The procedure is shown in Table 3.9.

In Table 3.9, in the column of water metering, the measured values of stage and discharge for the three dates with available measurements are placed. Q_{est} is the discharge derived by applying the rating curve for the data of the water-level gauge. h_{est} is the level resulting for the measured discharge, from the rating curve, i.e.,

$$Q = 22.20h^{1.840} \rightarrow h = \exp[(1/1.84)*(\ln Q - 3.1)]$$

Δ_h is the difference between the measured stage and the h_{est}. In places where there are no values of h_{est}, Δ_h is obtained by linear interpolation between the known values. h_{cor} is calculated by computing the difference of stage and Δ_h (in this case, the values are negative and added). Finally, Q_{cor} is obtained by applying the rating curve for a level equal to h_{cor}. It is observed that for the three dates for which discharge data are available, this is almost equal to the corrected calculated value.

3.6 RAINFALL–RUNOFF RELATIONSHIPS: EMPIRICAL METHODS

One of the most important concerns of flood hydrograph analysis is to estimate the (flood) peak flow. The early methods investigated for this purpose were empirical, but now more sophisticated methods are increasingly used mostly related to the theory of unit hydrograph (UH) and frequency analysis. Other key elements in the analysis of rainfall–runoff are the flood discharge volume for a single event or for a specified period of time and the rise time and the duration of the flood hydrograph. The choice of the method to use for this analysis in each case is based on certain criteria (Papazafiriou, 1983):

The available data: If there is a long-term hydrological observation record, statistical analysis procedures may be used, but the success of such methods is limited when there is lack of available data.

The basin area and other characteristics: These characteristics form the shape of the hydrograph in general and in particular the size and the time of occurrence of the peak. This part analyzes some representative empirical methods of flood peak assessment.

The intended purpose: The purpose is to design a hydraulic project based on flood peak or total flood volume. This paragraph refers to the estimation of flood peaks by using empirical methods described in the following.

3.6.1 Rational method for estimating flood peaks

This method is used to estimate the runoff peak of relatively small basins (<35 km^2). It is based on the principle that, for a rain with uniform intensity and distribution over the area, the maximum runoff occurs when the water from all parts of the basin reaches the outlet (see also Chapter 7). The runoff constitutes a certain percentage of the rain intensity which produces it. The rational method is expressed by the following equation:

$$Q_p = 0.278CIA \tag{3.23}$$

where

Q_p is the peak runoff in m^3/s
C is a dimensionless parameter known as the runoff coefficient
I is the rainfall intensity in mm/h
A is the area of the catchment in km^2

For the proper use of the method, it is necessary to clarify the limits of its application. At first, the method requires the rainfall intensity and spatial distribution to be uniform throughout the catchment. This condition is rarely met in nature and this assumption may only be valid for small catchments. Another factor which should be taken into account is the estimation of the rainfall intensity. To get the maximum peak, the water must reach the exit of the basin from all points, i.e. the rainfall duration with uniform intensity should be at least equal to the time of concentration t_c of the basin. This suggests that the method cannot be applied for rainfall duration less than t_c. But if the rain duration is greater than t_c, it cannot be used in the equation.

According to the process, the peak runoff per unit area of a basin caused by the rain of uniform intensity and of unlimited duration will be

$$q_p = \frac{Q_p}{A} = Ci \tag{3.24}$$

This suggests that the runoff factor C represents the ratio q_p/i. The runoff coefficient C should be selected based on the following factors: (1) the land use of the basin, (2) the extent and density of vegetation cover, (3) the slope of the basin and (4) the moisture content of the soil during the start of the rain. These factors suggest that the runoff coefficient is not constant even in the same catchment area, as it is a function of the moisture content of the soil and the rainfall intensity. Since these factors are difficult to evaluate, C is usually selected from tables taking into account all other factors, apart from the two. Such values given by the U.S. Army Corps of Engineers (USACE; 1948) are listed in Table 3.10 (see also Table 7.3).

As already mentioned earlier, the rational method can be applied for rainfall events whose duration is equal to or greater than the concentration time. It is, therefore, necessary to have ways to calculate this time.

In the case of hydrological basins, where the river path lengths are relatively large and the surfaces are uneven, several empirical relationships have been devised for the calculation of the concentration time. The following are the indicative equations devised by Kirpich (1940), Giandotti and the U.S. Soil Conservation Service (SCS):

The Kirpich equation is

$$t_c = 0.1947 \, L^{0.77} \, S^{-0.385} \tag{3.25}$$

where
  t_c is the concentration time (min)
  L is the maximum path length of water over the catchment (m)
  S is the slope equal to the ratio H/L, where H is the difference between the highest point of the basin and the outlet

Table 3.10 Indicative values of the coefficient C (rainfall for the return period of 5–10 years)

Type of basin	Runoff coefficient C
Permeable soil types (sandy)	0.10–0.20[a]
Medium permeability soil types	0.30–0.40[a]
Low permeability soil types	0.40–0.50[a]
Urban areas	0.70–0.90[a]
Industrial areas	0.50–0.90[a]
Forests	0.10–0.25
Roads (asphalt, concrete)	0.70–0.95
Rooftops	0.75–0.95

[a] The low values concern woodlands, while the highest values concern rural areas.

The Giandotti equation is

$$t_c = \frac{4\sqrt{A} + 1.5L}{0.8\sqrt{H_{av} - H_{out}}} \qquad (3.26)$$

where
A is the catchment area in km²
L is the distance of the mainstream to the outlet of the basin in km
H is the difference between the average elevation of the basin from the elevation at the outlet of the basin in m

The SCS equation is

$$t_c = \frac{L^{1.15}}{7700H^{0.38}} \qquad (3.27)$$

where
t_c is the time of concentration of the basin in h
L is the length of the main stream in m
H is the elevation difference between the farthest point and the outlet of the basin in m

Example 3.4

A bridge is planned to be constructed at the outlet of a basin, and therefore, it is necessary to estimate the maximum peak flood. At this location, there is no water gauge. The basin upstream of this location is 62 km², the maximum length of the main stream is 10 km, and the distance from the basin outlet to the nearest place in the river basin centroid is 3.75 km. It has also been estimated that the average altitude of the basin is 252 m, the basin outlet altitude is 121 m, and the altitude at the farthest point of the basin is 310 m. The basin runoff coefficient is equal to 0.2.

Calculate the maximum flood peak making use of the rational method when the concentration time is calculated by the Giandotti and SCS methods. The idf curve of the basin for a return period of $T = 20$ years is $h = 28.6t^{0.416}$, where h in mm and t in h

Solution

1. Giandotti

The concentration time is given by

$$t_c = \frac{4\sqrt{A} + 1.5L}{0.8\sqrt{\Delta H}} = 5.07 \text{ h}$$

where H_{av} and H_{out} are the average basin and outlet altitude in m, respectively.
The rainfall results from the idf curves

$$h_{cr} = 28.6 \, t_c^{0.416} = 28.6 \times 5.07^{0.416} = 56.18 \text{ mm}$$

and the corresponding intensity will be

$$i_{cr} = h_{cr}/t_c = 56.18/5.07 = 11.08 \text{ mm/h}$$

The peak discharge is calculated, using the rational method:

$$Q = 0.278CIA = 0.278 \times 0.2 \times 11.08 \times 62 = 38.19 \text{ m}^3/\text{s}$$

2. SCS

The concentration time is given by

$$t_c = \frac{L^{1.15}}{7700 H^{0.38}} = 1.76 \text{ h}$$

where

  L is the length of the main stream in feet
  H is the altitude difference of the farthest point of the basin and the elevation at the outlet of the basin

From the *idf* curves, the following rainfall is obtained:

$$h_{cr} = 28.6 t^{0.416} = 28.6 \times 1.76^{0.416} = 36.18 \text{ mm}$$

and the corresponding intensity will be

$$i_{cr} = h_{cr}/t_c = 36.18/1.76 = 20.55 \text{ mm/h}$$

The peak discharge is calculated with the aid of rational method as follows:

$$Q = 0.278 CIA = 0.278 \times 0.2 \times 20.55 \times 62 = 70.84 \text{ m}^3/\text{s}$$

3.6.2 Other empirical methods for calculating peak runoff

Apart from the rational method, several other empirical methods have been investigated to estimate flood peaks, which are used in cases where hydrological data for an area are inadequate. The difficulties encountered in the implementation of these methods are not due to the fact that these are empirical but to the lack of knowledge of the specific conditions under which each can be applied.

The most common and simplest methods use the catchment area as a parameter and have the following general forms (Papazafiriou, 1983):

$$Q_p = CA^n \tag{3.28}$$

$$Q_p = \frac{CA}{(a+bA)^m} dA \tag{3.29}$$

where

  Q_p is the discharge during the peak
  A is the catchment area
  C, a, b, d, n and m are empirical coefficients which must be evaluated for each particular case

An equation which was developed by the U.S. SCS (1957) uses, apart from the basin area A, the amount of excess rainfall P_r and the rise time t_p, as follows:

$$Q_p = 0.210 A P_r t_p^{-1} \tag{3.30}$$

where Q_p is in m³/s when A is in km², P_r in mm and t_p in h. Finally, Kinnison (1945) developed an equation which, apart from the basin area A, also uses the average basin elevation above its outlet point h, the average path needed to make the water reach the outlet L_m and the percentage of the basin a covered by natural or artificial ponds. The equation is expressed as follows:

$$Q_p = \frac{(0.000107\, h^{2.4} + 2.122)\, A^{0.9}}{a^{0.04} L_m^{0.7}} \tag{3.31}$$

where Q_p is in m³/s when A is in km², L_m in km and a is the percentage. The equation may be used only in the case where part of the basin is occupied by lakes; otherwise, $a = 0$ and Q_p gets an infinite value. It should also be clarified that the equation provides the maximum peak of the basin.

3.7 RAINFALL–RUNOFF RELATIONSHIPS: THE UH

The computation of a flood hydrograph for any rainfall can be based on the UH theory, which was first introduced in hydrological analysis by Sherman (1932). According to Sherman, a UH is the hydrograph caused by a unit rainfall excess (i.e. in the metric system equal to 1.0 cm), which is evenly distributed throughout the catchment area and has a uniform intensity. In general, the UH is a runoff hydrograph caused by a rainfall excess equal to 10 mm and of a definite duration.

3.7.1 Basic assumptions of UH

The theory of the UH is based on the following assumptions:

1. In a particular catchment area, rainfalls of equal duration, which cause runoff, derive direct runoff hydrographs of the same time base regardless of rainfall intensity.
2. At a given drainage area, the direct runoff caused by a specific rainfall is independent of the previous or next rainfalls.
3. The basin characteristics remain unchanged with time.

The aforementioned conditions are validated approximately only in natural catchments (Wilson, 1990). Regarding the first criterion, it is easy to observe that the capacity of streams increases as the water-level rises. Thus, for rainfalls with the same duration, the more the water stored in the streams, the greater the intensity of the rain. The stored water will continue draining after the stop of the rainfall, resulting in a prolonged direct runoff. Therefore, it is expected that there will be some variation in the runoff time depending on the rainfall intensity, and the definition of a single time base should be a result of assumptions.

Regarding the second condition, rainfall events preceding the flood affect the entire hydrograph and substantially the baseflow. For this reason, the UH cannot be applied to the total runoff but only to the direct runoff. But even in this case, this criterion is ambiguous to be applied, since the direct runoff depends on the soil moisture level and the filling of the soil cavities before the storm event.

Finally, the third condition can be considered valid because there are no significant changes in the state of hydrologic basins within reasonable periods of time, and there is no human intervention with construction of projects or alteration of vegetation cover in the basin.

After such interventions, it is expected that the hydrological behaviour of the basin will change accordingly, so the UHs to be investigated will be different from those investigated before the intervention.

Theoretically, an infinite number of UHs could be investigated due to variations in the duration and distribution of rainfall events. However, it is only necessary to determine the duration of rainfall for each river basin, for which the UH will be investigated in conjunction with the direct runoff, which will be representative for this basin. In the past, there have been various proposals for estimating the duration of the typical rainfall. Sherman (1949) recommends the following: for basins with an area larger than 2500 km², the typical rainfall duration should be from 12 to 24 h; for basins ranging between 250 and 2500 km², the duration from 6 to 12 h; for basins ranging from 50 to 250 km², the duration from 2 to 6 h and for smaller basins, the duration must be selected between the 1/3 or 1/4 of the time of concentration of the catchment area. Linsley et al. (1949) concluded that the duration of the typical rainfall should be about 1/4 of the time lag of the basin: this has been defined as the time interval between the centroid of the excess rainfall and the peak of the hydrograph. Moreover, the USACE (1948) proposed that in basins with an area less than 250 km², the typical duration of the rainfall should be half of the time lag.

Except for selecting the most appropriate rainfall duration for each basin, it is useful to investigate UHs of other durations less or greater than the typical ones. For classification purposes, the investigated UHs are characterized by the duration of the excess rainfall from which they originate. Thus, for example, a 6 h UH is the direct runoff hydrograph resulting from a rainfall excess of 6 h duration (and depth 1 cm).

The UH theory is based on two basic principles: the principle of proportionality and the principle of superposition.

3.7.1.1 Principle of proportionality

According to the principle of proportionality, two rainfall excesses of the same duration but with different intensities derive hydrographs with the same time base, but with ordinates proportional to the intensities, i.e. a rainfall excess with double intensity would result in a hydrograph with double discharges. The principle of proportionality is demonstrated in Figure 3.27. This principle is dependent upon the linearity of the basin, where the flood volume is directly proportional to the rainfall volume (double rainfall provides double runoff volume).

3.7.1.2 Principle of superposition

According to the principle of superposition, the total hydrograph resulting from individual rainfall events, is the hydrograph with ordinates at a given time the sum of the ordinates at that time of the individual hydrographs from each event. The principle of superposition is demonstrated in Figure 3.28, where the total hydrograph is the sum of individual hydrographs corresponding to three events of rainfall excess. The start of the individual hydrographs, which are summed, concurs with the start of the respective excess rainfall events.

3.7.2 Derivation of the UH from a single rainfall event

The necessary data for the derivation of a river basin UH are simultaneous observations of rainfall and discharge for a period of time, usually several years. From these time series, 4–5 intense rainfall events of the same duration are selected, as uniformly distributed as

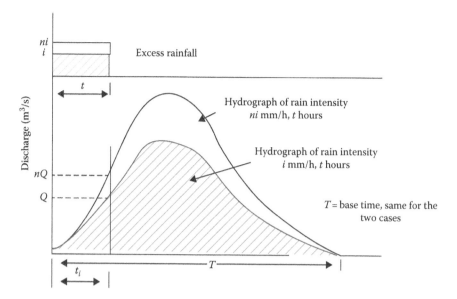

Figure 3.27 The principle of proportionality as applied to the unit hydrograph.

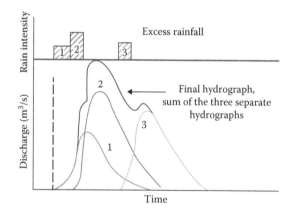

Figure 3.28 The principle of superposition as applied to the unit hydrograph.

possible over the catchment and with uniform intensity in time. From the measured hydrographs of these rainfall events, UHs are derived, which are used to extract a representative average UH of the basin.

With the assumption that a rainfall with considerable intensity, evenly distributed over the catchment and in time, and with representative duration for the basin is selected, in accordance to what has been said earlier, the procedure for the construction of the UH follows the following steps:

1. The total hydrograph is separated into direct runoff and baseflow.
2. The direct runoff hydrograph is plotted.
3. The total volume of direct runoff (m^3) and the rainfall excess depth (mm) is computed.

Table 3.11 Flood hydrograph at the outlet of basin *A*

Time (h)	0	1	2	3	4	5	6	7	8	9	10	11	12	13	14
Q (m³/s)	102	102	102	221	520	612	583	446	338	208	146	107	102	102	102

Table 3.12 Computation of UH

Time (h)	Direct runoff	UH
0	0	0
1	119	99.17
2	418	348.33
3	510	425.00
4	481	400.83
5	344	286.67
6	236	196.67
7	106	88.33
8	44	36.67
9	5	4.17
10	0	0

4. Based on the principle of proportionality, the ordinates of the direct runoff hydrograph are divided by the corresponding rainfall excess depth, calculated in the previous step. The values resulting from this process are the ordinates of the desired UH.

5. The duration of the rainfall excess, which is equal in magnitude to the known direct runoff, is determined.

Example 3.5

The rainfall excess of 12 mm and of 1h duration resulted in a flood hydrograph at the outlet of basin *A*, which is shown in Table 3.11 which follows. Find the 1 h duration UH of the basin.

Solution

First, the direct *flood hydrograph* will be calculated by the separation of the baseflow, which, in this case, seems constant and equal to 98 m³/s. Since the desired *direct hydrograph* and UH have the same duration, the latter will result from the principle of proportionality by simply multiplying the discharges of the direct runoff hydrograph by the ratio of rainfall excess depths:

$$Q_{UH}/Q_{DH} = i_{UH}/i_{DH} \rightarrow Q_{UH} = Q_{DH} \times (10/12)$$

The process of calculating the UH is shown in Table 3.12.

3.7.3 Mathematical determination of the UH of composite rainfall

The linear basin response at any precipitation is proportional to the UH, as referred earlier. This ratio can be generalized in composite storms and not only in a specific uniform event (Jones, 1997).

Figure 3.29 shows a composite storm, which extends, for example, in three equal time periods of 1 h each. The UH corresponding to a time period of excess rainfall can be used

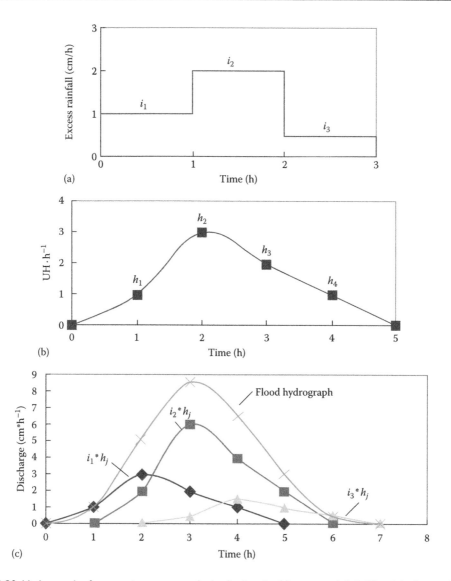

Figure 3.29 Hydrograph of composite storm, as derived using the (a) excess rainfall, (b) unit hydrograph and (c) flood hydrograph.

for the calculation of the flood hydrograph. Based on Figure 3.29, it is obvious that the following equations apply for the ordinates of the hydrograph (principles of proportionality and superposition):

$$i_1 h_1 = Q_1 \tag{3.32}$$

$$i_2 h_1 + i_1 h_2 = Q_2 \tag{3.33}$$

$$i_3 h_1 + i_2 h_2 + i_1 h_3 = Q_3 \tag{3.34}$$

$$i_3 h_2 + i_2 h_3 + i_1 h_4 = Q_4 \tag{3.35}$$

$$i_3 h_3 + i_2 h_4 + 0 = Q_5 \tag{3.36}$$

$$i_3 h_4 + 0 + 0 = Q_6 \tag{3.37}$$

This system of equations must be solved each time in order to estimate the UH. The general form of these equations can be expressed in the following matrix form (Bras, 1990):

$$I \cdot H = Q \tag{3.38}$$

where

$$I = \begin{bmatrix} i_1 & 0 & 0 & 0 \\ i_2 & i_1 & 0 & 0 \\ i_3 & i_2 & i_1 & 0 \\ 0 & i_3 & i_2 & i_1 \\ 0 & 0 & i_3 & i_2 \\ 0 & 0 & 0 & i_3 \end{bmatrix} \cdot H = \begin{bmatrix} h_1 \\ h_2 \\ h_3 \\ h_4 \end{bmatrix} \quad \text{and} \quad Q = \begin{bmatrix} Q_1 \\ Q_2 \\ Q_3 \\ Q_4 \\ Q_5 \end{bmatrix} \tag{3.39}$$

It should be noted that the number of the ordinates k of the flood hydrograph is calculated by

$$k = n + j - 1 \tag{3.40}$$

where
   n is the number of the UH ordinates
   j is the number of the excess rainfall ordinates

The general form of the equation which describes the method of calculating the hydrograph is

$$Q_i = \sum_{j=1}^{i} i_j h_{i-j+1} \tag{3.41}$$

This is also known as convolution integral.

Example 3.6

Excess rainfall of 2 h duration (shown in Table 3.13) resulted at the outlet of Basin A in the flood hydrograph that is given in Table 3.14. The baseflow is 96.23 m³/s. Derive the 1-hour UH.

Table 3.13 Analysis of rainfall

Time (h)	1	2
Rainfall (mm/h)	30	7.5

Table 3.14 Flood hydrograph at the outlet of the basin

Time (h)	0	1	2	3	4	5	6
Discharge (m³/s)	96.23	342.28	311.29	248.65	174.84	110.07	96.23

Solution

First, the direct runoff is calculated by abstracting the baseflow.

Time (h)	Flood hydrograph (m³/s)	Direct runoff (m³/s)
0	96.23	0
1	342.28	246.05
2	311.29	215.06
3	248.65	152.42
4	174.84	78.61
5	110.07	13.84
6	96.23	0

The direct flood hydrograph results from the superposition of two rainfall events. Suppose that the required ordinates of UH are U_0 (at the time $t = 0$) to U_6 (for $t = 6$ h). Applying the proportionality principle and the principle of superposition, the following system of equations is determined:

t (h)	UH	Displacement	Flood hydrograph
0	U_0		$3 \cdot U_0 = 0$
1	U_1	U_0	$3 \cdot U_1 + 0.75 \cdot U_0 = 246.05$
2	U_2	U_1	$3 \cdot U_2 + 0.75 \cdot U_1 = 215.06$
3	U_3	U_2	$3 \cdot U_3 + 0.75 \cdot U_2 = 152.42$
4	U_4	U_3	$3 \cdot U_4 + 0.75 \cdot U_3 = 78.61$
5	U_5	U_4	$3 \cdot U_5 + 0.75 \cdot U_4 = 13.84$
6	U_6	U_5	$3 \cdot U_6 + 0.75 \cdot U_5 = 0$

Solving the system of equations (line by line) gives the required ordinates of UH:

U_0	U_1	U_2	U_3	U_4	U_5	U_6
0	82.02	51.18	38.01	16.70	0.4	0

3.7.4 Determination of UH of a certain duration from known UH of a different duration: S-curve

The calculation of a UH with a duration that is an integer multiple of the duration of a known UH is relatively simple. For example, from a UH of 6 h, another UH of 12 h should be calculated. The UH of 6 h is added to itself with a 6 h time lag. The resulting hydrograph will last 12 h but will have a runoff volume equivalent to 2.0 cm depth, since each UH of duration of 6 h has a volume equivalent to 1.0 cm depth. If the ordinates of this hydrograph are divided by 2, the resulting hydrograph will have equivalent volume of 1.0 cm and will be the basin UH of 12 h duration. This method is known as the lagged hydrograph method.

So, it is very simple to derive UHs with duration a multiple of the duration of other known UHs. Difficulties arise when it is necessary to shorten or extend the UH duration by a fraction. In this case, the S-curve is used.

The S-curve is a hydrograph resulting at the basin outlet from infinite successive storms of a given duration. According to the UH theory, the S-hydrograph is the sum of the ordinates of infinite number of UHs of the given duration, as shown in Figure 3.30. The S-curve of D-hour duration is obtained by adding the corresponding UHs of D-hours, each one lagged by time D. It is understood that all these UHs are equal and each one results

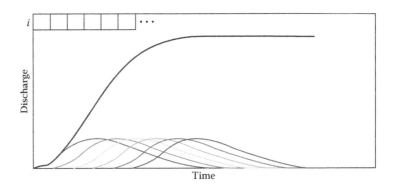

Figure 3.30 Calculation of S-curve from the unit hydrograph.

from a storm with duration $i = 1/D$ (cm/h) (Figure 3.30). The generated S-shaped curve takes its maximum value and as the rainfall excess continues, it becomes parallel to the axis of times, after the sum of T/D UHs, where T is the time base of each UH of D-hours. Since, as mentioned, the rainfall excess intensity of the hypothetical storm of uniform volume should be $1/D$, the S-curve is maximized and stabilized at the rate $Q = 1/D$, in units of discharge per surface area, or A/D in units of discharge, where A is the surface area of the basin.

When the S-curve of D-hour duration is available, it can be used to derive the UH of any other duration t, using the linearity property, according to the following procedure:

1. Lag the S-curve by t hours.
2. Subtract the ordinates of the lagged curve from the original S-curve. The result is a hydrograph of rainfall excess intensity t/D.
3. The resulting hydrograph from the previous step is converted to unit volume by multiplying its ordinates by D/t.

This process, shown in Figure 3.31, results in a UH of t hours.

During the process of calculating an S-curve of a given UH, an asymmetrical fluctuation of values around the mean value (at the part where the S-curve is supposed to be a straight line

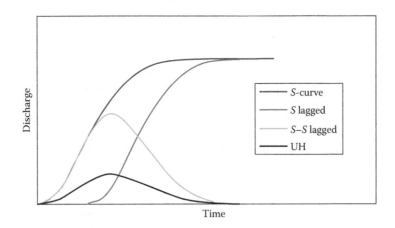

Figure 3.31 The use of the S-curve to derive a unit hydrograph of different durations.

parallel to the time axis) is usually observed, which becomes more intense as it approaches the end. This variation is mainly due to the lack of precision in the choice of the UH duration: the real duration of the UH differs somewhat from that used in the calculations. In this case, a normalization of the S curve is needed in order to be adjusted to the normal values.

Example 3.7

Derive the S-curve and use it to derive the UH of 3 h based on the UH of 1 h shown in Table 3.15.

Solution

The S-curve of t hours is the sum of infinite UH lagged by t h. In this case, while the S-curve base time is 1 h, it is calculated more simply by summing up at each time interval the ordinates of all the aforementioned UH. Then, the S-curve is shifted by time equal to the base time of the requested UH, i.e. 3 h. Finally, the difference is calculated at each

Table 3.15 UH ordinates of 1 h

T (h)	UH of 1 h
0	0.00
1	101.63
2	344.86
3	434.83
4	405.67
5	295.72
6	198.25
7	93.3
8	43.32
9	5.83
10	0

Table 3.16 Process of calculation of UH of 3 h

t (h)	UH of 1 h	S-curve	S displacement	Difference	UH of 3 h
0	0.00	0		0.00	0
1	101.63	101.63		101.63	33.88
2	344.86	446.49		446.49	148.83
3	434.83	881.32	0.00	881.32	293.77
4	405.67	1286.99	101.63	1185.36	395.12
5	295.72	1582.71	446.49	1136.22	378.74
6	198.25	1780.96	881.32	899.96	299.88
7	93.3	1874.26	1286.99	587.27	195.76
8	43.32	1917.58	1582.71	334.87	111.62
9	5.83	1923.41	1780.96	142.45	47.48
10	0	1923.41	1874.26	49.15	16.38
11		1923.41	1917.58	5.83	1.94
12		1923.41	1923.41	0.00	0
13			1923.41		
14			1923.41		

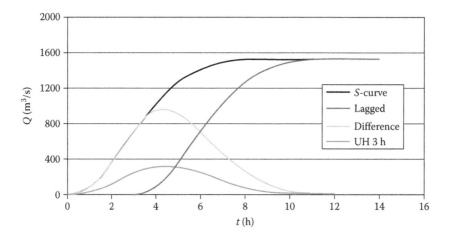

Figure 3.32 The S-curve and the unit hydrograph of 3 h.

time interval (Table 3.16; Figure 3.32). The desired UH is derived by the implementation of the multiplicative property, as follows:

$$\frac{Q_{UH}}{Q_{dif}} = \frac{i_{UH}}{i_{dif}} \Rightarrow \frac{Q_{UH}}{Q_{dif}} = \frac{10/3}{10} \Rightarrow Q_{UH} = \frac{Q_{dif}}{3}$$

Example 3.8

Based on the UH of 2 h shown in Table 3.17 and using the method of the S-curve, calculate the UH of 3 h of the same basin.

Solution

The 2 h UH is the measured runoff at the outlet of the basin resulting from a storm that lasted 2 h and had excess intensity 5 mm/h. The S-curve, which corresponds to the UH (curve S-2 h), is the runoff at the outlet of the basin, resulting from a hypothetical rainfall event which has infinite duration and intensity 5 mm/h. It is, thus, clear that the S-curve is calculated as the sum of infinite events similar to those producing the 2 h UH, shifted by 2 h relative to each other. In this case, 3–4 aggregations are sufficient to stabilize the S-curve at the maximum value. The algorithmic process of shifting the individual events and summing them, in order to construct the S-curve, is shown in Table 3.18.

For the calculation of 3 h UH, the S-curve should first be shifted by 3 h and the ordinates have to be subtracted from the ordinates of the initial S-curve. This subtraction produces an intermediate hydrograph corresponding to the rainfall excess of 3 h and intensity 5 mm/h, equal to the intensity of the S-curve. This direct runoff hydrograph of rainfall excess has the same duration with the desired UH (3 h) but a different intensity (5 mm/h for the intermediate hydrograph and 10/3 = 3.33 mm/h for the 3 h UH). Multiplying the direct hydrograph ordinates by 3.33/5 = 2/3 gives the requested UH results. The corresponding algorithmic procedure is presented in Table 3.19.

Table 3.17 UH of 2 h

t (h)	0	1	2	3	4	5	6
Q (m³/s)	0	50	120	95	50	25	0

Table 3.18 Construction of S-curve of UH 2 h

t (h)	First rainfall	Second rainfall	Third rainfall	Fourth rainfall	S-curve (2 h)
0	0				0
1	50				50
2	120	0			120
3	95	50			145
4	50	120	0		170
5	25	95	50		170
6	0	50	120	0	170
7		25	95	50	170
8		0	50	120	170

Table 3.19 Derivation 3 h UH from the 2 h S-curve

t (h)	S – 2 h	S – 2 h shift/Dt.	Difference	UH 3 h
0	0		0	0
1	50		50	33.3
2	120		120	79.92
3	145	0	145	96.57
4	170	50	120	79.92
5	170	120	50	33.3
6	170	145	25	16.65
7	170	170	0	0

3.7.5 Estimation of flood hydrograph using the UH

Once the UH of a basin is calculated (from simultaneous observations of rain and runoff), then it can be used for the derivation of direct runoff for any storm event, so that, the existing runoff data of a catchment can be used to cover periods for which there are observations of rainfall but not runoff data. It is necessary at this point to highlight the restrictions which affect the hydrologic analysis based on the UH.

The UH, as noted earlier, is determined on the basis of a particular rainfall. The distribution of this rainfall over the catchment affects the shape of the hydrograph. Thus, the uniformity of the distribution of the rainfall imposes an upper limit on the size of the basin, to which this method could be applied. What should be the upper limit of this size is a question that until now has not been answered. In general, it can be said that for the appropriate choice, the type of precipitation should be taken into account, such as the land use, the average slope and the shape of the basin. Thus, if in a region orographic rainfalls prevail, the maximum extent of catchments should be less than that of another area which is dominated by frontal or cyclonic rains. Indicatively, it can be said that in the first case, the area of the catchment should not exceed 5,000 km², while in the second, the limit can reach 10,000–12,000 km².

If the observations of the rainfall are recorded in 24 h, the area of the basin should be large enough so that the time of concentration of the water is greater than 24 h. In this case, the area of the basin, on the average, must be at least in the magnitude of 2000 km², and in certain conditions (basin shaped long and narrow, small slopes of embankments, stream slopes, etc.), the threshold can be lowered to 1000 km².

The hydrologic analysis based on the UH can be applied only when the characteristics of the streams of the basin do not change with time. Moreover, this can be applied to areas which have no significant ability of retaining water on their surface, because the ratio between the rate and the volume of runoff, required by the theory of the UH, implies a linear equation between the stream discharge and the stored water on the surface of the basin. This condition is violated when inside the basin, there are artificial or natural lakes which can hold a substantial volume of water or when, in the plains of the catchment, the water of the main stream is flooding the respective areas, so the effect is the same as in the case of an artificial lake storing water.

Finally, several problems arise when part or all the runoff water is derived from snow melting. In such cases, significant attention is required, and before applying the method, it is necessary to solve some specific problems regarding the area and the rate of melting, the distribution rates of runoff derived from rain and snow melting, etc.

If the conditions for the implementation of the UH are defined, the investigated hydrograph for a catchment area can be used to produce direct runoff hydrographs of this basin for any rainfall.

The construction of such hydrographs can be done by the following procedure:

1. Calculate the rainfall excess with one of the known methods (Φ, W, etc.).
2. Calculate the direct runoff hydrograph for each period having excess rainfall, by multiplying with the depth of excess rainfall of that period the ordinates of the appropriate UH, i.e. the UH of the same duration to the period.
3. The ordinates of the direct runoff hydrographs of each period are superimposed linearly, after making the proper time shift.

Example 3.9

Determine the total flood hydrograph caused by the following rainfall excess based on the UH of Example 3.5.

	Rainfall excess			
T (h)	1	2	3	4
I (mm/h)	0.4	1.2	1.9	1.6

Table 3.20 Calculation of flood hydrograph

t (h)	UH	First hour	Second hour	Third hour	Fourth hour	Direct runoff	Flood hydrograph
0	0.00	0.00				0.00	98.00
1	101.63	4.07	0.00			4.07	102.07
2	344.86	13.79	12.20	0		25.99	123.99
3	434.83	17.39	41.38	19.31	0	78.09	176.09
4	405.67	16.23	52.18	65.52	16.26	150.19	248.19
5	295.72	11.83	48.68	82.62	55.18	198.30	296.30
6	198.25	7.93	35.49	77.08	69.57	190.07	288.07
7	93.3	3.73	23.79	56.19	64.91	148.62	246.62
8	43.32	1.73	11.20	37.67	47.32	97.91	195.91
9	5.83	0.23	5.20	17.73	31.72	54.88	152.88
10	0	0	0.7	8.23	14.93	23.86	121.86
			0	1.11	6.93	8.04	106.04
				0	0.93	0.93	98.93
					0	0	98.00

Solution

Applying the principle of proportionality for each individual rainfall of 1 h duration, it is observed that each individual rainfall event is shifted in time from the previous one by 1 h. The direct runoff hydrograph results by horizontal sum while for the total flood hydrograph, the baseflow is added to each ordinate, according to Table 3.20.

3.7.6 Synthetic UHs

The UH method is one of the most popular for flood analysis; however, it requires streamflow data in order to derive the UH. Because many streams have no historical records, researchers have developed procedures to derive synthetic UHs. The procedure is to determine three characteristics of the synthetic UH – the peak time, the peak flow and the base time. With these three characteristics, a synthetic UH can be created. The shape of the synthetic UH is formulated such that the area under the curve equals 1 cm of runoff. It is important in these methods to calibrate the parameters among the peak flow and the lag time.

3.7.6.1 Snyder's UH

The most well-known and used method for synthetic UH derivation is Snyder's method, which was developed based on the analysis of many rainfall events in the Appalachian Mountain in North America.

This method defines the lag time t_p, the peak Q_p, the time base T and the widths of the UH W_{50} and W_{75} at discharges that correspond to 50% and 75% of the peak (McCuen, 1998).

The equations used to derive Snyder's UH are

$$t_p = C_t (L_{ca}L)^{0.3} \text{ (h)}$$

$$Q_p = C_p \frac{640 \cdot A}{t_p} \text{ (ft}^3/\text{s)} \tag{3.42}$$

$$T = 3 + 3 \cdot \left(\frac{t_p}{24} \right) \text{(days)}$$

where
  L_{ca} is the distance between the outlet and the centroid of the basin, measured along the main channel of the basin (mi)
  L is the length of the main stream from the outlet to the headwater (mi)
  C_t is a constant representing topographical and soil characteristics of the watershed normally ranging between 1.8 and 2.2. For steep basins, C_t tends to the lower value.
  C_p is a coefficient that depends on the applied unit system in the equation and by basin characteristics (ranging between 0.56 and 0.69)
  A is the area of the basin (mi²).
  T is the base time of the hydrograph with a minimum value of 3 days (Figure 3.33)

It is noted that t_p is the lag time (time between the peak of the hydrograph and the centroid of excess rainfall).

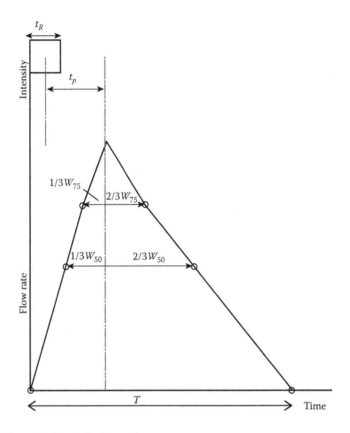

Figure 3.33 Snyder's synthetic unit hydrograph.

t_R is connected with t_p through the equation

$$t_R = \frac{t_p}{5.5} \tag{3.43}$$

If the required UH of $t_{R'} > t_R$, then $t_{R'}$ is adjusted by the equation

$$t_p' = t_p + \frac{(t_R' - t_R)}{4} \tag{3.44}$$

This adjusted value $t_{R'}$ must be replaced in Equations 3.42 in order to calculate the values of Q_p' and T'.

The widths of the UH in 50% and 75% of the peak, W_{50} and W_{75}, are estimated by the following equations:

$$W_{50} = \frac{830}{q_p^{1.1}}$$
$$\tag{3.45}$$
$$W_{75} = \frac{470}{q_p^{1.1}}$$

where $q_p = Q_p/A$ is the peak discharge divided by the area of the basin A.

After calculation of the aforementioned values, seven points of the synthetic UH graph are determined (including the first and the last of it), and by joining them the final hydrograph is plotted.

Example 3.10

At a river basin outlet with small infiltration rates, a bridge is planned to be constructed. Estimate the peak discharge of the steam. Discharge measurements in this position are not available. The total basin area is 38 km², the maximum mainstream length is 15 km, and the distance between the outlet and the nearest point of the main stream at the centre of the basin is 4.25 km.

Use Snyder's synthetic hydrograph method to calculate the UH of 3 h rainfall duration. C_t and C_p values are 2 and 0.65, respectively.

Solution

1 ft = 0.3048 m
1 mi = 1.609 km
t_p is given by

$$t_p = C_t \cdot (L_{ca} L)^{0.3} = 2 \cdot (2.641 \cdot 9.322)^{0.3} \ h = 5.22 \ h$$

with

L_{ca} = 4.25 km = 2.641 mi
L = 15 km = 9.322 mi
C_t = 2

The base time of excess rainfall in h is

$$t_R = t_p/5.5 = 5.22/5.5 \ h = 0.94 \ h < 3 \ h$$

Since the requested UH has a time of $(t'_R = 3 \ h) t_R$, *adjusted* t'_p is calculated:

$$t'_p = t_p + \frac{(t'_R - t_R)}{4} = 5.22 + \frac{(3 - 0.94)}{4} = 6.25 \ h$$

The peak discharge of the UH of $t_{R'}$ divided per mi² is given by

$$q_p = C_p \frac{640}{t'_p} = 0.65 \cdot \frac{640}{5.22} = 79.693 \ ft^3/s/mi^2$$

So, the final peak discharge is

$$Q_p = q_p \cdot A = 79.693 \cdot 14.67 = 1169.096 \ ft^3/s$$

where A = 14.67 mi², the area of the basin
The graph widths in 50% and 75% of the peak are, respectively,

$$W_{50} = 830/q_p^{1.1} = 830/(79.693)^{1.1} = 6.72 \ h$$

$$W_{75} = 470/q_p^{1.1} = 470/(79.693)^{1.1} = 3.806 \ h$$

These ordinates extend by 1/3 to the left and 2/3 to the right of the x ordinate of the peak (Figure 3.33).
Finally, the time base of the hydrograph is

$$T = 3 + 3 \cdot \left(\frac{t'_p}{24} \right) = T = 3 + 3 \cdot \left(\frac{6.25}{24} \right) = 3.781 \ days = 91 \ h$$

3.7.6.2 SCS dimensionless hydrograph method

This is an empirical method based on the analysis of a large number of UHs by the Soil (Natural Recourses now) Conservation Service (1971). The y-axis is in units h/h_p and for the volume–time diagram the units are V_a/V (Figure 3.34).

The volume of the rising limb of the hydrograph includes 37.5% approximately of the total volume.

3.7.6.3 Triangular SCS hydrograph method

SCS also proposed the use of a triangular hydrograph with the same percentage of volume on the rising limb (37.5%) (Figure 3.35).

t_b is given by the empirical relationships such as $t_b = 0.1021 \, (L/S^{0.5})^{0.968}$ or $t_b = 0.1L/S^{0.5}$ for simplicity.

This method is a simple version of SCS dimensionless hydrograph method.

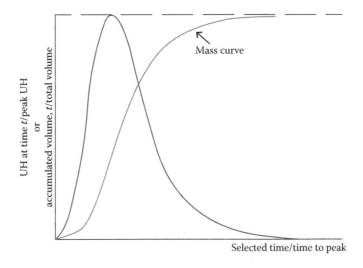

Figure 3.34 The unit hydrograph (UH) at time t/peak UH and accumulated volume/total volume.

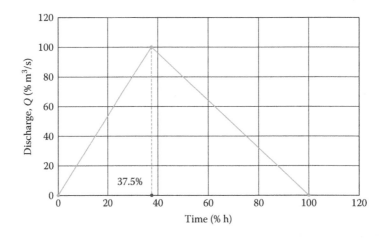

Figure 3.35 The triangular Soil Conservation Service hydrograph.

3.7.7 Instantaneous UH

In order to remove the restriction of uniform distribution of rainfall in time, a case which is difficult to match with the collected actual data of excess rainfall (ER), for the derivation of a UH of t_r-time, a more advanced tool, which is the instantaneous UH (IUH), has to be used.

IUH is a direct runoff hydrograph produced by a unit rainfall (P_{net} = 1 cm) instantaneously over a catchment. Therefore, the IUH is independent of the direct runoff (DR) duration, and it is considered an *impulse* response of the system.

For the derivation of IUH, four methods are applicable.

3.7.7.1 By an S-curve

If an S-curve is produced through the successive lagging of a UH by a time step and then the given UHs are summed up, in order to produce S-curve ordinates, the next method is implied: the curve is lagged by a small time of t_r'. Then, the t_r'-h UGO at time t can be expressed as $U(t_r', t) = (S_t - S_{t'})t_r/t_{r'}$, where S_t is an ordinate in the S-curve at any given time t and $S_{t'}$ is an ordinate after the lag by t_r'-h (Figure 3.36).

As can be seen in Figure 3.1, through deferential analysis when $t_r' \to 0$, then the ordinate of a IUH $U(0, t) = dS_t/dt$ = slope of S-curve.

The method is approximate since the S-curve derived from the rainfall–runoff data is not exact. Nevertheless, it is useful for theoretical purposes.

3.7.7.2 By using a convolution integral

Convolution integral (or Duhamel integral) is the defined function:

$$Q(t) = \int_0^{t'} u(t - \tau) \cdot i(\tau) d\tau$$

$$t' = t, \quad \text{when } t \leq t_0$$
$$t' = t_0, \quad \text{when } t \geq t_0$$

(3.46)

This method is based on the infinite division of a UG in infinitesimal elements of P_{net}.

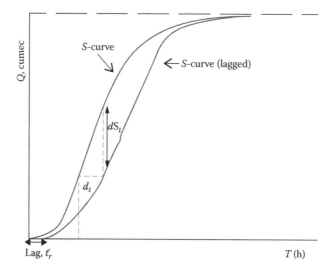

Figure 3.36 The S-curve and the lagged S-curve.

3.7.7.3 By the use of various conceptual models

One model which is commonly applicable is the one proposed by Nash (1957). The concept under this model is that a drainage basin consists of a series of linear reservoirs. Lineal reservoirs are fictitious reservoirs in which the storage is directly proportional to the outflow:

$$S = KO \tag{3.47}$$

Also, the inflow and the outflow of a system can be connected through the principle of continuity as

$$I - O = ds/dt \tag{3.48}$$

Using the condition $O = 0$, at $t = 0$, it is concluded that

$$u(t) = \frac{1}{k\Gamma n} e^{-t/k} \left(\frac{t}{k}\right)^{n-1}, \quad \Gamma n = (n-1)! \tag{3.49}$$

Two parameters in Nash model have significant importance: (1) n, considered a shape parameter, is a measure of the catchment channel storage, which defines the shape of IUH, and (2) K is a scale parameter (small K-value represents lower peak time of the runoff hydrograph and higher K-value the opposite). It is considered that by using a larger number of reservoirs (increasing n), the storage time (k) decreases reversely. Therefore, it is possible to have different sets of these two parameters which give similar results.

3.7.7.4 Routing time–area curve of basins

The principle which underlies this method is the division of a catchment into a series of sub-areas, each contributing inflow into drainage channels (which have storage) due to a flash storm. These sub-areas are called isochrones when the rain falling in any sub-area has the same time of travel to the outflow point.

Clark (1945) used time–area diagrams (TADs) for the calculation of IUH. Two parameters other than TADs are necessary for this method: the time of concentration for the basin t_c and K a storage coefficient. Clark's method is as follows (Figure 3.37):

Any ordinate following O_2, provided the first O_1, is given by the relationship

$$O_2 = C'I + C_2 O_1, \quad \text{where } C' = t/(K + 0.5t), \quad C_2 = 1 - C', \quad O_1 = Q_1$$

where

t is the Δt_c between successive isochrones
K is the storage coefficient
I is the inflow to the isochrone sub-area

K can be determined from an observed hydrograph as follows.

3.7.8 CEH-UK: Revitalized flood hydrograph

The Revitalized Flood Studies Report (ReFSR) of 2005, presented by the Centre of Ecology and Hydrology (CEH, UK) as a replacement of an older method from Flood Engineering Handbook—1999 (FEH), introduced a new approach in the development of event-based rainfall–runoff models concerning hydrologic design. This method was applied since an

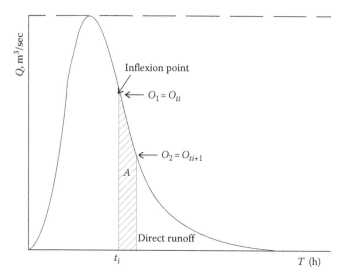

Figure 3.37 The relation of the discharge in the falling limb.

analysis of a large number of observed flood events (1488) and model parameter estimation from 143 gauged catchments in the United Kingdom were carried out. The model parameters were linked to basin characteristics using multivariate linear regression providing a useful tool applying the rainfall–runoff model at ungauged catchments throughout the United Kingdom.

The components which shape the revitalized flood hydrograph (ReFH) are loss, routing and baseflow models based on PDM Moore (1985) model, time to peak t_p parameter adopted from the triangular-shaped IUH and linear reservoir concept, respectively. This method is implemented both in gauged and in ungauged catchments but with different methodologies in each one. This paragraph will focus on the ungauged sites routing model and the estimation of t_p parameter.

The ReFH model uses the triangular IUH concept for routing the direct runoff but goes further with a new shape, i.e. *twisted/bended triangle* form, as shown in Figure 3.38. The new shape is described by a timescale factor t_p and two dimensionless parameters Q_p and Q_k, affecting the height and the twisting/bending, respectively, of the IUH. Q_k is a multiplier to the Q_c ordinate of a non-bended triangular IUH.

In the ungauged sites, the optimal combination of catchment descriptors adopted by Houghton–Carr (1999) helped to predict time to peak by using the FEH methodology. As a result, for the ReFH routing model,

$$t_p = \frac{DPLBAR^{0.60}}{1.56\, PROPWET^{1.09} \left(1 + URBTEXT_{1990}\right)^{3.34} DPSBAR^{0.23}} \tag{3.50}$$

where
  $DPLBAR$ is the mean drainage path length (km)
  $DPSBAR$ is the mean drainage path slope (m/km)
  $PROPWET$ is the proportion of time when soil moisture deficit (mm) was less than or
    equal to 6 mm
  $URBTEXT_{1990}$ is the extent of urban and suburban land cover (year 1990)

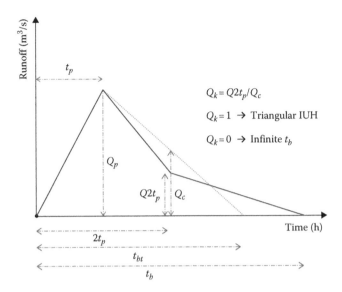

Figure 3.38 Instantaneous dimensionless hydrograph adopted in revitalized flood hydrograph.

From the new geometry of the suggested UH, it is acquired that

$$Q_c = Q_p \frac{t_{bt} - 2t_p}{t_{bt} - t_p} \tag{3.51}$$

where $t_{bt} = 2(t_p/Q_p)$ ensuring unit area under the non-bended triangular UH. Other useful equations completing this new approach are

$$t_b = t_p\left(1 + 2\frac{1 - Q_p}{Q_k Q_c}\right) \tag{3.52}$$

where $Q_p = 0.65$ and $Q_k = 0.8$, which are designated as average values after unsuccessful attempts to relate these parameters. Due to the new approach, the ReFH has a lower peak flow value and longer base time than the triangular IUH. In order to convert the dimensionless IUH into required units of m³/s/mm, a scaling factor is applied. This factor is equal to $AREA/(3.6\ t_p)$, where $AREA$ is in km² and t_p in h.

3.7.9 Shortcomings of the UH and the UH-based models

An important restriction in the use of the UH and consequently in UH-based models is that it is a prerequisite to basin linearity, or in other words, the volume of the runoff has to be proportional to the rain depth falling over the basin. This case is invalid in non-linear basins where depending on the extent of non-linearity, the runoff in the exit of the basin is not proportional to the precipitation over the basin. In order to overcome this issue, a number of rainfall–runoff models in a semi-distributed or fully distributed mode have been developed for the calculation of the runoff hydrographs in non-linear basins as well.

3.7.10 General overview of hydrological models: Specific rainfall–runoff models

Hydrological models refer to a wide range of mathematical transformations using field data and logical assumptions about physical processes. The goal of these models is to make quantitative estimates of different hydrological variables. Such models are, for example, water balance models using simulation tools which represent the physical processes at the scale of a river basin or rainfall–runoff models estimating, e.g. the flood peak, flood volume and flood duration.

The hydrological models are categorized according to the field of application, the time resolution, the spatial scale and the mathematical structure.

The time resolution of a hydrological model depends on the specific application of the model. Water management models need a daily or monthly scale, while flood models adopt finer scales and, in some cases, an hour or less time step.

According to the spatial scale, the models can be divided as follows:

1. *Lumped* models are the models whose parameters are the same over the study area.
2. *Semi-distributed* models are the models which can be applied with differentiation in their parameters in river basin which are divided into discrete areas corresponding to natural sub-basins with the same hydrological and geomorphological characteristics for each one.
3. *Semi-lumped* models are an intermediate form between the lumped and semi-distributed schematization, which considers discrete spatial areas with different characteristics but common parameters for the entire modelling area.
4. *Distributed* models are the models which can be applied in a fully discretized area through GIS applications (using, for example, the method of grids) with different set of characteristics and parameters for each grid.

Depending on the parametric or mathematical structure, the hydrological models can be categorized as follows:

1. *Physical* models are detailed models which use a large number of equations with limited application at a river basin scale due to high uncertainty.
2. *Conceptual* models with parametric relations which represent the hydrological processes on a river basin.
3. *Deterministic* with deterministic description of the variables and processes *and statistical (probabilistic–stochastic)* models which simulate the natural behaviour of the observed measurements by using statistical tools. The probabilistic models use the probability theory, while the stochastic ones include a stochastic component and moreover the time evolution of the processes (time series).
4. *Black-box* models which only take into account water balance equations and consider only the input–output of the models without taking into account the intermediate systems (e.g. the characteristics of the basin).

In the website http://hydrologicmodels.tamu.edu/models.htm, a substantial effort has been undertaken to list, according to pre-defined categories, all the available commercial software of hydrological models, accompanied with the respective links. In the category of the distributed models, MIKE SHE, *Storm Water Management Model* and *Central Valley Groundwater and Surface Water Model* are the most famous, while that of the lumped models, MIKE 11 RR, *Hydrological Model and Forecasting System* (WATFLOOD) and SCS-CN are commonly used. A monthly water balance model with many applications is the *Water Balance Simulation Model*. In the category of GIS applications in hydrology and hydraulics, the frequently used models are BASINS, HEC-GeoRAS, HEC-GeoHMS and *Soil and Water Assessment Tool*, among others. An example of global hydrology model is the *Global Hydrologic Evaluation Model*, while a stochastic model is the *Stochastic Analysis, Modelling and Simulation*.

In the following, the HEC-HMS model is presented as a case study. This model is a UH-based conceptual rainfall–runoff model and is applied to an experimental semi-urban basin in the greater Athens area.

3.8 CASE STUDY USING THE HEC-HMS HYDROLOGICAL MODEL

3.8.1 General information for HEC-HMS

HEC-HMS is a product of the Hydrologic Engineering Center of the USACE. HEC-HMS is an open-source software which can be freely downloaded from the website of USACE (http://www.hec.usace.army.mil/software/hec-hms/). HEC-HMS is designed to simulate the precipitation–runoff processes of watershed systems. It is applicable in a wide range of geographic areas for solving the widest possible range of problems. This includes large river basin water supply and flood hydrology and small urban or natural watershed runoff. Hydrographs produced by the program are used directly or in conjunction with other software.

This model offers the option to import a geographical file with the study area, its sub-basins and their properties, created in GIS environment. One of the options which can be used is the extension HEC-GeoHMS. Particularly, it is a geospatial hydrology toolkit, which can be downloaded from the website of USACE as well. The program allows the user to visualize spatial information, document watershed characteristics, perform spatial analysis and delineate sub-basins and streams. Generally, it creates hydrological inputs for HEC-HMS.

3.8.1.1 Components of HEC-HMS

The analysis is based on formulation of models for the computation of the following *hydrological parameters*: hydrological loss, direct runoff, baseflow and channel flow routing. Once the methods for those models are selected, a *meteorological model* and a *control specification manager* are designed. Then, a list of representative events is drawn, and these events are simulated by the model. Some events need to be used for calibration, and the rest of them can be used for validation purposes. In order to compute each one of the aforementioned models, the user may choose between a wide range of relevant options. The main criteria for the selection of the appropriate method for each model are the availability of data required by each method and the experience of the modeller, based on which the modeller may exclude some methods for specific applications. In the following sections, indicative methods for each model of each element (basin elements, reach elements) are included.

3.8.1.1.1 Basin elements

3.8.1.1.1.1 HYDROLOGICAL LOSS

Hydrological losses include inter alia infiltration, subsurface loss and retention. All these processes and their interaction are simulating in a unified way by HEC-HMS through a *loss method* which needs to be defined for every sub-basin. HEC-HMS provides 12 different loss methods. The *Deficit and constant loss* method, which uses a single soil layer to account for continuous changes in moisture content and is combined with a meteorological model for the evapotranspiration estimation. The *Exponential loss* method which represents incremental infiltration as a logarithmically decreasing function of accumulated infiltration. There is an option for increased initial infiltration, when the soil is particularly dry before the arrival of a storm. However, it is not suitable for continuous simulation. The *Green and Ampt loss* method assumes the soil is initially at uniform moisture content and infiltration takes place with the *piston displacement*. This method automatically accounts for ponding on the surface. In correspondence with the above, the *Gridded deficit constant loss* method implements the deficit constant method on a grid cell by grid cell basis, where each grid cell receives separate precipitation and potential evapotranspiration from the meteorological model. The *gridded Green and Ampt loss* method implements the Green and Ampt method on a grid cell by grid cell basis. The *Gridded SCS curve number loss* method implements the SCS curve number loss method on a grid cell by grid cell basis, as well as, the *gridded soil moisture accounting* with the soil moisture accounting method. The *initial and constant loss* method is simplier and more appropriate for watersheds that lack detailed soil information. The *SCS curve number loss* computes incremental precipitation during a storm by recalculating the infiltration volume at the end of each time interval. The *Smith–Parlange loss* approximates Richard's equation for infiltration into soil by assuming the wetting front can be represented with an exponential scaling of the saturated conductivity. The *soil moisture accounting loss* method uses three layers to represent the dynamics of water movement in the soil and is usually used in conjunction with a canopy and surface method. Some of the methods are designed for simulating events, while others are intended for continuous simulation.

3.8.1.1.1.2 DIRECT RUNOFF

The calculations for direct surface runoff are performed using a transformation method. HEC-HMS provides seven different transform methods. The *Clark UH transform*, which is a synthetic UH method, in the sence that, the user is not required to develop a UH through the analysis of the earlier observed hydrographs. The *kinematic wave transform* is a conceptual model that includes one or two representative planes (pervious-impervious planes) and is applied either in urban areas, or in undeveloped regions. The *ModClark transform* is a linear, quasi-distributed transform that is based on the Clark conceptual UH. This represents the sub-basin as a collection of grid cells. The *SCS UH transform*, it uses observed data collected in small, agricultural watersheds, that are generalized as dimensionless hydrographs and a best approximate hydrograph is developed for general application. The *Snyder UH transform* is a synthetic UH method, where the original data only supported the computation of the peak flow as a result of a unit of precipitation. The *User-specified S-graph transform* method is not synthetic and uses a summation UH to represent the response of a sub-basin to a unit precipitation. The s-graph is defined in terms of percentage of unit flow versus percentage of time lag. The *User-specified UH transform* is, also, not synthetic and a separate UH must be computed for each sub-basin. Usually, these UHs are developed from multiple storm observations when precipitation and flow have been measured at the same time interval.

3.8.1.1.1.3 BASEFLOW

Baseflow method is used for the estimation of the subsurface processes. HEC-HMS provides four different methods. The *Bounded recession baseflow* method is intended primarily for real-time forecasting operation and is very similar to the recession method. However, the monthly baseflow limits can be specified. The baseflow is computed according to the recession methodology and the monthly limits are imposed. After a storm event, this method does not reset the baseflow. The *constant monthly baseflow* does not conserve mass within the sub-basin and is intended primarily for continuous simulation in sub-basins where the baseflow is approximated by a constant flow for each month. The *linear reservoir baseflow* uses a linear reservoir to model the recession of baseflow after a storm event, while it conserves mass within the sub-basin. Infiltration computed by the loss method is connected as the inflow to the linear reservoir. The *non-linear Boussinesq baseflow* is designed to approximate the typical behaviour observed in watersheds when channel flow recedes after an event. This method is intended primarily for event simulation. It may be used for any continuous simulation, as it has the ability to automatically reset after each storm event. The *recession baseflow* is designed to approximate the typical behaviour observed in watersheds when channel flow recedes exponentially after an event. It is intended primarily for event simulation, but, because of its ability to automatically reset after each storm event, it may be used for continuous simulation.

3.8.1.1.2 Reach elements

3.8.1.1.2.1 CHANNEL FLOW ROUTING

Channel flow routing is applied to reach elements that represent stream segments. HEC-HMS provides six different methods to define channel flow routing. Each of these methods included in the model provides a different level of detail and not all methods are equally adept at representing a particular stream. The *kinematic wave routing* method approximates the full unsteady flow equations by ignoring inertial and pressure forces. This method is best suited to fairly steep streams, as it is assumed that the energy slope is equal to the bed slope. The *lag routing method* represents the translation of flood waves. It does not include any representation of attenuation or diffusion processes and it is suitable for short stream segments with a predictable travel time that does not vary with flow depth. The *modified Puls routing* is often called storage routing or level pool routing and uses conservation of mass and a relationship between storage and discharge to route flow through the stream reach. The *Muskingum routing* uses a simple conservation of mass approach to route flow through the stream reach. The *Muskingum-Cunge routing* is based on the combination of the conservation of mass and the diffusion representation of the conservation of momentum. It represents attenuation of flood waves and can be used in reaches with a small slope. The *Straddle stagger routing*, which is the last one, uses empirical representations of translation and attenuation processes to route water through a reach.

3.8.1.1.2.2 LOSS/GAIN METHOD

It is also applied to reach elements, and it represents the modelling of interactions with the subsurface. Particularly, it represents losses from the channel, additions to the channel from groundwater or bidirectional water movements depending on the specific implementation of a method. HEC-HMS provides two different methods. The *Constant loss/gain* method, which uses an empirical relationship to calculate channel loss using a fixed flow rate reduction and a ratio of the flow, and the *Percolation loss/gain* method, which uses a constant infiltration rate in combination with the inundated area in the reach to compute channel loss.

3.8.1.1.3 Meteorological model

The meteorological model is one of the main components of HEC-HMS. Its purpose is to prepare meteorologic boundary conditions for sub-basins. It includes precipitation, evapotranspiration and snowmelt. Six different historical and synthetic precipitation methods are included. There are many methods to define the meteorological model.

3.8.1.1.3.1 PRECIPITATION

There are many precipitation methods to choose from or the user can choose to have no precipitation at all. If the basin model contains sub-basins, the precipitation method is necessary. In the case when the basin model does not contain sub-basins, the precipitation method cannot be used. HEC-HMS provides seven different methods. The *frequency storm* method is designed to produce a synthetic storm from statistical precipitation data. However, this method uses the same parameter data for all sub-basins. The *gauge weights* method, such as the Thiessen polygons method, is designed to work with recording and non-recording precipitation gauges. The *gridded precipitation* is based on the ModClark gridded transform, but it can be used with other transform methods. The most common use of this method is to utilize radar-based precipitation. The *inverse distance* is designed for application in real-time forecasting systems and can use recording gauges that report at regular intervals like 15 min or 1 h. The *SCS storm* implements the design storm developed by the Natural Resources Conservation Service. This method was developed mainly for agricultural applications, but it can be used for other applications as well. The *specified hyetograph* gives the opportunity to user to specify the exact time series to use the hyetograph at sub-basins and it is useful when precipitation data are processed externally by the program and imported without alteration. The *standard project* storm is no longer frequently used, but it is included for projects where it is still necessary.

3.8.1.1.3.2 EVAPOTRANSPIRATION

For the evapotransiration factor, the user can use either one of the three different evapotranspiration methods proposed by HEC-HMS, or no evapotranspiration in the case that basin model does not include sub-basins.

The available methods are: the *gridded Priestley–Taylor* which can be used when ModClark transform method is selected, the *monthly average* which is used when measured pan evaporation data are available and which uses separate parameter data for each sub-basin in the meteorological model.

3.8.1.1.3.3 SNOWMELT

For the snowmelt factor, the user may choose to either use one of the two different methods proposed by HEC-HMS, or not to model snowmelt at all.

Generally, using the temperature determines whether the precipitation previously computed was liquid rain or frozen snow. The *gridded temperature index* is designed to work with ModClark transform method and is the same as the regular temperature index method. The main difference is that the equations for simulating the snowpack are computed separately for each grid cell with separate precipitation and temperature boundary conditions. The *temperature index* is an extension of the degree-day approach to modelling a snowpack. A typical approach to the degree day is to have a fixed amount of snowmelt for each degree above freezing.

3.8.1.1.4 Control specification manager

The final step is the creation of a simulation time window, which is referred to as 'control specification manager'. In this window, the time span and the time interval are set. There is the possibility to have several simulation time windows by adding new ones or copying the already existing ones.

3.8.1.1.4.1 SIMULATION RUNS

Simulation run is the main component which can compute results. Each run is composed of one basin model, one meteorological model and one control specification. Results can be visualized as graphs, summary tables and time series tables. The format of the results will be shown in the example which will be presented in the following.

3.8.1.1.4.2 OPTIMIZATION TRIALS

Having completed the simulation runs, the model gives the opportunity of an automatic optimization. For this process, an observed outflow in a particular position is necessary. Particularly, to complete optimization trial and acquire results, the user should choose a simulation run and an observed outflow. For the optimization trial, the user should define the objective function, the method that minimizes the objective function and the controlling search tolerance. Many parameters can be optimized at the same time.

3.8.1.1.4.3 OBJECTIVE FUNCTION

The objective function measures the goodness of fit between the computed outflow and observed streamflow at the selected element. Seven different functions are provided:

1. Peak-weighted RMS error
2. Percent error peak
3. Percent error volume
4. RMS log error
5. Sum absolute residuals
6. Sum squared residuals
7. Time-weighted error

3.8.1.1.4.4 SEARCH METHOD

Two search methods are available for minimizing the objective function and finding optimal parameter values. The first one is the *univariate gradient method* which evaluates and adjusts one parameter at a time while keeping the other parameters constant. The other one is the *Nelder–Mead method* which uses a downhill simplex to evaluate all parameters simultaneously and determine which parameter to adjust.

3.8.1.1.4.5 CONTROLLING SEARCH TOLERANCE

Two methods are provided for controlling the search process with the univariate gradient or Nelder–Mead methods. The tolerance determines the change in the objective function value which will terminate the search.

3.8.2 Case study

3.8.2.1 Study area

The area which is used for the implementation of the hydrologic model is Rafina Basin. It is a periurban area in the greater southeast Mesogeia region in eastern Attica, Greece. The area covers 126 km² and geographically extends east of Ymittos Mountain to the coastline of Evoikos Gulf. The mean altitude of this region is approximately 227 m, with the maximum value being 909 m and the minimum 0 m. Regarding the ground slope, it ranges from 0% to 37.8% and the mean value is estimated as 7.5% (Figure 3.39).

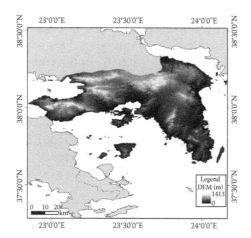

Figure 3.39 The DEM of Attica region and the boundaries of the study area.

3.8.2.2 HEC-HMS environment

3.8.2.2.1 Basin model

The basin model in this study was created through the use of HEC-GeoHMS in ArcGIS 9.3 environment (Figure 3.40).

3.8.2.2.2 Definition of hydrological parameters

3.8.2.2.2.1 HYDROLOGICAL LOSS

The method adopted in this study was *SCS curve number loss*. For the application of this method, it is necessary to define some parameters. The first one is the initial abstraction. This parameter is optional. In case the user leaves it 'blank', it is automatically calculated

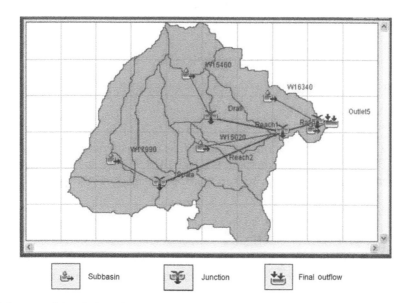

Figure 3.40 Basin model with the sub-basins and the elements.

as 0.2 times the potential retention, which is calculated from the curve number. The second parameter is the curve number (CN) which quantifies the impact of soil properties and land use on the sub-basin. The last one is imperviousness. This percentage can be adjusted to the CN parameter or can be calculated separately. In the current study, the percentage of imperviousness was included in the estimation of CN parameter.

3.8.2.2.2.2 DIRECT RUNOFF

The method adopted for the estimation of direct runoff was *Snyder's synthetic UH*. To define this method, lag time (h) and peaking coefficient are computed. The value of peaking coefficient ranges from 0.4 to 0.8, and it is estimated using the best judgement of the user according to the watershed physical features.

3.8.2.2.2.3 BASEFLOW

The method which was adopted in the current study for the estimation of the contribution of baseflow to the sub-basin outflow was *recession baseflow method*, a method intended for event simulation. At first, the initial discharge type needs to be defined. There are two options: initial discharge and initial discharge per area. The first was selected. The other parameters which have to be estimated for this option are the initial discharge, which is the initial baseflow as a discharge, the recession constant which describes the rate at which baseflow recedes and the threshold type which is the method determining how to reset the baseflow during the event.

3.8.2.2.2.4 CHANNEL FLOW ROUTING

Channel flow routing is applied to reach elements. In the current study, the Muskingum method was applied. For this method, the travel time through the reach (Muskingum K [h]), the weighting between inflow and outflow influence (Muskingum X) and the number of subreaches need to be imported. Regarding the parameter Muskingum X, it ranges from 0 to 0.5. The value 0 corresponds to maximum attenuation, while the value 0.5 corresponds to no attenuation.

3.8.2.2.2.5 ESTIMATION OF INITIAL VALUES OF PARAMETERS

The estimated values for all the earlier methods are presented in Table 3.21.

3.8.2.2.3 *Meteorological model*

In this study, a precipitation method was selected, as the basin model contains sub-basins. Particularly, the gauge weights method was adopted. These weights were calculated using Thiessen polygons. This procedure was performed in GIS environment using ArcMap 9.3.

3.8.2.2.4 *Control specification manager*

In this section, the start and the end of the simulation run were defined, which were the same as the start and the end of the rainfall event that was used. Particularly, the rainfall event of 3 February 2011 was selected. The format of the control specification is clear in Figure 3.41.

Table 3.21 Initial estimated parameters

	Loss (SCS curve number)		Transform (Snyder's UH)	
	CN	Initial abstraction (mm)	Lag time	C_p
W15020	50	5	7.865	0.55
W17990	50	5	5.575	0.45
W15460	40	7	4.36	0.74
W16890	75	5	2.67	0.5
W16340	50	5	4.847	0.6

	Recession baseflow			
	Initial discharge (m^3/s)	Recession constant	Threshold type	
W15020	0.668	0.5	Ratio to peak	0.3
W17990	0.2	0.06	Threshold discharge	0.03 (m^3/s)
W15460	0.05	0.8	Threshold discharge	0.05 (m^3/s)
W16890	0.668	0.5	Ratio to peak	0.3
W16340	0.5	0.06	Ratio to peak	0.03

	Muskingum routing		
	K (h)	X	Subreaches
Reach 1	2	0.2	6
Reach 2	2.5	0.35	15
Reach 3	0.6	0.25	4
Reach 4	0.7	0.25	5

Figure 3.41 Format of the control specification manager.

3.8.2.2.5 Simulation run and results

Having estimated all the aforementioned parameters, the simulation run was accomplished. The results which are presented in the following deal with the positions of Rafina and Drafi (Figure 3.42), in which there were measurements.

3.8.2.2.6 Results

A flood hydrograph has occurred in each of the positions of Rafina and Drafi. Observing Figures 3.43 and 3.44, the gray colour represents the simulated result, while the black colour represents the observed one.

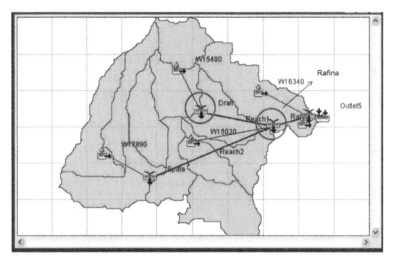

Figure 3.42 Position in which the results are presented.

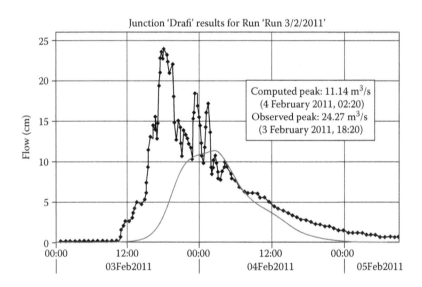

Figure 3.43 Computed and observed flood hydrograph at Drafi station.

3.8.2.2.6.1 RAFINA POSITION

Observing the Figure 3.44, the continuous line represents the simulated result, while the line with points represents the observed one.

3.8.2.2.7 Calibration of model

To get the best results, calibration is a necessary process. The parameter which was calibrated in this study is the CN parameter and the parameters of Snyder's UH. Particularly, with regard to Snyder's UH, a reduction in the lag time (C_t) is implemented to bring the peaks closer and to increase the peaking coefficient (C_p). Finally, a reduction in the values of the

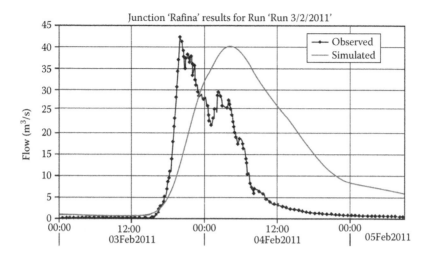

Figure 3.44 Observed and simulated flood hydrograph at Rafina station.

Table 3.22 Calibrated values (C_t, C_p)

	Snyder's UH	
	Lag time (h)	C_p
W15020	2	0.7
W17990	2	0.6
W15460	0.8	0.8
W16890	2	0.6
W16340	3	0.7

Table 3.23 Calibrated CN values

	Loss (SCS curve number)
	CN
W15020	40
W17990	40
W15460	45
W16890	55
W16340	40

CN parameter was realized in order to get better results. The exact values of the calibrated parameters are presented in Tables 3.22 and 3.23.

3.8.2.2.8 Results after calibration

For Drafi, the difference between the observed and the simulated flow is significantly smaller and flow peaks are also captured. For Rafina, the simulated peak flow and the discharge volume are slightly bigger than the observed ones (Figures 3.45 and 3.46).

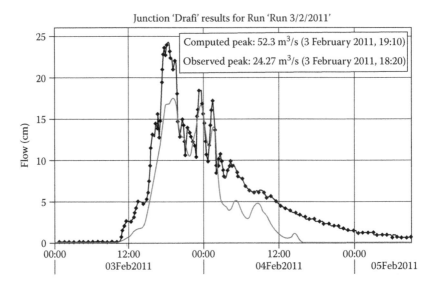

Figure 3.45 Drafi position.

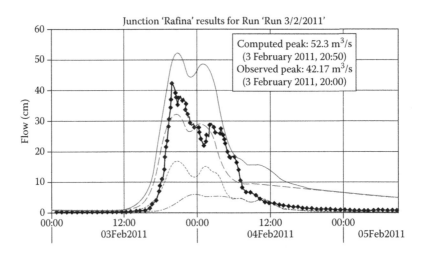

Figure 3.46 Rafina position.

REFERENCES

Barnes, B.S., 1939, The structure of discharge recession curves, *Transactions of the American Geophysical Union*, 20, 721–725.

Bedient, P.B. and Huber, W.C., 1992, *Hydrology and Floodplain Analysis*, 2nd edn. Published by Addison-Wesley.

Bras, R.L., 1990, *An Introduction to Hydrologic Science*, Published by Addison-Wesley.

Butler, S.S., 1957, *Engineering Hydrology*, Prentice-Hall, Englewood Cliffs, NJ.

Clark, C.O., 1945, Storage and the unit hydrograph, *Transactions of the American Society of Civil Engineers*, 110, 1419–1446, Paper No. 2261.

Giandotti, M., 1934, Previsione delle piene e delle magre dei corsi d'acqua. Ministero LL.PP., Memorie e studi idrografici, Vol. 8, Rep. No. 2, Servizio Idrografico Italiano, Rome, Italy (in Italian).

Horner, W.W. and Flynt, F.L., 1936, Relations between rainfall and runoff from small urban areas, *Transactions of the American Society of Civil Engineers*, 101, 140–206.

Horton, R.E., 1945, Erosional development of streams and their drainage basins: Hydrophysical approach to quantitative morphology, *Bulletin of the Geological Society of America*, 56, 275–370.

Jones, G.V., 1997, A synoptic climatological assessment of viticultural phenology, Dissertation, Department of Environmental Sciences, University of Virginia, Charlottesville, VA, 394pp.

Kinnison, H.B. and Colby, B.R., 1945, Flood formulas based on drainage basin characteristics, *Proceedings of the American Society of the Civil Engineers*, 69, 849–876.

Kirpich, Z.P., 1940, Time of concentration in small agricultural watersheds, *Civil Engineering*, 10(6), 362.

Kjeldsen, Thomas Rodding. 2007. The revitalised FSR/FEH rainfall-runoff method. Wallingford, NERC/Centre for Ecology & Hydrology, 68pp, ISBN 0 903741 15 7. (Flood Estimation Handbook, Supplementary Report No. 1).

Linsley, R., Kohler, M., and Paulhus, J., 1982, *Hydrology for Engineers*. 3rd Edn, McGraw-Hill, New York.

Linsley, R.K., Jr., Kohler, M.A., and Paulhus, J.L.H., 1949, *Applied Hydrology*, McGraw-Hill Book Co., New York, 689pp.

McCuen, R.H., 1998, *Hydrologic Analysis and Design*, 2nd edn. Published by Prentice-Hall, Inc., NJ.

Moore, R.J., 1985, The probability-distributed principle and runoff production at point and basin scales, *Hydrological Sciences Journal*, 30, 273–297.

Mimikou, M., Baltas, E., Zobanakis, G., Michaelides, S., Spanides, N., and Gikas, A., 2001, National Data Bank of Hydrological and Meteorological Information (NDBHMI), Athens, Greece, http://ndbhmi.chi.civil.ntua.gr/.

Nash, J.E., 1957, The form of the instantaneous unit hydrograph, *IAHS AISH Publications*, 42, 114–118.

Papazafiriou, Z., 1983, *Surface Hydrology*, Aristotle University of Thessaloniki, Thessaloniki, Greece (in Greek).

SCS, 1957/1971, Hydrology, Engineering Handbook, Supplement A, Section 4, U.S. Department of Agriculture, Washington, DC.

SCS, 1986, Urban hydrology for small watersheds, Technical Release TR-55, Soil Conservation Service, Hydrology Unit, U.S. Department of Agriculture, Washington, DC.

Sherman, L.K., 1932, Streamflow from rainfall by the unit-graph method, *Engineering News-Record*, 108, 501–505.

Sherman, L.K., 1949, The unit hydrograph method, in *Hydrology, pt. 9 of Physics of the Earth*, O.E. Meinzer (ed.), Dover Publications, New York, pp. 514–525.

Wilson, E.M., 1990, *Engineering Hydrology*. Published by Palgrave MacMillan, U.K.

Chapter 4

Probability and statistics in hydrology

4.1 GENERAL CONCEPTS AND DEFINITIONS

The design of hydraulic structures often requires hydrological information governed by the laws of probability and statistics. This results in the necessity of using the statistical and probabilistic analyses of hydrological data. Deterministic hydrology deals with causes and assesses hydrological variables and processes with complete certainty; thus, it cannot cover the random component of the hydrological variables, something significantly impacting both simulation and forecast. This has led to the development of the field of statistical hydrology. This branch of hydrology, which deals with the analysis of hydrological variables, is divided into probabilistic hydrology and stochastic hydrology. Probabilistic hydrology analyzes and synthesizes hydrological events without taking into account their temporal sequence, i.e. the probability P of the variable to have a specific value at the spatiotemporal point of zero. The theory of probability provides the framework for modelling processes that we cannot determine precisely. Stochastic hydrology takes into account the time sequence and considers the probability, previously reported equal to ε, where ε is a positive number between 0 and 1. In hydrological variables, there are two categories of uncertainty: (1) natural randomness and (2) imprecise knowledge of reality.

Each feature which presents variability is called a variable. Variables are divided into continuous, when their values are the result of continuous measurement, for example rainfall depth in mm, and discontinuous or discrete, when the values are enumerated, for example number of rainfall days. The values of the variables are the data. With the term 'population', we imply a set of values of a variable which we want to study. A population is finite when we can count all the members, while it is infinite when its members cannot be enumerated. In an infinite population or even in a finite population, technical and economic reasons make in most cases the study of the entire population impossible. Therefore, we must limit ourselves to studying a part of the population, called sample. A sample is called random when each member of the population has the same probability of being included therein.

Some basic concepts of probabilistic analysis are as follows:

- *Experiment*: Conditions under which we observe a random variable, for example the rainy days of a month and the daily rainfall depth
- *Outcome or sample point*: The result of such observation, for example 10 days of rainfall per month and 30 mm of rainfall on the 10th day
- *Sample space*: The concentration of all possible outcomes of an experiment, for example the rainfall days of a month $S_1 = (0, 1, 2, 3, 4,..., 31)$ and the rainfall depth greater than zero $S_2 = (x/x \geq 0)$

- *Event*: The concentration of possible outcomes or the subset of the sample space, for example the event E_1 = 8 or fewer days of rainfall in the month and the event of the rainfall between 5 and 15 mm, $E_2 = (x/5 \leq x \leq 15)$

The main task of statistics is to draw conclusions from the sample for the population in a way which allows the calculation of the uncertainty of the conclusions drawn. This uncertainty is quantified through the theory of probability which plays a central role in statistics.

The classical definition of probability was formulated by Laplace as follows:

The probability of an event A is the ratio of the number m of cases favourable to it to the number n of equally possible cases. The probability of A is defined as $P(A) = m/n$.

4.1.1 Experiments and sample spaces

We are often interested in sets of points of the sample space. Each set is called incident or event and is a subset of the basic set. The basic properties of these sets are given in the following.

Intersection of two events A and B is the set of all objects which are members of both A and B. It is denoted by $A \cap B$ (Figure 4.1).

The union of the sets A and B is the set of all objects which are a member of A or B or both. It is denoted by $A \cup B$ (Figure 4.2).

The difference of set B from set A is the set of all members of B which are not members of A (Figure 4.3).

Two events A and B of the same sample space Ω are called incompatible or mutually exclusive if the realization of one excludes the other. The relationship for incompatible events is

$$A \cap B = \varnothing$$

where $\varnothing$ is the empty set. The event which is symbolized this way is considered impossible.

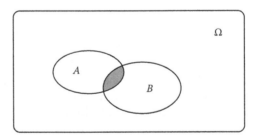

Figure 4.1 Intersection of A and B (A ∩ B).

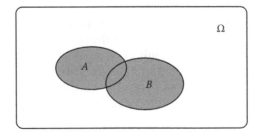

Figure 4.2 Union of the sets A and B.

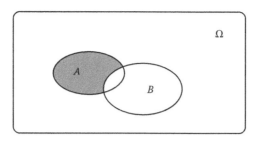

Figure 4.3 Difference of set *B* from set *A*.

A set of events $[A_1,..., A_m]$ is considered to include any mutually exclusive events, if $Ai \cap Aj = \varnothing$ for every i, j.

Two events A and B of the same sample space Ω are called complementary if the realization of one excludes the realization of the other and their sum (union) gives a certain event, the entire sample space Ω. The complementary event of A is denoted by A^C.

A set of events $[A_1,..., A_m]$ is said to exhaust the sample space if $\bigcup\limits_{i=1}^{m} A_i = \Omega$.

Properties

- Commutative: $A \cup B = B \cup A, \quad A \cap B = B \cap A$
- Combinational: $A \cup (B \cup C) = (A \cup B) \cup C, \quad A \cap (B \cap C) = (A \cap B) \cap C$
- Distributive: $A \cap (B \cup C) = (A \cap B) \cup (A \cap C), \quad A \cup (B \cap C) = (A \cup B) \cap (A \cup C)$

Furthermore,

$$\left(A^C\right)^C = A$$

$$A \cap \Omega = A, \quad A \cup \Omega = \Omega, \quad A \cap \varnothing = \varnothing, \quad A \cup \varnothing = A$$

$$A \cap A^C = \varnothing, \quad A \cup A^C = \Omega, \quad A \cap A = A, \quad A \cup A = A$$

$$(A \cup B)^C = A^C \cap B^C, \quad (A \cap B)^C = A^C \cup B^C$$

$$\left(\bigcup\limits_{i=1}^{N} Ai\right)^c = \bigcap\limits_{i=1}^{N} Ai^C, \quad \left(\bigcap\limits_{i=1}^{N} Ai\right)^c = \left(\bigcup\limits_{i=1}^{N} Ai\right)^c = \bigcap\limits_{i=1}^{N} Ai^C$$

$$A - B = A \cap B^C$$

The commutative, combinational and distributive properties do not apply to the difference:

$$(A - B) \cup B = A \cup B \neq A \cup B - B \cup B = A - B$$

4.1.2 Probability function

A probability function $P[\cdot]$ is a function which gives a value (real number) in each event, satisfying the following relations:

$$0 \leq P[A] \leq 1$$

$$P[\Omega] = 1$$

If $A_1, A_2, \ldots, A_N$ is a set of mutually excluded events, then the following relation applies:

$$P\left[\bigcup_{i=1}^{\infty} A_i\right] = \sum_{i=1}^{\infty} P[A_i]$$

The following are the properties of a probability function:

$$P[\varnothing] = 0$$

$$P[A_1 \cup A_2 \cup \cdots \cup A_N] = \sum_{i=1}^{N} P[A_i]$$

$$P[A^C] = 1 - P[A]$$

For any two events A and B, the following applies:

$$P[A \cup B] = P[A] + P[B] - P[A \cap B] \text{ (additive law)}$$

$$P[A] = P[A \cap B] + P[A \cap B^C]$$

$$P[A - B] = P[A \cap B^C] = P[A] - P[A \cap B]$$

If $A \subset B$
   then $P[A] \leq P[B]$
(The symbol $\subset$ denotes that the event A is a subset of event B.)

4.1.3 Conditional probability

The *conditional probability* of an event A is the probability that the event will occur given the knowledge that an event $B \neq \varnothing$ has already occurred. This probability is written as $P[A \mid B]$

$$P[A \mid B] = \frac{P[A \cap B]}{P[B]}, \quad \text{if } P[B] > 0$$

It is not defined if $P[B] = 0$.

The conditional probability is obtained by applying the general rule of probability multiplication as follows:

$$P[A \cap B] = P[A] \, P[B|A] \text{ if } P[A] > 0 \quad \text{and} \quad P[A \cap B] = P[B] \, P[A \mid B] \text{ if } P[B] > 0$$

The events A and B are *independent* when the probability of event A is not affected by the probability of event B. Then $P[A \mid B] = P[A]$ and $P[B \mid A] = P[B]$, so $P[A \cap B] = P[A] \, P[B]$.

The properties of conditional probability are

$$P[A \cap B] = P[A \mid B]P[B] = P[B \mid A]P[A]$$

$$0 \le P[A \mid B] \le 1$$

$$P[\Omega \mid B] = 1$$

$$P\left[\bigcup_{i=1}^{m} A_i \mid B \right] = \sum_{i=1}^{m} P[A_i \mid B]$$

$$P[A^C \mid B] = 1 - P[A \mid B]$$

$$P[A_1 \cup A_2 \mid B] = P[A_1 \mid B] + P[A_2 \mid B] - P[A_1 \cap A_2 \mid B]$$

If $A_1 \subset A_2$, then

$$P[A_1 \mid B] \le P[A_2 \mid B]$$

4.1.4 Total probability and Bayes' theorem

If $B_1, B_2, \ldots B_M$ is a set of mutually excluded events of the sample space, then for each event A, with $P[A] > 0$, the following relation applies:

$$P[A] = \sum_{i=1}^{m} P[A \mid B_i]P[B_i] = \sum_{i=1}^{m} P[B_i \mid A]P[A]$$

Bayes' theorem

If $B_1, B_2, \ldots B_M$ is a set of mutually excluded events of the sample space, then for each event A, with $P[A] > 0$, the following relation applies:

$$P[B_K \mid A] = \frac{P[B_K \cap A]}{P[A]} = \frac{P[A \mid B_K]P[B_K]}{\sum_{i=1}^{m} P[A \mid B_i]P[B_i]}$$

Example 4.1

A company is supplied with three types of automatic telemetric equipment for measuring rainfall depth, type A, type B and type C, at 50%, 30% and 20%, respectively. If the electronic equipment of types A, B and C are faulty by 2%, 3% and 10%, respectively, calculate the following:

1. The probability that the equipment is of type A and good
2. The probability that the equipment is faulty
3. The probability that the equipment is of type A given that it is faulty

Solution

1. The probability that the equipment is in good condition (K) given that it is of type A is complementary to the probability that the equipment is defective (E) and of type A, i.e.
$P[K \mid A] = 1 - P[E \mid A] = 1 - 0.02 = 0.98$. Thus, the probability that the equipment is of type A and good is $P(A \cap K) = P(A)*P(K \mid A) = 0.5*0.98 = 0.49$.
2. The probability that the equipment is faulty is given by the relationship

$$P(E) = P(E \mid A)*P(A) + P(E \mid B)*P(B) + P(E \mid C)*P(C) = 0.02*0.5 + 0.03*0.3 + 0.1*0.2$$
$$= 0.039$$

3. The probability that the equipment is of type A given that it is faulty is given by

$$P[A \mid E] = \frac{P[A \cap E]}{P[E]} = \frac{P[E \mid A]*P[A]}{P[E]} = \frac{0.02*0.5}{0.039} \cong 0.256$$

Example 4.2

Two automatic rain gauges A and B operate independently. If the percentage of operating time is 0.85 for A and 0.70 for B, what is the probability that at a given point of time?

1. Both of them operate
2. None of them operates

Solution

1. The probability that both of them operate is

$$P(A \cap B) = P(A)*P(B) = 0.85*0.70 = 0.595$$

2. The probability that none of them operate is

$$P[\bar{A} \cap \bar{B}] = P(\bar{A})*P(\bar{B}) = 0.15*0.30 = 0.045$$

where
The probability of operation failure of A is

$$P(\bar{A}) = 1 - P(A) = 1 - 0.85 = 0.15$$

The probability of operation failure of B is

$$P(\bar{B}) = 1 - P(B) = 1 - 0.70 = 0.30$$

4.2 RANDOM VARIABLE

Each outcome of a random experiment is usually characterized by a numeric value. The characterization of each outcome with a number is equivalent to the definition of a function which takes a certain numerical value for each outcome of the experiment. That function is called random variable. The function of the random variable is also a random variable. If X is a random variable and $f(x)$ is a function of X, then the following applies:

1. Mean (expected) value $E[X]$ or μ_x of the random variable X
   This is given by the following relationship:

$$\mu_x = E[X] = \begin{cases} \displaystyle\sum_j x_j f_X(x_j) \text{ (discrete)} \\[2ex] \displaystyle\int_{-\infty}^{+\infty} x f_X(x) dx \text{ (continuous)} \end{cases}$$

   $E[X]$ defines the centre of weight of $f_X(\cdot)$ (Figure 4.4).

2. Other measurements of the central tendency
   For (median) med (X):

$$\int_{-\infty}^{\text{med}(X)} f_X(x) dx = \frac{1}{2} \text{ (continuous)}$$

$$P[X \le \text{med}(x)] \ge \frac{1}{2} \quad \text{and} \quad P[X \ge \text{med}(X)] \ge \frac{1}{2} \text{ (generally)}$$

Assuming that the N values are ranked in order, the median is the value on either side of which there is an equal number of values. If the number N is even, the median is taken as the average of the two middle values. The most probable mode, mod (X), is given as

$$\text{mod}(X) = \arg\left\{\max_x f_X(x)\right\}$$

The most probable mean of a parameter is the value which has the greatest frequency (Figure 4.5).

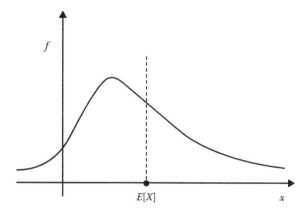

Figure 4.4 Mean (expected) value.

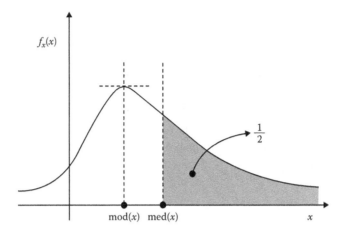

Figure 4.5 Median and most probable mode.

3. Expected value of a function of random variable
   If $g(\cdot)$ is a real function of the random variable X, then

$$E[g(X)] = \begin{cases} \displaystyle\sum_j g(x_j)f_X(x_j) \text{ (discrete)} \\[2em] \displaystyle\int_{-\infty}^{+\infty} g(x)f_X(x)dx \text{ (continuous)} \end{cases}$$

4. Properties of the mean
   a. $E[C] = C$ for every constant C
   b. $E[Cg(X)] = CE[g(X)]$ for every constant C and real function $g(\cdot)$
   c. $E[C_1g_1(X) + C_2g_2(X)] = C_1E[g_1(X)] + C_2E[g_2(X)]$
   (mean is linear)
5. Variance, Var $[X]$

$$\text{Var}[X] = E[(X - \mu_x)^2] = \begin{cases} \displaystyle\sum_j (x_j - \mu_x)^2 f_X(x_j) \text{ (discrete)} \\[2em] \displaystyle\int_{-\infty}^{+\infty} (x - \mu_x)^2 f_X(x)dx \text{ (continuous)} \end{cases}$$

The variance Var$[X]$ is the moment of inertia of the probability density function (pdf) around the axis which passes through the mean. It is a measure of the opening of the pdf (Figure 4.6).

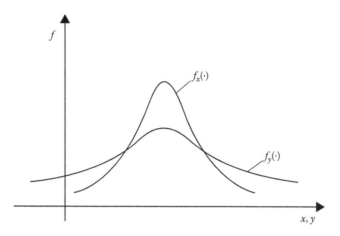

Figure 4.6 Variance, Var[X] < Var[Y].

6. Standard deviation

$$\sigma_x = +\sqrt{\text{Var}[X]}$$

7. Variation coefficient
   The coefficient of variation is a dimensionless parameter of dispersion, which is defined as the ratio of standard deviation to the mean and is given by $V_x = \sigma_x/\mu_x$.
8. Variation properties
   a. Var $[c] = 0$ for every constant c
   b. Var $[cX] = c^2\text{Var}[X]$
   c. Var $[a + bX] = b^2\text{Var}[X]$
   d. Var $[X] = E[X^2] - (E[X])^2$

   Proof
   $$\begin{aligned}
   \text{Var } [X] &= E[(X-E[X])^2] = E[X^2-2XE[X]+(E[X])^2] \\
   &= E[X^2]-2(E[X])^2+(E[X])^2 \\
   &= E[X^2]-(E[X])^2
   \end{aligned}$$
9. Asymmetry α_x

   $$\alpha_x = E[(X - \mu_x)^3]$$

   α_x is a measure of the symmetry of the pdf (Figure 4.7).
   Coefficient of asymmetry: $\gamma_x = \alpha_x/\sigma_x^3$
10. Kurtosis K_x:

    $$K_x = \frac{E[(X - \mu_x)^4]}{\sigma_x^4}$$

    K_x is a measure of the kurtosis of the pdf (Figure 4.8).

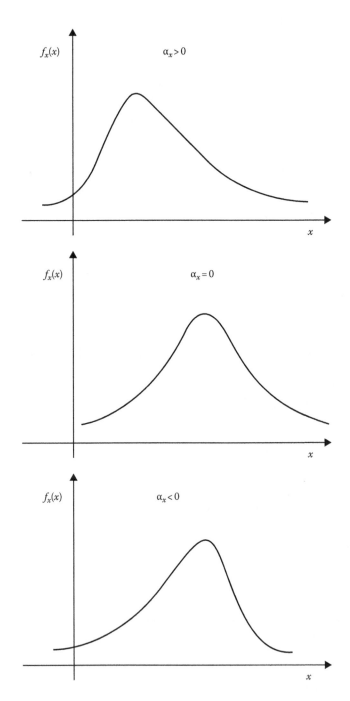

Figure 4.7 Asymmetry.

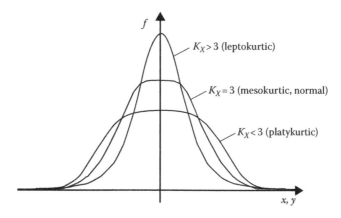

Figure 4.8 Kurtosis.

11. General functions of statistical moments:
    a. Central moments (around the mean)

$$\mu_x^{(r)} = E[(X - \mu_x)^r]$$

    ($r = 2$, dispersion; $r = 3$, asymmetry)
    b. Moments around the origin of the axes

$$m_x^{(r)} = E[X^r]$$

    ($r = 1$, mean value)
12. Function for the creation of moments $m_x(t)$:

$$m_x(t) = F[e^{tX}] = \begin{cases} \displaystyle\sum_{x_j} e^{tx_j} f_X(x_j) \text{ (discrete)} \\ \displaystyle\int_{-\infty}^{+\infty} e^{tx} f_X(x)dx \text{ (continuous)} \end{cases}$$

$$\frac{d^r m_x(0)}{dt^r} = E[X^r] = m_x^{(r)}$$

13. Standard error:
    One of the most important theorems in statistics is the central limit theorem, according to which the mean value of n independent variables with mean m and standard deviation σ also follows the normal distribution with mean *m* and standard deviation $\sigma/\sqrt{n}$ or $s/\sqrt{n}$ in the case of sample. The term $s/\sqrt{n}$ is called standard error.

14. Frequency distributions:

In the science of hydrology, depending on the time step used for data analysis, the amount of data may be too large; thus, their management is difficult. In such cases, the data are divided into classes. The selection of space in classes affects the appearance of the histogram. Several scientists have derived various relationships on the number of classes. Spiegel (1961) proposed that the number of classes should be from 5 to 20, Steel and Torrie (1960) suggested that space should be between ¼ and ½ of the standard deviation of the data, and Sturges (1926) presented the equation $m = 1 + 3.3$ $\log N$, where m is the number of classes and N the number of data. After determining the number of classes and their respective limits, the number of values in each class is calculated. This number is called the absolute frequency of the class.

Relative frequency is the percentage ratio of the absolute frequency of the class to the absolute frequency of all classes. Cumulative frequency is the sum of the frequencies of all classes up to the limit, while relative cumulative frequency is the sum of the relative frequencies up to the limit. Based on these reasons, the respective polygons of frequency or diagrams giving information on the data can be generated. The histograms are used for the graphic representation of frequency distributions of continuous data. The same applies for frequency polygons.

4.3 DISTRIBUTIONS

From the plethora of theoretical distributions available in the literature, this chapter focuses on those which find applications in hydrology. The distributions described in the following are divided into discrete or continuous, depending on whether the variable is discrete or continuous. The study and understanding of the theoretical distributions are necessary to approximate the empirical distributions of historical data.

4.3.1 Normal distribution

It is the most important distribution from the theoretical and practical points of view. It arises in cases where a random variable is the sum of a large number of independent random variables, each contributing a small amount in the total score (central limit theorem). The normal distribution is characterized by a symmetric pdf in the shape of a bell (also known as Gaussian distribution). It has two parameters, the mean μ and standard deviation σ. The pdf of the normal distribution is given by the relationship

$$f_X(x; \mu_x, \sigma_x) = \frac{1}{\sqrt{2\pi}\sigma_x} e^{-\frac{1}{2}\left[\frac{x-\mu_x}{\sigma_x}\right]^2}, \quad -\infty < x < +\infty$$

The mean value of the normal distribution is denoted as $E[X] = \mu_x$, the mod is $\text{mod}(X) = \arg\left\{\max_x f_X(x)\right\}$, and the variance is denoted as $\text{Var}[X] = \sigma^2_x$

$f(x) \to 0$ as $x \to \pm\infty$ and $f(x)$ is maximum at $x = \mu$. Therefore, $f(x)$ is symmetric around μ. The central moment r of the distribution is given by

$$\mu_x^{(r)} = E[(X - \mu_x)^r] = \begin{cases} 0, & \text{if } r = 1,3,5,\ldots \\ -1 \cdot 3 \cdot 5 \cdots (r-1) \cdot \sigma_x^r, & \text{if } r = 2,4,6,\ldots \end{cases}$$

The distribution function of the normal distribution is given by the relationship

$$F(x) = \frac{1}{\sqrt{2\pi}} \int_{-\infty}^{x} e^{-\frac{1}{2}\left(\frac{x-\mu}{\sigma}\right)^2} x dx$$

The values of $F(x)$ represent the area under the curve of the normal distribution. To facilitate the calculation of the values of pdf and the distribution function, the standard normal distribution variable $Z = (x - \mu)/\sigma$ is used, which follows the normal distribution and has a zero mean and a standard deviation of 1. It appears that the pdf for the variable Z is given by

$$f(Z) = \frac{1}{\sqrt{2\pi}} \exp\left(\frac{-Z^2}{2}\right)$$

and the cumulative distribution function is

$$F(Z) = \frac{1}{\sqrt{2\pi}} \int_{-\infty}^{z} \exp\left(\frac{-Z^2}{2}\right) dZ$$

The value of Z is taken from tables, where it is related to the values of $F(Z)$, and then x is calculated from the relationship

$$x = \mu + \sigma^* Z$$

for any value of mean (μ) and standard deviation (σ).

Table 4.1 shows $F(Z)$ in relation to Z for $0 \leq Z \leq 3.09$ and Table 4.2 for $-3.09 \leq Z \leq -0.10$.

In a specially designed graph, known as a normal distribution graph or a Gaussian graph, the distribution function is a straight line. Specifically, in this graph, the pairs of points are designed by plotting the value of the standard normal variable Z and the respective value of the variable x. The chart is designed in such a way that by joining these points a straight line is obtained. A normal distribution diagram is given in Figure 4.9.

All these calculations concern a sample of the population. Thus, the average value corresponds to the average value of the representative sample. In practice, it is useful to set limits around the mean of the sample, within which is the mean of the population with some probability. These limits are known as confidence limits and for the normal distribution are given by the following equations:

$$X(T)_{,max} = x(T) + Z_{(1+\alpha)/2}^* \sigma_T \quad \text{and} \quad X(T)_{,min} = x(T) - Z_{(1+\alpha)/2}^* \sigma_T$$

$$\text{with } \sigma_T = \delta \frac{\sigma}{\sqrt{N}}, \delta = \sqrt{1 + \frac{K(T)^2}{2}} \quad \text{and} \quad K(T) = Z_{(1-1/T)}$$

where

$Z_{(1+\alpha)/2}$ is the variable of the standardized normal distribution for confidence level α%
σ_T is the standard deviation of $X(T)$
σ is the standard deviation of the sample
N is the number of observations of the sample

Table 4.1 Cumulative distribution function F(Z) for $0 \leq Z \leq 3.09$

	0	0.01	0.02	0.03	0.04	0.05	0.06	0.07	0.08	0.09
0	50.00	50.40	50.80	51.20	51.60	51.99	52.39	52.79	53.19	53.59
0.1	53.98	54.38	54.78	55.17	55.57	55.96	56.36	56.75	57.14	57.53
0.2	57.93	58.32	58.71	59.10	59.48	59.87	60.26	60.64	61.03	61.41
0.3	61.79	62.17	62.55	62.93	63.31	63.68	64.06	64.43	64.80	65.17
0.4	65.54	65.91	66.28	66.64	67.00	67.36	67.72	68.08	68.44	68.79
0.5	69.15	69.50	69.85	70.19	70.54	70.88	71.23	71.57	71.90	72.24
0.6	72.57	72.91	73.24	73.57	73.89	74.22	74.54	74.86	75.17	75.49
0.7	75.80	76.11	76.42	76.73	77.04	77.34	77.64	77.94	78.23	78.52
0.8	78.81	79.10	79.39	79.67	79.95	80.23	80.51	80.78	81.06	81.33
0.9	81.59	81.86	82.12	82.38	82.64	82.89	83.15	83.40	83.65	83.89
1	84.13	84.38	84.61	84.85	85.08	85.31	85.54	85.77	85.99	86.21
1.1	86.43	86.65	86.86	87.08	87.29	87.49	87.70	87.90	88.10	88.30
1.2	88.49	88.69	88.88	89.07	89.25	89.44	89.62	89.80	89.97	90.15
1.3	90.32	90.49	90.66	90.82	90.99	91.15	91.31	91.47	91.62	91.77
1.4	91.92	92.07	92.22	92.36	92.51	92.65	92.79	92.92	93.06	93.19
1.5	93.32	93.45	93.57	93.70	93.82	93.94	94.06	94.18	94.29	94.41
1.6	94.52	94.63	94.74	94.84	94.95	95.05	95.15	95.25	95.35	95.45
1.7	95.54	95.64	95.73	95.82	95.91	95.99	96.08	96.16	96.25	96.33
1.8	96.41	96.49	96.56	96.64	96.71	96.78	96.86	96.93	96.99	97.06
1.9	97.13	97.19	97.26	97.32	97.38	97.44	97.50	97.56	97.61	97.67
2	97.72	97.78	97.83	97.88	97.93	97.98	98.03	98.08	98.12	98.17
2.1	98.21	98.26	98.30	98.34	98.38	98.42	98.46	98.50	98.54	98.57
2.2	98.61	98.64	98.68	98.71	98.75	98.78	98.81	98.84	98.87	98.90
2.3	98.93	98.96	98.98	99.01	99.04	99.06	99.09	99.11	99.13	99.16
2.4	99.18	99.20	99.22	99.25	99.27	99.29	99.31	99.32	99.34	99.36
2.5	99.38	99.40	99.41	99.43	99.45	99.46	99.48	99.49	99.51	99.52
2.6	99.53	99.55	99.56	99.57	99.59	99.60	99.61	99.62	99.63	99.64
2.7	99.65	99.66	99.67	99.68	99.69	99.70	99.71	99.72	99.73	99.74
2.8	99.74	99.75	99.76	99.77	99.77	99.78	99.79	99.79	99.80	99.81
2.9	99.81	99.82	99.82	99.83	99.84	99.84	99.85	99.85	99.86	99.86
3	99.87	99.87	99.87	99.88	99.88	99.89	99.89	99.89	99.90	99.90

The normal distribution is the most widely used distribution. It is used for the analysis of variance, the hypothesis testing, the estimation of random errors of hydrological measurements, the comparison of distributions and the generation of random numbers. A random variable is expected to follow a normal distribution if it is the sum of independent effects.

Remark 1: The tables of normal distribution calculate the distribution of the standardized normal variable $Z = (X-\mu_x)/\sigma_x$, which has zero mean value and unit standard deviation. Then, the following applies:

$$f_U(Z;0,1) = \frac{1}{\sqrt{2\pi}} e^{(-1/2)Z^2} \quad -\infty < Z < +\infty$$

$$f_X(x) = \frac{1}{\sigma_x} f_z\left(\frac{x-\mu_x}{\sigma_x}\right)$$

Table 4.2 Cumulative distribution function F(Z) for −3.09 ≤ Z ≤ −0.10

	0	0.01	0.02	0.03	0.04	0.05	0.06	0.07	0.08	0.09
−3	0.13	0.13	0.13	0.12	0.12	0.11	0.11	0.11	0.10	0.10
−2.9	0.19	0.18	0.18	0.17	0.16	0.16	0.15	0.15	0.14	0.14
−2.8	0.26	0.25	0.24	0.23	0.23	0.22	0.21	0.21	0.20	0.19
−2.7	0.35	0.34	0.33	0.32	0.31	0.30	0.29	0.28	0.27	0.26
−2.6	0.47	0.45	0.44	0.43	0.41	0.40	0.39	0.38	0.37	0.36
−2.5	0.62	0.60	0.59	0.57	0.55	0.54	0.52	0.51	0.49	0.48
−2.4	0.82	0.80	0.78	0.75	0.73	0.71	0.69	0.68	0.66	0.64
−2.3	1.07	1.04	1.02	0.99	0.96	0.94	0.91	0.89	0.87	0.84
−2.2	1.39	1.36	1.32	1.29	1.25	1.22	1.19	1.16	1.13	1.10
−2.1	1.79	1.74	1.70	1.66	1.62	1.58	1.54	1.50	1.46	1.43
−2	2.28	2.22	2.17	2.12	2.07	2.02	1.97	1.92	1.88	1.83
−1.9	2.87	2.81	2.74	2.68	2.62	2.56	2.50	2.44	2.39	2.33
−1.8	3.59	3.51	3.44	3.36	3.29	3.22	3.14	3.07	3.01	2.94
−1.7	4.46	4.36	4.27	4.18	4.09	4.01	3.92	3.84	3.75	3.67
−1.6	5.48	5.37	5.26	5.16	5.05	4.95	4.85	4.75	4.65	4.55
−1.5	6.68	6.55	6.43	6.30	6.18	6.06	5.94	5.82	5.71	5.59
−1.4	8.08	7.93	7.78	7.64	7.49	7.35	7.21	7.08	6.94	6.81
−1.3	9.68	9.51	9.34	9.18	9.01	8.85	8.69	8.53	8.38	8.23
−1.2	11.51	11.31	11.12	10.93	10.75	10.56	10.38	10.20	10.03	9.85
−1.1	13.57	13.35	13.14	12.92	12.71	12.51	12.30	12.10	11.90	11.70
−1	15.87	15.62	15.39	15.15	14.92	14.69	14.46	14.23	14.01	13.79
−0.9	18.41	18.14	17.88	17.62	17.36	17.11	16.85	16.60	16.35	16.11
−0.8	21.19	20.90	20.61	20.33	20.05	19.77	19.49	19.22	18.94	18.67
−0.7	24.20	23.89	23.58	23.27	22.96	22.66	22.36	22.06	21.77	21.48
−0.6	27.43	27.09	26.76	26.43	26.11	25.78	25.46	25.14	24.83	24.51
−0.5	30.85	30.50	30.15	29.81	29.46	29.12	28.77	28.43	28.10	27.76
−0.4	34.46	34.09	33.72	33.36	33.00	32.64	32.28	31.92	31.56	31.21
−0.3	38.21	37.83	37.45	37.07	36.69	36.32	35.94	35.57	35.20	34.83
−0.2	42.07	41.68	41.29	40.90	40.52	40.13	39.74	39.36	38.97	38.59
−0.1	46.02	45.62	45.22	44.83	44.43	44.04	43.64	43.25	42.86	42.47

$$F_X(x) = P[X \le x] = P\left[\frac{X - \mu_x}{\sigma_x} \le \frac{x - \mu_x}{\sigma_x}\right] = P\left[z \le \frac{x - \mu_x}{\sigma_x}\right]$$

$$= F_z\left(\frac{x - \mu_x}{\sigma_x}\right) = \frac{1}{\sqrt{2\pi}} \int_{-\infty}^{(x-\mu_x)/\sigma_x} e^{-(1/2)z^2} \, dz$$

Remark 2: $F_X(x) = 1 - F_X(-x)$ symmetrical

Remark 3: If $X_1 \sim N(m_{x_1}, \sigma_{x_1})$ and $X_2 \sim N(m_{x_2}, \sigma_{x_2})$ are two independent normal random variables, then $X = X_1 + X_2$ is also normal with $N\left(m_{x_1} + m_{x_2}, \sqrt{\sigma_{x_1}^2 + \sigma_{x_2}^2}\right)$.

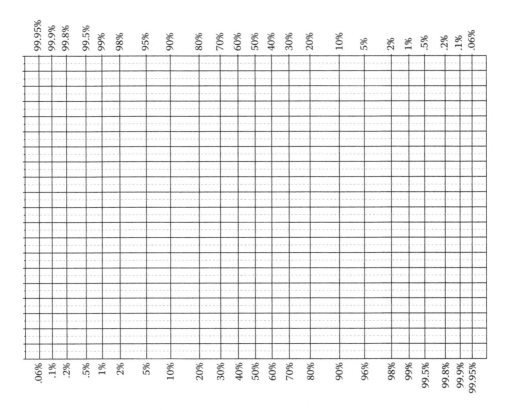

Figure 4.9 Normal distribution diagram.

4.3.2 Log-normal distribution

A log-normal distribution occurs when a random variable is the product of many random variables, e.g. in sediment transport, where the volume of particles is the multiplier effect of many conflicts such as the failure of structures or machines:

$$X = X_1 X_2 \ldots X_N \rightarrow \ln X = \ln X_1 + \ln X_2 + \cdots + \ln X_N$$

From the central limit theorem, given that the number N is relatively large, $\ln X$ will be normally distributed. Let $\mu_{\ln X}$, $\sigma_{\ln X}$ be the mean and standard deviation of $\ln X$. Then, the pdf of the log-normal distribution is

$$f_X(x; \mu_{\ln X}, \sigma_{\ln X}) = \begin{cases} \dfrac{1}{x\sigma_{\ln X}\sqrt{2\pi}} \exp\left\{-\dfrac{1}{2}\left[\dfrac{\ln x - \mu_{\ln x}}{\sigma_{\ln X}}\right]^2\right\}, & x \geq 0 \\ 0, & \text{else} \end{cases}$$

or

$$f_x(x; \text{med}(X), \sigma_{\ln X}) = \begin{cases} \dfrac{1}{x\sigma_{\ln X}\sqrt{2\pi}} \exp\left\{-\dfrac{1}{2}\left[\dfrac{1}{\sigma_{\ln X}}\ln\left(\dfrac{x}{\text{med}(X)}\right)\right]^2\right\}, & \text{if } x \geq 0 \\ 0, & \text{else} \end{cases}$$

$$E[X] = \text{med}(X) \exp\left\{\frac{1}{2}\sigma_{\ln X}^2\right\} = \exp\left\{\mu_{\ln X} + \frac{1}{2}\sigma_{\ln X}^2\right\} = \mu_x$$

$$\text{Var}[X] = (E[X])^2 \left(\exp\left\{\sigma_{\ln X}^2\right\} - 1\right)$$

$$\text{med}(X) = E[X] \exp\left\{-\frac{1}{2}\sigma_{\ln X}^2\right\} = \exp\left\{\mu_{\ln X}\right\}$$

The variation coefficient V_x and asymmetry C_s are given from the following relations:

$$V_X^2 = \left(\frac{\sigma_x}{\mu_x}\right)^2 = \exp\left\{\sigma_{\ln X}^2\right\} - 1$$

$$C_s = 3V_x + V_x^2$$

$$\mu_{\ln X} = \ln\left(\text{med}(X)\right)$$

$$\sigma_{\ln X}^2 = \ln\left\{V_X^2 + 1\right\}$$

The confidence limits for confidence level α are given as follows:

$$X(T)_{,\max} = x(T) + Z_{(1+\alpha)/2}{}^*\sigma_T \quad \text{and} \quad X(T)_{,\min} = x(T) - Z_{(1+\alpha)/2}{}^*\sigma_T$$

$$\text{with } \sigma_T = \delta\frac{\sigma}{\sqrt{N}}, \delta = \sqrt{1 + \frac{K(T)^2}{2}} \quad \text{and} \quad K(T) = Z_{(1-1/T)}$$

where
$Z_{(1+\alpha)/2}$ is the variable of the standard normal distribution for confidence level $\alpha\%$
σ_T is the standard deviation of $X(T)$
σ is the standard deviation of the logarithms of the sample
N is the number of observations of the sample

Remark 1: If X follows log-normal distribution, then the same applies for $Y = aX^b$ where

$$\mu_{\ln Y} = \ln a + b\mu_{\ln X}, \quad \sigma_{\ln Y}^2 = b^2\sigma_{\ln X}^2$$

Remark 2: If $\{X_i\}_{i=1}^N$ follow log-normal distribution and are independent, then the same applies for $Y = aX_1^{b_1} X_2^{b_2} \ldots X_N^{b_N}$ and

$$\mu_{\ln Y} = \ln a + \sum_{i=1}^N b_i\mu_{\ln X_i}, \quad \sigma_{\ln Y}^2 = \sum_{i=1}^N b_i^2\sigma_{\ln X_i}^2$$

Remark 3: Using the tables of normal distribution (Tables 4.1 and 4.2),

$$f_X\left(x;\sigma_{\ln X},\mathrm{med}(X)\right) = \frac{1}{x\sigma_{\ln X}}f_z\left(\frac{\ln\left(x/\mathrm{med}(X)\right)}{\sigma_{\ln X}}\right)$$

$$F_X(x;\sigma_{\ln X},\mathrm{med}(X)) = F_z\left(\frac{\ln\left(x/\mathrm{med}(X)\right)}{\sigma_{\ln X}}\right)$$

where f_U and F_U are the pdf and the cumulative probability function of the standard normal variable.

4.4 SOME IMPORTANT DISCRETE DISTRIBUTIONS

4.4.1 Bernoulli trials and the Bernoulli distribution

Bernoulli distribution gives the probability of an event. The variable x is defined as follows:

$$X = \begin{cases} 1 & \text{if } S\text{th occurs with probability } p \\ 0 & \text{it does not occur with probability } 1-p \end{cases}$$

The distribution is given by the following relationship:

$$f_X(x;p) = \begin{cases} p^x(1-p)^x, & \text{if } x = 0,1,\dots \\ 0, & \text{else} \end{cases}$$

where

$$E[X] = p$$

$$\mathrm{Var}[X] = p(1-p)$$

$$m_x(t) = (1-p) + pe^t$$

4.4.2 Binomial distribution

Let us define N Bernoulli variables $[X_1, X_2,\dots, X_N]$ which take on independent values. If $X = X_1 + X_2 + \cdots + X_N$, then

$$f_X(x;p) = \begin{cases} (N/x)p^x(1-p)^{N-x}, & \text{if } x = 0,1,2,\dots,N \\ 0, & \text{else} \end{cases}$$

where $0 \leq p \leq 1$ and N is an integer. The parameters are given by the following relationships:

$$E[X] = Np$$

$$\text{Var}[X] = Np(1-p)$$

$$m_x(t) = [(1-p)+pe^t]^N$$

4.4.3 Geometric distribution

Let $[X_1, X_2,..., X_N]$ be N Bernoulli variables of probability p. If X = the number of experiments, where an event occurs for the first time, then

$$f_X(x;p) = \begin{cases} p(1-p)^{x-1}, & \text{if } x = 0,1,... \\ 0, & \text{else} \end{cases}$$

where $0 \leq p \leq 1$ and

$$E[X] = \frac{1}{p}$$

$$\text{Var}[X] = \frac{1-p}{p^2}$$

$$m_x[t] = \frac{pe^t}{1-(1-p)e^t}$$

4.4.4 Poisson distribution

Let p be the probability of the occurrence of a storm in a period Δt_j, $\Sigma \Delta t_j = t$. The assumptions are the following: (1) the probability of the occurrence of more than one storm is negligible, (2) storms are independent events and (3) the duration of the storm is very short compared to the time periods. Let x be the number of storms which occurred during the time t of an experiment in which an event occurs for the first time. Then, from the binomial distribution,

$$f_X(x;p) = \begin{cases} \binom{N}{x}p^x(1-p)^{N-x}, & \text{if } x = 0,1,2,...,N \\ 0, & \text{else} \end{cases}$$

where
  $0 \leq p \leq 1$ and N is an integer
  $E[X] = Np$ is the expected number of storms in time t
  If $\Delta t_j \to 0$, then $E[X] = Np = v$.

In this case, $f(x)$ follows the Poisson distribution

$$f_X(x;v) = \begin{cases} \dfrac{v^x e^{-v}}{x!}, & \text{if } x = 1,2,\ldots \\ 0, & \text{else} \end{cases}$$

$$E[X] = v$$

$$\text{Var}[X] = v$$

$$m_x(t) = e^{v(e^t - 1)}$$

The parameter $v = \lambda t$, where λ is the mean, is equal to the mean value of the events during the period t. Then,

$$f_X(x;\lambda t) = \begin{cases} \dfrac{(\lambda t)^x e^{-\lambda t}}{x!}, & \text{if } x = 1,2,\ldots \\ 0, & \text{else} \end{cases}$$

Also, if X_1, X_2 are two variables which follow the Poisson distribution with parameters v_1, v_2, then the sum $X = X_1 + X_2$ also follows the Poisson with parameter $v = v_1 + v_2$.

4.4.5 Uniform distribution

It is the simplest distribution. All results have the same probability of occurrence and are defined as follows:

$$f_X(x;N) = \begin{cases} \dfrac{1}{N}, & \text{if } x = 1,2,\ldots,N \\ 0, & \text{else} \end{cases}$$

The parameters of the distribution are given by the following relationships:

$$E[X] = \frac{N+1}{2}$$

$$\text{Var}[X] = \frac{N^2 - 1}{12}$$

$$m_x(t) = \left(\frac{1}{N}\right)\sum_{j=1}^{N} e^{jt}$$

4.5 SOME IMPORTANT CONTINUOUS DISTRIBUTIONS

4.5.1 Uniform distribution

The pdf of the uniform distribution is

$$f_X(x; a, b) = \begin{cases} \dfrac{1}{b-a}, & \text{if } a \le x \le b \\ 0, & \text{else} \end{cases}$$

$$E[X] = \frac{a+b}{2}$$

$$\text{Var}[X] = \frac{(b-a)^2}{12}$$

$$m_x(t) = \frac{(e^{bt} - e^{at})}{(b-a)t}$$

4.5.2 Exponential distribution

The pdf of the exponential distribution is

$$f_X(x; \lambda) = \begin{cases} \lambda e^{-\lambda x}, & \text{if } x \ge 0 \\ 0, & \text{else} \end{cases}$$

where $\lambda > 0$ and

$$E[X] = \frac{1}{\lambda}$$

$$\text{Var}[X] = \frac{1}{\lambda^2}$$

$$m_x(t) = \frac{\lambda}{\lambda - t}, \quad t < \lambda$$

4.5.3 Gamma distribution

If $X_1, \ldots, X_K$ are K independent, exponentially distributed random variables, with $f_{X_i}(x_i; \lambda)$ and $X = \sum_{i=1}^{K} X_i$, then the pdf of gamma distribution is

$$f_X(x; \kappa, \lambda) = \begin{cases} \dfrac{\lambda (\lambda x)^{\kappa - 1} e^{-\lambda x}}{(\kappa - 1)!}, & \text{if } x \ge 0 \\ 0, & \text{else} \end{cases}$$

$$E[X] = \frac{\kappa}{\lambda}$$

$$\mathrm{Var}[X] = \frac{\kappa}{\lambda^2}$$

$$m_x(t) = \left(\frac{\lambda}{\lambda - t}\right)^{\kappa}, \gamma\iota\alpha \quad t < \lambda$$

Generally,

$$f_X(x; k, \lambda) = \begin{cases} \dfrac{\lambda(\lambda x)^{\kappa-1} e^{-\lambda x}}{\Gamma(\kappa)}, & \text{if } x \geq 0 \\ 0, & \text{else} \end{cases}$$

where

$$\Gamma(\kappa) = \begin{cases} (k-1)!, & \text{if } \kappa \text{ integer} \\ \displaystyle\int_0^{\infty} e^{-u} u^{\kappa-1} du, & \text{else} \end{cases}$$

The limits of confidence for confidence level $a\%$ are given by the following relationship:

$$X(T) = \left(\dot{x} + k(T)s \pm Z_{(1+a)/2} \frac{s}{\sqrt{N}} \sqrt{1 + 2Ck(T) + \frac{1}{2}(1 + 3C_v^2)k(T)^2} \right)$$

where
  $Z_{(1+\alpha)/2}$ is the variable of the normal distribution for confidence level $\alpha\%$
  x is the mean of the sample
  s is the standard deviation of $X(T)$
  $k(T) = Z_{(1-1/T)}$
  C is the variability coefficient
  N is the size of the sample

A gamma distribution paper for each function cannot be manufactured, except for the case when the parameter k has a specific value.

Remark: If X_1, X_2 are independent gamma variables, with parameters (κ_1, λ), (κ_2, λ), then $X = X_1 + X_2$ is also a gamma variable with parameters $(\kappa_1 + \kappa_2, \lambda)$.

4.5.3.1 Gamma distribution of three parameters (Pearson type III)

Adding a constant c to the two-parameter gamma distribution, the result is the three-parameter gamma distribution, which is widely used in hydrology. The pdf is written as follows:

$$f_X(x; \kappa, \lambda) = \begin{cases} \dfrac{\lambda^x (x-c)^{\kappa-1} e^{-\lambda(x-c)}}{\Gamma(\kappa)}, & \text{if } x \succ c \\ 0, & \text{else} \end{cases}$$

where

$$\Gamma(\kappa) = \begin{cases} (k-1)!, & \text{if } k \text{ integer} \\ \displaystyle\int_0^\infty e^{-u} u^{\kappa-1} du, & \text{else} \end{cases}$$

The distribution function is $F_X(x) = \displaystyle\int_c^x f_x(s)ds$, and the parameters are estimated as follows:

$$E[X] = c + \frac{\kappa}{\lambda}$$

$$\mathrm{Var}[X] = \frac{\kappa}{\lambda^2}$$

$$m_x(t) = \left(\frac{\lambda}{\lambda-t}\right)^\kappa, \quad t < \lambda$$

The estimation of the confidence limits is a complex procedure and the gamma distribution paper can linearize every gamma distribution function for a given k.

4.5.4 Log-Pearson distribution

It has been widely used in the United States since 1967 after the formation of the Water Resources Council for adoption as the official flood frequency distribution to all services. If the random variable $Y = \ln X$ follows the distribution Pearson type III, then X follows the distribution log-Pearson type III, with pdf:

$$f_X(x; \kappa, \lambda) = \begin{cases} \dfrac{\lambda^x (\ln x - c)^{\kappa-1} e^{-\lambda(\ln x - c)}}{x\Gamma(\kappa)}, & \text{if } x \succ e^c, \kappa \succ 0, \lambda \succ 0 \\ 0, & \text{else} \end{cases}$$

where

$$\Gamma(\kappa) = \begin{cases} (k-1)!, & \text{if } k \text{ integer} \\ \displaystyle\int_0^\infty e^{-u} u^{\kappa-1} du, & \text{else} \end{cases}$$

The distribution function is given by the relationship $F_X(x) = \int_{e^c}^{x} f_x(s)ds$, and the parameters are estimated as follows:

$$E[X] = e^c \left(\frac{\lambda}{\lambda - 1} \right)^\kappa$$

$$\text{Var}[X] = e^{2c} \left[\left(\frac{\lambda}{\lambda - 2} \right)^\kappa - \left(\frac{\lambda}{\lambda - 1} \right)^{2\kappa} \right]$$

$$m_x(t) = e^{tc} \left(\frac{\lambda}{\lambda - 1} \right)^\kappa \quad \text{for } t < \lambda$$

The estimation of the confidence limits is a complex procedure, and the gamma distribution paper can linearize every gamma distribution function for a given k.

4.6 STATISTICAL ANALYSIS OF EXTREMES

Consider a sample of N independent and identical distributed random variables $Y_1, ..., Y_N$. If X is defined as $X = \max\{Y_1, ..., Y_N\}$, then

$$F(x) = P[X \leq x] = P[(Y_1 \leq x) \cap (Y_2 \leq x) \cap \cdots \cap (Y_N \leq x)]$$

$$= P[(Y_1 \leq x)] \cdots P[(Y_N \leq x)]$$

$$= F_{Y_1}(x) \cdots F_{Y_N}(x) = \left(F_Y(x) \right)^N$$

$$f_X(x) = \frac{dF_X(x)}{dx} = N(F_Y(x))^{N-1} \frac{dF_Y(x)}{dx} = N(F_Y(x))^{N-1} f_Y(x)$$

Consequently, the distribution of X depends on N and the parent distribution $F_Y(\cdot)$ and $f_Y(\cdot)$. Under certain conditions and for a variety of parent distributions when N becomes sufficiently large, the extreme value distribution $f_x(\cdot)$ tends to characterize the following limit distributions:

1. Type I: The paternal distribution is not limited in the direction of the specific limit and has finite moments (e.g. exponential, normal, log-normal and gamma).
2. Type II: The parent distribution is the same as in type I, but all moments are not necessarily finite.
3. Type III: The parent distribution is limited to the direction of the desired edge (e.g. exponential, log-normal and gamma).

Type II has no important applications in hydrology. Types I and III are often used for the simulation of maximum or minimum runoff, rainfall depth, etc.

4.6.1 Point frequency analysis

Frequency analysis is carried out to determine the frequency of hydrological events. The hydrological data for frequency analysis should be processed according to the purpose of analysis, the length of the recording period, the completeness of data and the homogeneity. The period length should be greater than 25 years, so the results of the distribution are acceptable. The missing data should be estimated using regional analysis or by correlation with other hydrological data in the region. The hydrological data are presented in chronological order. These data constitute the full series. The frequency distributions may be combined with data using two methods: the graphical method and the method of frequency factors.

4.6.1.1 Graphical method

This method involves the combination of hypothetical probability distribution data and observed data. The data are allocated in ascending or descending order of size. In descending ranking, the unit is given to the highest value. The last value is set to n, where n is the number of data. This classification gives an estimate of the probability of exceedance, i.e. the probability of a value to be equal to or greater than the value of the parameter. If the classification is done in ascending order, then we have an estimate of the probability of non-exceedance, i.e. the probability of a value to be equal to or less than the given value. The point data are designed with the help of an equation. One of the most common in hydrology is

$$P_m = \frac{m}{n+1}$$

where P_m is the probability of exceedance of the m point data of the sample, which is ranked in descending order. The return period of the m point data is

$$T_m = \frac{N+1}{m}$$

The observed data and their exceedance probabilities are designed depending on the hypothetical distribution of probability.

4.6.1.2 Method of frequency factor

Chow (1964) proposed the use of a frequency factor in hydrological frequency analysis. If a hydrological variable X is designed chronologically, then an x value will consist of two parts: the average x and the difference Δx from the average, i.e.

$$x = \bar{x} + \Delta x$$

The value of Δx may be positive or negative and may be expressed by multiplying the standard deviation S with the frequency factor K. In this case,

$$x = \bar{x} + SK$$

where K depends on the return period T and the probability distribution of x. K is the number of standard deviations above and below the average. For distributions of two parameters, the value of K varies with the probability or the return period T. For asymmetric distributions, the value of K changes with the asymmetry coefficient and is sensitive to the length of the recording period (Singh, 1992).

4.6.2 Gumbel maximum distribution

Consider a set of N observations of a random variable, where N is relatively large. If the series is divided into n subsamples of size m, so $N = nm$, each sample contains a large and a small value, which are the extremes. Gumbel (1958) showed that the n higher values of the samples asymptotically follow an extreme value distribution (Type I). The pdf and cumulative distribution function are given by the following relations:

$$f(x; a, u) = a \exp\{-a(x - u) - \exp[-a(x - u)]\}$$

where $-\infty < x < \infty$, $-\infty < u < \infty$, $\alpha > 0$

$$F(x; a, u) = \exp\{-\exp[-a(x - u)]\}$$

Gumbel (1954) was the first who used the extreme value theory for the analysis of flood frequency. The distribution is often referred to as the Gumbel distribution or the double negative exponential distribution. It is one of the most widely used distributions for the analysis of flood frequency, maximum values of rainfall, etc.:

$$E[X] = \left\{ u + \frac{0.577}{a}, \quad \text{max} \right.$$

$$\text{Var}[X] = \frac{\pi^2}{\sigma\alpha^2} \cong \frac{1.645}{\alpha^2}$$

$$E[(X - \mu_x)^3] = \{1.1396, \quad \text{max}$$

In line with the confidence limits of the normal distribution, the confidence limits of the Gumbel distribution are given by the following relations:

$$X(T)_{,\max} = x(T) + Z_{(1+a)/2} {}^* \sigma_T \quad \text{and} \quad X(T)_{,\min} = x(T) - Z_{(1+a)/2} {}^* \sigma_T$$

with $X(T) = \mu + K(T)^* \sigma$, $\sigma_T = \delta \dfrac{\sigma}{\sqrt{N}}$, $\delta = \sqrt{1 + 1.1396 * K(T) + 1.1 * K(T)^2}$

and $K(T) = -0.45 - 0.7797 * \ln(-\ln(1 - 1/T))$

Remark 1: The tables give the distribution of the standard variable $W = (X - u)a$. The following applies:

$$\left. \begin{aligned} f_X(x) &= a f_W((x - u)a) \\ F_X(x) &= F_W((x - u)a) \end{aligned} \right\} \text{max}$$

The probability of exceedance F_1 is the inverse of the return period T, while the probability of non-exceedance is $1 - F_1$. The Gumbel graph paper is constructed based on the relation (W, T). Different values of T are selected, such as 1.01, 1.5, 2, 5, 10, 20, 50, 100, 200 and 250, and the corresponding values of W are calculated. The plotting of the pair (X, T) of the maximum values of a variable on the Gumbel graph paper is a test of how well the Gumbel distribution fits on the data. However, based on a relation, there is a correspondence between the maximum value x and the return period T. Different methods such as the method of frequency coefficients, maximum likelihood and maximum entropy have been applied in the estimation of the Gumbel parameters. The one presented in this chapter and widely used is the method of moments.

4.6.3 Gumbel minimum distribution

Consider a set of N observations of a random variable, where N is relatively large. If the series is divided into n subsamples of size m, so $N = nm$, each sample contains a large and a small value, which are the extremes. Gumbel (1958) showed that the n minimum values of the samples asymptotically follow an extreme value distribution (Type I). The pdf and cumulative distribution function are given by the following relations:

$$f(x) = a \exp\{a(x - u) - \exp[a(x - u)]\}$$

where

$$-\infty < x < \infty, \ -\infty < u < \infty, \ \alpha > 0$$

$$F(x; a, u) = 1 - \exp\{-\exp[-a(x - u)]\}$$

$$E[X] = \left\{ u - \frac{0.577}{a}, \quad \text{min} \right.$$

$$\text{Var}[X] = \frac{\pi^2}{\sigma\alpha^2} \cong \frac{1.645}{\alpha^2}$$

$$E[(X - \mu_x)^3] = \{-1.1396, \quad \text{min}$$

In line with the confidence limits of the normal distribution, the confidence limits of the Gumbel distribution are given by the following relations:

$$X(T)_{,\text{max}} = x(T) + Z_{(1+a)/2}{}^*\sigma_T \quad \text{and} \quad X(T)_{,\text{min}} = x(T) - Z_{(1+a)/2}{}^*\sigma_T$$

with

$$X(T) = \mu + K(T)^*\sigma$$

$$\sigma_T = \delta \frac{\sigma}{\sqrt{N}}$$

$$\delta = \sqrt{1 + 1.1396 * K(T) + 1.1 * K(T)^2}$$

$$K(T) = -0.45 - 0.7797 * \ln(-\ln(1 - 1/T))$$

Remark 1: The tables give the distribution of the standard variable $W = (X - u)a$. The following applies:

$$\left. \begin{aligned} f_X(x) &= af_W\left(-(x - u)a\right) \\ F_X(x) &= 1 - F_W\left(-(x - u)a\right) \end{aligned} \right\} \text{ min}$$

4.6.4 Weibull distribution

A Type III distribution in its full form is a three-parameter distribution. It describes the distribution of independent variables in the range of values $(c, +\infty)$ with distribution function

$$F_Y(y) = \rho(y - c)^a,$$

where ρ and a are positive constants.

The minimum distribution function of the type III extreme value distribution is

$$F_X(x; a, b) = 1 - e^{-\left(\frac{x-c}{b-c}\right)^a}$$

In the special case where $c = 0$, we obtain the simplified, two-parameter distribution known as the Weibull distribution. The Weibull distribution is widely applicable in hydrology, since it describes the distribution of low flows in streams. The pdf of the Weibull distribution is

$$f_X(x; a, b) = ax^{a-1}b^{-a}e^{-(x/b)^a}, \quad \text{for } x \geq 0 \text{ and } a, b > 0$$

The distribution function of the Weibull distribution is

$$F_X(x; a, b) = 1 - e^{-(x/b)^a}$$

The mean and the standard deviation are

$$E(X) = b * \Gamma\left(1 + \frac{1}{a}\right)$$

$$\text{Var}(X) = b^2 * \left[\Gamma\left(1 + \frac{2}{a}\right) - \Gamma^2\left(1 + \frac{1}{a}\right)\right]$$

The estimation of the parameters of the distribution is based on the method of moments:

$$\frac{\Gamma\left(1 + \frac{2}{a}\right)}{\Gamma^2\left(1 + \frac{1}{a}\right)} = \frac{s_x^2}{\overline{x}^2} + 1$$

$$b = \frac{\overline{x}}{\Gamma\left(1 + \frac{1}{a}\right)}$$

where the parameter a can be estimated only by the numerical solution of the respective corresponding equation.

In the case when the lower bound on the parent distribution is not zero, a displacement parameter is added and the distribution function becomes

$$F_X(x; a, b, c) = 1 - e^{-\left(\frac{x-c}{b-c}\right)^a}$$

By using the transformation,

$$y = \left[\frac{(x-c)}{(b-c)}\right]^a$$

Tables of e^{-y} can be used to calculate the probability P.

The mean and the standard deviation become (Haan, 1985)

$$E(X) = c + (b-c)\Gamma\left(1 + \frac{1}{a}\right)$$

$$\text{Var}(X) = (b-c)^2 * \left[\Gamma\left(1 + \frac{2}{a}\right) - \Gamma^2\left(1 + \frac{1}{a}\right)\right]$$

Through the transformation of y and the use of equations of mean and variance, the following equations are obtained (Haan, 1985):

$$b = E(X) + \text{Var}(X)^{0.5}A(a)$$

$$c = E(X) - \text{Var}(X)^{0.5}B(a)$$

where

$$A(a) = \left(1 - \Gamma\left(1 + \frac{1}{a}\right)\right)B(a)$$

$$B(a) = \left(\Gamma\left(1 + \frac{2}{\alpha}\right) - \Gamma^2\left(1 + \frac{1}{\alpha}\right)\right)^{0.5}$$

The distribution function of the Weibull distribution is a straight line when drawn on the special graph of the Weibull distribution. The pairs on the graph are designed by plotting the value of the quantity ln [−ln (1−F)] on the horizontal axis and the value of lnX on the vertical axis.

4.7 TESTING OF THE DISTRIBUTIONS

The successful statistical analysis of the hydrological data depends on the fit of the distribution to the empirical distribution of the sample. The question is how well the theoretical distribution is fitted to the empirical distribution. The following describes two important tests for the adjustment of the sample data, which are widespread in hydrology: the X^2 and the Kolmogorov–Smirnov tests.

4.7.1 Test X^2

X^2 test is suitable for the test of the fit of a distribution of a pdf $f_X(x; \Theta_1, ..., \Theta_p)$ in n points. It is summarized in the following steps:

1. The parameters $\{\Theta_1, ..., \Theta_p\}$ of the pdf are calculated.
2. The sample data are classified from the larger sample ($m = 1$) to the smallest ($m = N$).
3. k different classes $((x_0, x_1], ..., (x_{k-1}, x_k])$ are defined such that the expected number of points of each class is greater or equal to S and approximately the same for all classes, where the expected number of points of class i $(x_{i-1}, x_i]$ is

$$\ell_i = N \int_{x_{i-1}}^{x_i} f_X(x; \Theta_i, ..., \Theta_p) dx = N \left[F_X(x_i) - F_X(x_{i-1}) \right]$$

and $f_X(.;...)$ is the selected and calculated pdf.

4. The actual number of points n_i in each class i is measured, and the following statistical term is calculated:

$$x_c^2 = \sum_{i=1}^{\kappa} \frac{(n_i - \ell_i)^2}{\ell_i}$$

x_c^2 is distributed according to the distribution X^2 with $\nu = k - 1 - p$ degrees of freedom, where p is the number of parameters calculated from the data.

If $F_{X^2_\nu}(x_c^2) > 0.05$, the distribution is appropriate.
If $F_{X^2_\nu}(x_c^2) < 0.01$, the distribution is not appropriate.

If $0.01 < F_{X^2_\nu}(x_c^2) < 0.05$, the conclusion cannot be drawn regarding the appropriateness of the distribution. The test X^2 should be repeated after the collection of more data.

4.7.2 Kolmogorov–Smirnov test for the appropriateness of a distribution

To test the fit of the distribution of a pdf $f_X(x; \Theta_1, ..., \Theta_p)$ for a group of n points, in accordance with the Kolmogorov–Smirnov test, the difference between the cumulative distribution function $F_X(X_i)$ and the observed cumulative histogram $F^*(X)$ is calculated. The observed cumulative histogram is given by $F^*(X_i) = i/n$, where i is the biggest observed value from a sample of size n. Based on the sample data, the statistical parameter D is calculated:

$$D = \max_{i=1}^{n}[|\, F^*(X_i) - F_X(X_i)\,|] = \max_{i=1}^{n}[|\, \tfrac{i}{n} - F_X(X_i)\,|]$$

The distribution is acceptable if $D < c$, where the values of parameter c are given in tables depending on the sample size and the desired significance level, a. The significance level a is defined as the probability of making type I error, i.e. rejecting the null hypothesis while it is correct. It is described in detail in test-related statistical books (Eadie et al., 1971; Hollander and Wolfe, 1973).

Example 4.3

Calculate the mean and the standard deviation of the random variable Y:

$$Y = \frac{1}{N}(X_1 + X_2 + X_3 + \cdots + X_N)$$

where X_1, X_2, ..., X_N are random variables, which are independent and have the same mean, μ, and variance, σ^2.

Solution

Calculation of the mean μ_Y:

$$E(Y) = E\left(\frac{1}{N}(X_1 + X_2 + X_3 + \cdots + X_N)\right) = \frac{1}{N} * E(X_1 + X_2 + X_3 + \cdots + X_N)$$

$$= \frac{1}{N} * (E(X_1) + E(X_2) + E(X_3) + \cdots + E(X_N)) = \frac{1}{N} * (\mu + \mu + \mu + \cdots + \mu) = \frac{1}{N} * (N * \mu) = \mu$$

Calculation of the standard deviation σ_Y:

$$\text{Var}(Y) = \text{Var}\left(\frac{1}{N}(X_1 + X_2 + X_3 + \cdots + X_N)\right) = \frac{1}{N^2} * \text{Var}(X_1 + X_2 + X_3 + \cdots + X_N)$$

$$= \frac{1}{N^2} * (\text{Var}(X_1) + \text{Var}(X_2) + \text{Var}(X_3) + \cdots + \text{Var}(X_N)) = \frac{1}{N^2} * (\sigma^2 + \sigma^2 + \sigma^2 + \cdots + \sigma^2)$$

$$= \frac{1}{N^2} * (N * \sigma^2) = \frac{\sigma^2}{N}.$$

Thus, the standard deviation of the random variable Y is $\sigma_Y = \sigma/N^{1/2}$.

Example 4.4

For the standard variable $U = (x - m_x)/\sigma_x$, prove that

1. $E[(X - m_x)/\sigma_x] = 0$ and $\text{Var}[(X - m_x)/\sigma_x] = 1$
2. The correlation coefficient between the variables X and Z is the covariance between the respective standard variables $[(X - m_x)/\sigma_x]$ and $[(Z - m_Z)/\sigma_Z]$

Solution

1. $E[(X - m_x)/\sigma_x] = 0$

Proof:

$$E[(X - m_x)/\sigma_x] = E(X - m_x)/\sigma_x = 1/\sigma [E(X) - E(m_x)]$$

The expected value of the mean is the mean $\mu = E(X)$.

Thus,

$$E\left[(X - m_x)/\sigma_\chi\right] = 1/\sigma\ \overline{[E(X)]} - E(X)] = 0$$

$$\mathrm{Var}[(X - m_x)/\sigma_\chi] = 1$$

Proof:

First way: $\mathrm{Var}[(X - m_x)/\sigma_\chi] = \mathrm{Var}(1/\sigma_x X - 1/\sigma_\chi m_x) = 1/\sigma_x^2 \mathrm{Var}(X) = 1/\sigma_x^2 \sigma_x^2 = 1$

Second way: $\mathrm{Var}[(X - m_x)/\sigma_\chi] = E[(X - m_x)/\sigma_\chi]^2 - (E[(X - m_x)/\sigma_\chi])^2$

$$E[(X - m_x)/\sigma_\chi] = 0$$

Thus,

$$
\begin{aligned}
\mathrm{Var}[(X - m_x)/\sigma_\chi] &= E[(X - m_x)/\sigma_\chi]^2 = E[(X^2 - 2Xm_x + m_x^2)/\sigma_\chi^2] \\
&= 1/\sigma_\chi^2[E(X^2) - 2m_x E(X) + E(m_x^2)] = 1/\sigma_\chi^2[E(X^2) - 2m_x^2 + m_x^2] \\
&= 1/\sigma_\chi^2[E(X^2) - m_x^2] = 1/\sigma_\chi^2[E(X^2) - E(X)^2] = 1/\sigma_\chi^2 \mathrm{Var}(X) = \sigma_\chi^2/\sigma_\chi^2 = 1
\end{aligned}
$$

2. $\mathrm{Cov}[(X - m_x)/\sigma_\chi, (Z - m_Z)/\sigma_Z] = \rho_{x,z}$

Proof:

We know that

$$\mathrm{Cov}(X,Z) = E(XZ) - E(X)E(Z)$$

and

$$\rho_{x,z} = \mathrm{Cov}(X,Z)/(\sigma_x \sigma_z)$$

$$\mathrm{Cov}[(X - m_x)/\sigma_\chi, (Z - m_Z)/\sigma_Z] = E((X - m_x)/\sigma_\chi\ (Z - m_Z)/\sigma_Z) - E((X - m_x)/\sigma_\chi) E((Z - m_Z)/\sigma_Z)$$

where $E((X - m_x)/\sigma_\chi = 0$, $E((Z - m_Z)/\sigma_Z = 0$; thus,

$$
\begin{aligned}
\mathrm{Cov}[(X - m_x)/\sigma_\chi, (Z - m_Z)/\sigma_Z] &= E((X - m_x)/\sigma_\chi (Z - m_Z)/\sigma_Z) \\
&= E((XZ - Xm_Z - m_x Z + m_x m_Z)/(\sigma_\chi \sigma_Z)) \\
&= 1/(\sigma_\chi \sigma_Z)[E(XZ) - E(Xm_Z) - E(m_x Z) + E(m_x m_Z)] \\
&= 1/(\sigma_\chi \sigma_Z)[E(XZ) - m_Z E(X) - m_x E(Z) + E(m_x m_Z)] \\
&= 1/(\sigma_\chi \sigma_Z)[E(XZ) - m_Z m_x - m_x m_Z + m_x m_Z] \\
&= 1/(\sigma_\chi \sigma_Z)[E(XZ) - m_Z m_x] = \mathrm{Cov}(X,Z)/(\sigma_\chi \sigma_Z) = \rho_{x,z}
\end{aligned}
$$

Example 4.5

A flood control reservoir was designed for the N-year flood, i.e. its capacity will be exceeded by the N-year or greater flood. The size of the N-year flood is defined as that which has the probability $1/N$ to be exceeded each year. We assume that the successive annual floods are independent.

1. What is the probability that a flood equal to or greater than the 50-year flood will occur in 50 years?
2. What is the probability that three floods equal to or greater than the 50-year flood will occur in 50 years?
3. What is the probability that one or more floods equal to or greater than the 50-year flood will occur in 50 years?
4. If a company constructs 20 independent systems – i.e. in 20 different areas – designed for the flood of 500 years, what is the distribution of the number of systems that will fail at least once in the first 50 years after the construction?
5. In 1958, the size of the 50-year flood was estimated. In the next 10 years, two floods were found to be greater than this size. If the initial assessment is correct, what is the probability of such a remark?

Solution

The random variable is the annual flood and follows a discrete distribution. For the calculation of probabilities, the binomial distribution is used:

$$f_X(x; p) = \begin{cases} \binom{N}{x} p^x (1-p)^{N-x} & \text{if } x = 0,1,\ldots,N \\ 0 & \text{else} \end{cases}$$

where

x is the number of floods equal to or greater than the flood with a return period of 50 years

N is the number of years for which the probability of occurrence or exceedance of the 50-year flood is examined

$p = 1/T$ is the probability of occurrence or exceedance of the 50-year flood at each time step (1 year)

1.

N (years)	T (years)	$p = 1/T$	x
50	50	0.02	1

Thus, the probability of a flood occurrence equal to or greater than the 50-year flood in a time period of 50 years is

$$P(X = 1) = f_x(1) = \binom{50}{1} * 0.02^1 * (1 - 0.02)^{50-1} = 0.372$$

2.

N (years)	T (years)	$p = 1/T$	x
50	50	0.02	3

Thus, the probability of occurrence of three floods equal to or greater than the 50-year flood in a time period of 50 years is

$$P(X = 3) = f_x(3) = \binom{50}{3} * 0.02^3 * (1-0.02)^{50-3} = 0.061$$

3. The probability of occurrence of one or more floods equal to or greater than the 50-year flood in a time period of 50 years is $P(X \geq 1) = 1-P(X = 0)$.

Thus, the probability of occurrence of no flood equal to or greater than the 50-year flood can be calculated.

N (years)	T (years)	$p = 1/T$	x
50	50	0.02	0

$$P(X = 0) = f_x(0) = \binom{50}{0} * 0.02^0 * (1-0.02)^{50-0} = 0.364$$

and

$$P(X \geq 1) = 1 - 0.364 = 0.636$$

4. Initially, the probability of failure of each system j is examined. The system is considered to fail in the case of occurrence or exceedance of the 500-year flood at least once in a 50-year time period. x is the number of floods equal to or greater than the 500-year flood.

N (years)	T (years)	$p = 1/T$	x
50	500	0.002	0

Thus,

$$P(X = 0) = f_x(0) = \binom{50}{0} * 0.02^0 * (1-0.002)^{50-0} = 0.905$$

Thus, the probability of the failure of any system j is

$$P(X \geq 1) = 1 - 0.905 = 0.095$$

The probability of failure of none ($X = 0$) to 20 ($X = 20$) systems in a 50-year time period with a failure probability of each system $p = 0.095$ is calculated as follows:

$$f_x = \binom{N}{x} * p^x * (1-p)^{N-x} \quad \text{for } x = 0, 1, 2, ..., 20$$

where
   x is the number of systems which fail
   N is the total number of systems
   $p = 1/T$ is the probability of failure of each system j in a 50-year time period

N (systems)	$p = 1/T$
20	0.095

and

$$P(X) = f_x = \binom{20}{x} * 0.095^x * (1 - 0.095)^{20-x}$$

Thus, the probability of failure of the systems for $x = 0, 1, 2,..., 20$ is given in Table 4.3.

We observe that the probability of failure of X systems decreases as X increases from 1 to 20.

5. In the next 10 years, the flood with the 50-year return period was exceeded two times. Thus,

N (years)	T (years)	$p = 1/T$	x
10	50	0.02	2

and

$$P(X = 2) = f_x(2) = \binom{10}{2} * 0.02^2 * (1 - 0.02)^{10-2} = 0.015$$

Table 4.3 Probability of failure

X	P(X)
0	0.136
1	0.285
2	0.284
3	0.179
4	0.080
5	0.027
6	0.007
7	0.001
8	2.5E−04
9	3.5E−05
10	4.1E−06
11	3.9E−07
12	3.1E−08
13	2.0E−09
14	1.0E−10
15	4.4E−12
16	1.4E−13
17	3.5E−15
18	6.2E−17
19	6.8E−19
20	3.6E−21

Example 4.6

The mean annual and the maximum daily runoff are given in Table 4.4. The values were estimated at a river cross section for a 30-year time period.

1. Calculate the statistical characteristics of both samples (mean, standard deviation, coefficients of variation, skewness and kurtosis, maximum and minimum values).
2. Fit the normal distribution (Gaussian) to the sample of the mean annual discharges.
3. Design the sample and the theoretical distribution on a normal distribution graph.
4. Estimate the values of the mean annual discharges, which correspond to return periods of 10, 50 and 200 years, based on the normal distribution.
5. If 75% of the mean annual discharge is sufficient to meet the water needs of an adjacent city, find the probability of failure of the complete coverage of the city's water needs during a year.
6. Adjust the distributions of Gumbel and log-normal to the sample of the maximum daily discharge. Check the appropriateness of the Gumbel distribution with the test X^2.
7. Design the sample of the max daily runoff and the theoretical Gumbel distribution on a Gumbel distribution graph.
8. Estimate the maximum daily runoff values corresponding to return periods of 10, 20 and 1000 years, using both fit distributions.
9. Calculate the 95% confidence limits of Gaussian distribution for the values of question 4 and Gumbel distribution for the values of question 8.

Solution

1. The statistical characteristics of both samples are given in Table 4.5.
2. Fit of normal distribution

The statistical characteristics of the sample of mean annual discharge, which are necessary for normal distribution, are the mean and the standard deviation (calculated in question 1). Having calculated these characteristics, the sample is ranked in descending order and the sample values are numbered. Then the return period is calculated using the Weibull relationship $(T = (N + 1)/m)$, where N is the number of the sample values and m is the rank. Then, the probability of non-exceedance F is calculated (from $F = 1 - 1/T$). For each value of F, the standard variable Z is calculated (from Table 4.6). Based on the values of Z, X is calculated (from $X = \mu + Z*\sigma$) which corresponds to each probability. The results are shown in Table 4.6.

Table 4.4 Mean annual and maximum daily runoff

| | Discharge (m³/s) | | | Discharge (m³/s) | | | Discharge (m³/s) | |
Year	Mean annual	Maximum daily	Year	Mean annual	Maximum daily	Year	Mean annual	Maximum daily
1	27.7	751	11	36.4	387	21	27.0	695
2	24.3	320	12	35.2	520	22	25.3	647
3	27.3	207	13	23.5	419	23	21.9	290
4	24.6	349	14	31.7	437	24	24.8	185
5	16.2	165	15	20.6	328	25	22.1	165
6	20.2	425	16	30.5	389	26	14.5	123
7	28.2	605	17	19.3	177	27	18.3	147
8	25.0	360	18	18.5	190	28	25.4	357
9	37.2	507	19	16.4	445	29	22.4	424
10	32.9	369	20	10.6	85	30	33.4	904

Table 4.5 Statistical characteristics of both samples

	Mean annual discharge (m³/s)	Maximum daily discharge (m³/s)
Mean	24.71	379.07
Standard deviation	6.63	198.58
Variation coefficient	0.27	0.52
Skewness coefficient	0.07	0.76
Kurtosis coefficient	−0.38	0.42
Min	10.60	85.00
Max	37.20	904.00

Table 4.6 Results of the fit of normal distribution

Year	Mean annual discharge (m³/s)	Descending order	T (Weibull)	F	Z	X
1	27.7	37.2	31.00	0.9677	1.8486	36.97
2	24.3	36.4	15.50	0.9355	1.5179	34.78
3	27.3	35.2	10.33	0.9032	1.3002	33.33
4	24.6	33.4	7.75	0.8710	1.1310	32.21
5	16.2	32.9	6.20	0.8387	0.9892	31.27
6	20.2	31.7	5.17	0.8065	0.8649	30.45
7	28.2	30.5	4.43	0.7742	0.7527	29.70
8	25	28.2	3.88	0.7419	0.6493	29.02
9	37.2	27.7	3.44	0.7097	0.5524	28.38
10	32.9	27.3	3.10	0.6774	0.4605	27.77
11	36.4	27	2.82	0.6452	0.3723	27.18
12	35.2	25.4	2.58	0.6129	0.2869	26.62
13	23.5	25.3	2.38	0.5806	0.2035	26.06
14	31.7	25	2.21	0.5484	0.1216	25.52
15	20.6	24.8	2.07	0.5161	0.0404	24.98
16	30.5	24.6	1.94	0.4839	−0.0404	24.45
17	19.3	24.3	1.82	0.4516	−0.1216	23.91
18	18.5	23.5	1.72	0.4194	−0.2035	23.36
19	16.4	22.4	1.63	0.3871	−0.2869	22.81
20	10.6	22.1	1.55	0.3548	−0.3723	22.25
21	27	21.9	1.48	0.3226	−0.4605	21.66
22	25.3	20.6	1.41	0.2903	−0.5524	21.05
23	21.9	20.2	1.35	0.2581	−0.6493	20.41
24	24.8	19.3	1.29	0.2258	−0.7527	19.72
25	22.1	18.5	1.24	0.1935	−0.8649	18.98
26	14.5	18.3	1.19	0.1613	−0.9892	18.16
27	18.3	16.4	1.15	0.1290	−1.1310	17.22
28	25.4	16.2	1.11	0.0968	−1.3002	16.09
29	22.4	14.5	1.07	0.0645	−1.5179	14.65
30	33.4	10.6	1.03	0.0323	−1.8486	12.46

3. In order to check the fit of the normal distribution to the sample of mean annual discharge, the sample and the theoretical distribution are designed on normal distribution graph. The values of the distribution function F are marked on the horizontal axis of the graph, and the values of discharge are marked on the vertical axis. The points (F, X) of the table of the preceding question are joined by a straight line. The closer the straight line is to the line joining the points (F, Q) (where Q is the time series with the descending values), the more representative is the normal distribution for the sample.

4. The mean ($\mu = 24.71$ m³/s) and the standard deviation ($\sigma = 6.63$ m³/s) of the sample of mean annual discharges were calculated in question 1. The probability of non-exceedance is calculated for return periods $T = 10$, 50 and 200 years using the relation $F = 1 - 1/T$, and then the standard variable Z is based on tabular data. The respective value of the mean annual runoff is obtained by applying the relationship $X = \mu + Z*\sigma$ for each Z, given that μ and σ are known from question 1. Thus, we obtain Table 4.7.

5. Suppose that the water needs satisfy the normal distribution. Then, the standardized variable Z is $Z = (18.54 - 24.71)/6.63 = -0.932$. The value of non-exceedance probability, F, which corresponds to Z is equal to 17.57%, as per Table 4.8. Thus, the probability of failure of the complete coverage of the city's water demand during any given year reflects the probability of exceedance and is given by $F' = 1 - F = 0.8243$.

6. Fit of the Gumbel distribution.

The statistical characteristics of the sample of maximum daily discharges, which are useful for the Gumbel distribution, are the mean, standard deviation calculated in question 1 and the parameters a and c, which are given as mean = 379.07 m³/s. Standard deviation = 198.58 m³/s, $a = 0.01$ and $c = 289.70$, respectively.

Once the statistical characteristics of the sample have been calculated, in order to fit the Gumbel distribution to the sample, the sample data values are ranked in descending order and the values are numbered. Then, the return period from the Weibull relationship $(T = (N + 1)/m)$ is calculated, where N is the number of the sample values) and the non-exceedance probability, $F = 1 - 1/T$, is calculated.

For each value of F, the quantity $\ln[\ln(T) - \ln(T - 1)]$ is calculated and then the respective value of X directly from the equation of the theoretical Gumbel distribution. The process has been tabulated, and the results are given in Table 4.8.

Fit of the log-normal distribution

The statistical characteristics of the sample of maximum daily discharges which are of interest for the log-normal distribution are the mean, standard deviation of the sample calculated in question 1 and the mean and standard deviation of the time series of logarithms of the sample:

Mean (sample) $\mu_x = 379.07$ m³/s
Standard deviation (sample) $\sigma_x = 198.58$ m³/s
Mean (sample logarithms) $\mu_y = 5.79$ m³/s
Standard deviation (sample logarithms) $\sigma_y = 0.57$ m³/s

Table 4.7 Probability of non-exceedance

Return period, T	Probability of not being exceeded, F	Standard variable, Z	Mean annual runoff (m³/s)
10	0.9	1.2816	33.21
50	0.98	2.0537	38.33
200	0.995	2.5758	41.79

Table 4.8 Results of the theoretical Gumbel distribution

Year	Maximum daily discharges, Q_2 (m³/s)	Descending order	T (Weibull)	F	ln(ln(T)−ln(T − 1))	X
1	751	904	31.00	0.9677	−3.4176	818.8776
2	320	751	15.50	0.9355	−2.7077	708.9505
3	207	695	10.33	0.9032	−2.2849	643.4913
4	349	647	7.75	0.8710	−1.9794	596.1884
5	165	605	6.20	0.8387	−1.7379	558.7924
6	425	520	5.17	0.8065	−1.5366	527.6249
7	605	507	4.43	0.7742	−1.3628	500.7204
8	360	445	3.88	0.7419	−1.2090	476.9020
9	507	437	3.44	0.7097	−1.0702	455.4072
10	369	425	3.10	0.6774	−0.9430	435.7114
11	387	424	2.82	0.6452	−0.8250	417.4365
12	520	419	2.58	0.6129	−0.7143	400.2989
13	419	389	2.38	0.5806	−0.6095	384.0784
14	437	387	2.21	0.5484	−0.5095	368.5984
15	328	369	2.07	0.5161	−0.4134	353.7128
16	389	360	1.94	0.4839	−0.3203	339.2965
17	177	357	1.82	0.4516	−0.2295	325.2389
18	190	349	1.72	0.4194	−0.1404	311.4379
19	445	328	1.63	0.3871	−0.0523	297.7957
20	85	320	1.55	0.3548	0.0355	284.2139
21	695	290	1.48	0.3226	0.1235	270.5880
22	647	207	1.41	0.2903	0.2125	256.8015
23	290	190	1.35	0.2581	0.3035	242.7162
24	185	185	1.29	0.2258	0.3975	228.1587
25	165	177	1.24	0.1935	0.4961	212.8967
26	123	165	1.19	0.1613	0.6013	196.5956
27	147	165	1.15	0.1290	0.7167	178.7305
28	357	147	1.11	0.0968	0.8482	158.3760
29	424	123	1.07	0.0645	1.0083	133.5879
30	904	85	1.03	0.0323	1.2337	98.6789

Once the statistical characteristics of the sample are calculated for the fit of the log-normal distribution to the sample, the samples are ranked in descending order and the values are numbered. Then, the return period from the Weibull relationship $T = (N + 1)/m$ is calculated, where N is the number of the sample values, and the non-exceedance probability, $F = 1 - 1/T$, is calculated.

The standard variable Z is calculated for each value of F (from Table 4.9), and then, the respective value of X which corresponds to each probability is calculated from the equation

$$X = e^{Z*\sigma_y + \mu_y}$$

The process has been tabulated, and the results are presented in Table 4.9.

Test X^2

The first step for the test X^2 is the calculation of the parameters of the distribution to be adjusted. The Gumbel distribution has two parameters ($r = 2$), and the parameter values were calculated at the beginning of question 6. Then, the sample is divided into k classes (intervals) of equal probability. The division of the sample into $k = 5$ classes was chosen.

Table 4.9 Results of the theoretical log-normal distribution

Year	Maximum daily discharges, Q_2 (m^3/s)	$ln(Q_2) = Y$	Descending order	T	F	Z	X	Exp (in descending order)
1	751	6.621	6.807	31.00	0.9677	1.8486	948.46	904.00
2	320	5.768	6.621	15.50	0.9355	1.5179	784.27	751.00
3	207	5.333	6.544	10.33	0.9032	1.3002	691.99	695.00
4	349	5.855	6.472	7.75	0.8710	1.1310	627.87	647.00
5	165	5.106	6.405	6.20	0.8387	0.9892	578.71	605.00
6	425	6.052	6.254	5.17	0.8065	0.8649	538.82	520.00
7	605	6.405	6.229	4.43	0.7742	0.7527	505.17	507.00
8	360	5.886	6.098	3.88	0.7419	0.6493	476.02	445.00
9	507	6.229	6.080	3.44	0.7097	0.5524	450.23	437.00
10	369	5.911	6.052	3.10	0.6774	0.4605	427.05	425.00
11	387	5.958	6.050	2.82	0.6452	0.3723	405.94	424.00
12	520	6.254	6.038	2.58	0.6129	0.2869	386.50	419.00
13	419	6.038	5.964	2.38	0.5806	0.2035	368.41	389.00
14	437	6.080	5.958	2.21	0.5484	0.1216	351.46	387.00
15	328	5.793	5.911	2.07	0.5161	0.0404	335.44	369.00
16	389	5.964	5.886	1.94	0.4839	−0.0404	320.20	360.00
17	177	5.176	5.878	1.82	0.4516	−0.1216	305.61	357.00
18	190	5.247	5.855	1.72	0.4194	−0.2035	291.55	349.00
19	445	6.098	5.793	1.63	0.3871	−0.2869	277.91	328.00
20	85	4.443	5.768	1.55	0.3548	−0.3723	264.59	320.00
21	695	6.544	5.670	1.48	0.3226	−0.4605	251.51	290.00
22	647	6.472	5.333	1.41	0.2903	−0.5524	238.56	207.00
23	290	5.670	5.247	1.35	0.2581	−0.6493	225.64	190.00
24	185	5.220	5.220	1.29	0.2258	−0.7527	212.62	185.00
25	165	5.106	5.176	1.24	0.1935	−0.8649	199.34	177.00
26	123	4.812	5.106	1.19	0.1613	−0.9892	185.60	165.00
27	147	4.990	5.106	1.15	0.1290	−1.1310	171.07	165.00
28	357	5.878	4.990	1.11	0.0968	−1.3002	155.22	147.00
29	424	6.050	4.812	1.07	0.0645	−1.5179	136.95	123.00
30	904	6.807	4.443	1.03	0.0323	−1.8486	113.25	85.00

The degree of freedom of the distribution is $v = k - r - 1 = 5 - 2 - 1 = 2$.

X is determined, which corresponds to the cumulative probability of each class and the limits of the classes.

The expected (theoretical) number of observations is calculated for each class, $E_i = N^* p_i$, where N is the size of the sample ($N = 30$).

The values N_i of the sample, which are within the limits of each class, are measured, and the statistical parameter D is calculated: $D = \Sigma[(N_i - E_i)^2/E_i]$. Thus, we obtain Table 4.10.

The value of parameter D is compared with the value from the tables of X^2 for the specific degree of freedom ($v = 2$) and significance level α. The usual values of α were considered: 1%, 5% and 10%.

For $v = 2$ and $\alpha = 1\%$, it results from the tables of X^2 that $X^2 = 9.2$
For $v = 2$ and $\alpha = 5\%$, it results from the tables of X^2 that $X^2 = 6$
For $v = 2$ and $\alpha = 10\%$, it results from the tables of X^2 that $X^2 = 4.6$

Table 4.10 Test X^2

Limits of classes	Expected probability, p_i	E_i	N_i	$(N_i - E_i)^2/E_i$
$0 < X < 215.988$	0.2	6	6	0
$215.988 < X < 303.246$	0.2	6	6	0
$303.246 < X < 393.756$	0.2	6	6	0
$393.756 < X < 522.047$	0.2	6	6	0
$522.047 < X < 1$	0.2	6	6	0
	D	0		

The result is that $D < X^2$ for the usual significance levels for which the sample was tested. Thus, the sample is considered to follow the Gumbel distribution for the usual significance levels.

7. To check the adjustment of the Gumbel distribution to the sample of daily maximum discharges, the sample and the theoretical distribution are designed on the Gumbel distribution graph. The values of X are marked on the vertical axis and the values of F on the horizontal axis, which results in a straight line. Then, the points are plotted with the values of the descending order of the table in question 6 on the vertical axis and the values of F on the horizontal axis. The shorter the distance of these points from the straight line, the more representative is the Gumbel distribution for the sample.
8. Gumbel distribution

Given the values of the return periods and the parameters of the Gumbel distribution, which were calculated in question 6, the non-exceedance probability F (= $1 - 1/T$) is calculated, and the respective discharge value is based on the relevant equations. The results are shown in Table 4.11.

Log-normal distribution

First, the non-exceedance probability F (=$1 - 1/T$) is calculated using the given return periods and the parameters of the log-normal distribution, as calculated in question 6. Then, the respective value of the standard variable Z is calculated, and the discharge is calculated based on the relationship $X = e^{Z \cdot \sigma_y + \mu_y}$.

The results are shown in Table 4.12.

Table 4.11 Non-exceedance probability using the Gumbel distribution

Return period, T	Probability, F	Discharge (m^3/s)
10	0.9	638.290
20	0.95	749.793
1000	0.999	1359.651

Table 4.12 Non-exceedance probability using the log-normal distribution

Return period, T	Probability of non-exceedance, F	Standard variable, Z	Discharge (m^3/s)
10	0.9	1.2816	684.63
20	0.95	1.6449	843.63
1000	0.999	3.0902	1936.37

Table 4.13 Non-exceedance probability using the normal distribution

Return period, T	F	Z_F	δ	S_T	X(T)	$X(T)_{max}$	$X(T)_{min}$
10	0.9	1.2816	1.350	1.633	33.21	36.410	30.007
50	0.98	2.0537	1.763	2.134	38.33	42.510	34.145
200	0.995	2.5758	2.078	2.515	41.79	46.717	36.859

Table 4.14 Calculation of the limits for α = 95%

Return period, T	F	K(T)	δ	S_T	X(T)	$X(T)_{max}$	$X(T)_{min}$
10	0.9	1.305	2.088	75.696	638.29	786.652	489.928
50	0.98	1.866	2.637	95.623	749.79	937.210	562.375
200	0.995	4.936	5.781	209.600	1359.65	1770.460	948.842

9. Normal distribution

For the return periods of question 4, the probability of non-exceedance ($F = 1 - 1/T$), the respective standard variable Z_F and the parameters δ and S_T are calculated. For $a = 95\%$, the value of $Z_{(1 + \alpha)/2} = 1.96$ (from Table 4.13) is calculated.

$$X(T) = \mu + Z_F^* \sigma \quad \text{and}$$

$$x(T)_{max} = x(T) + Z_{(1+\alpha)/2} S_T$$

$$x(T)_{min} = x(T) - Z_{(1+\alpha)/2} S_T$$

The calculations are shown in Table 4.13.

Gumbel distribution

For the return periods of question 4, the non-exceedance probability F ($F = 1 - 1/T$) and the parameters $K(T) = -0.45 - 0.7797 * \ln[-\ln(1 - 1/T)]$, δ and S_T are calculated. For α = 95%, the value of $Z_{(1+\alpha)/2} = 1.96$ (from Table 4.14).

$$X(T) = \mu + K(T)^* \sigma \quad \text{and}$$

$$x(T)_{max} = x(T) + Z_{(1+\alpha)/2} * \frac{S_X}{\sqrt{n}} \sqrt{1 + 1.1396 * K(T) + 1.1 * K(T)^2}$$

$$x(T)_{min} = x(T) - Z_{(1+\alpha)/2} * \frac{S_X}{\sqrt{n}} \sqrt{1 + 1.1396 * K(T) + 1.1 * K(T)^2}$$

The calculations are shown in Table 4.14.

Example 4.7

The minimum annual monthly discharges on a river are found to have an average of 150 m³/s, a standard deviation of 70 m³/s and a skew coefficient of 1.4. Using both the Weibull distribution and the Gumbell distribution, calculate the probability of an annual minimum flow, which is less than 100 m³/s.

Solution

1. The Weibull distribution

By solving the equations given in the Weibull distribution, we have
$a = 1.266$, $A(\alpha)=0.098$ and $B(\alpha) = 1.36$

$$b = 150 + 70(0.098) = 156.8$$
$$c = 156.8 - 70(1.36) = 61.6$$
$$P_{rob}(X \leq 100) = P_x(100) = 1 - \exp^{-y}$$

where

$$y = \left\{ \frac{(x - \hat{\varepsilon})}{(\hat{\beta} - \hat{\varepsilon})} \right\}^{\hat{a}} = \left\{ \frac{(100 - 61.66)}{(156.86 - 61.66)} \right\}^{1.266} = 0.316$$

$$P_x(100) = 0.271$$

2. The Gumbel distribution
   Using the method of moments, we calculate the two parameters α and u

$$\hat{a} = \frac{s}{1.282} = \frac{75}{1.282} = 58.50$$

$$u = \bar{x} + 0.45 \times s = 150 + 0.45 \times (75) = 183.75$$

$$P_{rob}(X \leq 100) = P_x(100) = 1 - \exp(-\exp^y)$$

where

$$y = \frac{(x - u)}{\hat{a}} = \frac{(100 - 183.75)}{58.50} = -1.432$$

$$P_x(100) = 1 - 0787 = 0.213$$

4.8 INTENSITY–DURATION–FREQUENCY CURVES

Several major water resource projects such as storm sewer networks, reservoirs and flood protection works are designed based on an analysis of rainfall intensity (i), duration (d) and frequency (f). Sometimes, rainfall intensity is replaced by the rainfall depth and the frequency by the return period. The rainfall intensity is usually the average intensity during a rainfall. A high-intensity rainfall occurs less frequently than a low-intensity rainfall.

The annual highest rainfall depth h (mm) and the intensity i (mm/h), observed at a rainfall monitoring station, for a particular rainfall duration t, during n years of measurements are ordered in descending order of magnitude. The result of this analysis is the construction of the relationship (i, t, T) or (h, t, T) between the rainfall intensity (or depth) and the duration and return period, which is known as the *idf* curve. *idf* curves are necessary for the hydrologic design of numerous hydrological works, such as the rainwater sewage network, the design of a water retention basin or a spillway. Based on them, the rainfall depth and intensity can be easily calculated for every rainfall duration and return period, and generally, one of the three variables (h or i, t, T) can be estimated when the other two are known. For example, under such estimation, the result can be the design storm of an infrastructure with a specific return period and duration, whose standards are most often used as design criteria (the researcher decides for the scope). The design storm properly distributed (during its duration) in time can be used with the response function of the basin (e.g. the unit hydrograph, if the basin is linear) at the site of the infrastructure (spillway), and in this way the design flood for the infrastructure is derived for a specific return period and risk, as explained in Chapter 6.

The process, following the already known frequency analysis of the ordered series of maxima rainfall depths h_i or rainfall intensities i_i for a certain duration t_1, consists of repeating the frequency analysis of maxima h_i or i_i for several different durations t_2, t_3,..., t_r durations. From the proper probability distribution (e.g. Gumbel) for every t_1, t_2,..., t_r durations, the value of $h_1, h_2,..., h_r$ or $i_1, i_2,..., i_r$ of the rainfall depth or intensity for a specific return period T is estimated. The analytical *idf* relationships for a specified return period T between rainfall depth h and rainfall intensity i with duration t are usually one of the following forms:

Simplified exponential:

$$h = kt^m \quad \text{or} \quad i = kt^{m-1}$$

where k and m are constants. These relationships are the simplest ones because they can easily become linear using logarithmic transformation, and the values of k and m are calculated using the least squares method.

Hyperbolic form:

$$h = \frac{kt}{t+b} \quad \text{or} \quad i = \frac{k}{t+b}$$

where b is an additional constant.

Mixed form:

$$h = \frac{kt}{(t+b)^m} \quad \text{or} \quad i = \frac{k}{(t+b)^m}$$

These relationships have better flexibility and generality as they contain the correction parameter b, which corrects the timescale as the scattering points around the *idf* curve are optimized. The acquisition of relationships (h or i, t) for different return periods leads to *idf* family of curves (h or i, t, T) of the form

$$i = kT^\alpha / (t+b)^m$$

where α is an additional parameter. In another approach, the *idf* curves are parallel in double logarithmic paper with T parameter and h or i and $(t+b)$ axes, following the form $\log i = \log(kT^\alpha) - m\log(t+b)$.

The setting of this equation begins with the calculation of the correction parameter b so that the scattering points around the *idf* curves is optimized (it can be resolved analytically as well as graphically in double logarithmic paper). Then, for every T duration of the analysis and by the use of the least squares method, $A_T = \log(kT^\alpha)$ and m can be calculated. Finally, based on the pair of values (A_T, $\log T$) which are already known, k and α are estimated with the use of the least squares method and the analytical equation $A_T = \log k + \alpha \log T$.

Therefore, all four parameters (b, m, k, α) used in the analytical expression of the *idf* curves for a specific site of a monitoring station are calculated. Usually, *idf* curves are useful when they represent a whole basin and not a specific site. In that case, for every $t_1, t_2, ..., t_r$ durations of the analyzed rainfalls, the annual maximum average values for the whole basin must be calculated (e.g. by the Thiessen method) followed by an analysis using the average values of the basin.

The acquisition of primary data, meaning the series h_i or i_i for every rainfall duration time, is in great importance for the frequency analysis of the rainfall maxima. Particularly, in order to record rainfall durations and multiples, even after 24 h, it is necessary a rain gauge to be installed at the analysis site in order to continuously record the rainfall. In recent years, fully automatic telemetric monitoring stations with a high temporal resolution are being used widely, from which it is easy to derive the rainfall maxima depth at different durations. From the gauge's tape (usually weekly or daily) and for every year the period t_1, $t_2, ..., t_r$, of hours with the highest annual depth is chosen. The rainfall being selected must be continuous or with minor gaps so that it is not considered as a different event. Therefore, every maximum should be referred to a single meteorological event and not to two consecutive events. Also, special attention should be given in the case where the maxima h_i or i_i of a t duration, but average for a basin, is calculated. First, the maximum value for a specific monitoring station is pointed out (usually for the one with the highest rainfall depth). Then, the values of the other stations, which are situated at the same basin and for the same rainfall event for the specific duration, are being pointed out. Finally, the average value is determined using these methods, e.g. the Thiessen method and the Kriging method. The average highest value of the basin for the specific duration must be verified by estimating the average highest value of the basin for this duration starting from the highest annual point value of the same duration in another station situated in the basin or by depleting practically the probability of the existence of another highest value in the same year.

Finally, the constructed *idf* curves in many countries have been encountered massively for extensive geographical areas. So, ready-to-use maps have been established for each region, so *idf* curves for any point can be easily retrieved, without the need of primary historical data. The U.S. Weather Bureau established such maps in 1961 (Viessman et al., 1989; Wanielista et al., 1997). These maps illustrate the curves of equal precipitation of maxima rainfall depths for different rainfall durations and different return periods.

4.8.1 Construction of the idf curves

The precipitation curves may be designed using the frequency analysis of recorded rainfall data. The following are the steps for their design:

1. The rainfall duration is selected, such as 5, 10, 20, 30 and 60 min or 2, 6, 12 and 24 h.
2. For the selected duration, the annual maximum rainfall intensity or rainfall depth is calculated for each year of records.
3. A proper frequency curve is adjusted to the values given earlier.
4. From the frequency curve, we obtain the values of rainfall intensity for the selected return periods (5, 10, 20, 30, 50, 80 and 100 years).
5. Steps 2 and 3 are repeated for different durations.
6. The data are redistributed as rainfall intensity versus duration for various return periods or frequencies.
7. For a selected frequency, the values of rainfall intensity are placed on the vertical axis and the duration on the horizontal axis of a logarithmic graph.
8. The same is repeated for other frequencies, resulting in a group of curves (Figure 4.10).

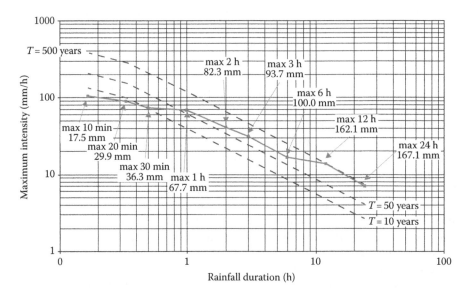

Figure 4.10 Precipitation curves. (From Singh, V.P., *Elementary Hydrology*, Prentice Hall, Englewood Cliffs, NJ, 1992.)

Example 4.8

The maximum annual rainfall depth for durations of 1, 2, 6, 12, 24 and 48 h are given in Table 4.15. The values have been recorded with the use of a rain gauge in Attica, Greece.

1. Assuming that the values of maximum rainfall depth of all durations follow the Gumbel distribution, calculate the precipitation curves for return periods $T = 5$, 20, 50 and 100 years.
2. The construction of a flood protection project to a nearby water stream is studied. The upstream basin with an area of 60 km² and average altitude of 350 m is hilly, with moderate vegetation and a small percentage of impermeable surfaces. The length of the main stream is 9.65 km, and the altitude at the position of the project is 120 m. Estimate the runoff coefficient and the concentration time of the basin.
3. Estimate the peak discharge of the stream for return periods of 5 and 100 years, applying the rational method.
4. Calculate the risk of the project for the aforementioned return periods, given that the useful lifetime is 50 years.

Solution

1. Based on the values of rainfall depth, the respective values of rainfall intensity are calculated for each rainfall duration. For each time series, the following are calculated: mean, standard deviation and the values of the parameters a and c of the Gumbel distribution. The results of the calculations are presented in Table 4.16.

   The non-exceedance probability F ($F = 1 - 1/T$) is calculated for return periods of 5, 20, 50 and 100 years and the parameter $\ln[-\ln(1 - 1/T)]$. Then, the values of X are calculated for each T and rainfall duration based on the statistical characteristics of the sample. The results of the calculations are presented in Table 4.17.

   For the calculation of the precipitation curves for the given return periods, the parameters a and b of the equation $i = aD^b$ will be estimated. If the values of

Table 4.15 Maximum rainfall depth

Hydrological year	1 h	2 h	6 h	12 h	24 h	48 h
1965–1966	2.8	4.7	6.9	7.3	10.2	10.2
1966–1967	13.1	23.4	35.6	41.9	41.9	41.9
1967–1968	1.8	2.3	4.6	4.8	4.8	4.8
1968–1969	8.9	12.8	20.4	30.0	30.0	30.0
1969–1970	1.8	2.7	3.0	3.0	5.2	7.2
1970–1971	25.0	25.0	69.0	39.7	58.8	59.0
1971–1972	9.9	12.4	12.5	12.5	12.5	12.5
1972–1973	8.6	10.0	12.6	13.2	13.7	13.7
1973–1974	11.1	12.9	19.0	24.3	34.1	34.1
1974–1975	8.9	12.7	30.8	36.9	36.9	36.9
1975–1976	27.8	31.1	37.0	37.0	43.1	47.4
1976–1977	8.9	10.3	19.9	20.1	38.0	38.0
1977–1978	9.1	13.8	22.4	37.7	38.5	45.4
1978–1979	14.4	14.7	21.0	35.4	50.8	71.9
1979–1980	23.4	36.2	43.1	44.8	46.0	46.0
1980–1981	13.0	18.7	40.3	62.7	68.2	70.4
1981–1982	6.1	10.9	25.2	42.1	50.9	50.9
1982–1983	16.9	24.9	32.6	38.1	38.2	38.2
1983–1984	11.4	18.8	41.9	56.3	77.0	77.0
1984–1985	5.5	7.8	11.1	12.4	12.4	12.4
1985–1986	30.0	39.0	50.5	73.5	79.7	80.0
1986–1987	12.8	14.2	14.8	20.2	23.5	23.5
1987–1988	18.8	21.5	23.3	39.7	46.5	51.4
1988–1989	5.0	9.3	15.6	20.6	39.3	47.1
1989–1990	5.7	9.0	20.9	34.7	42.4	42.6
1990–1991	12.9	22.2	50.6	82.4	92.2	92.2
1991–1992	8.9	16.4	30.9	33.3	34.6	34.6
1992–1993	11.4	16.7	29.1	36.2	47.5	49.0
1993–1994	7.3	11.2	18.7	21.7	25.4	25.4
1994–1995	14.3	26.3	66.0	78.9	79.9	79.9
1995–1996	13.8	22.6	35.2	36.0	38.2	38.2
1996–1997	30.4	39.0	64.9	69.1	105.3	116.3
1997–1998	20.0	23.6	29.8	47.9	54.3	59.3

rainfall intensity are plotted on normal graph in relation to the duration (1, 2, 6, 12, 24 and 48 h), then Graph 1 is obtained, while if the values of rainfall intensity are plotted on a logarithmic graph in relation to the duration, then Graph 2 is obtained (Figures 4.11 and 4.12).

The quantities $\ln i$ and $\ln D$ are related linearly through an equation of the form $\ln i = c_1 + c_2 \ln D$. Thus, for the determination of the precipitation curve, the calculation of the parameters c_1 and c_2 is sufficient for the calculation of the parameters a and b.

For each return period, T, the logarithms of intensity and duration of Table 4.17 are calculated. $\ln i$ and $\ln D$ are linearly related ($\ln i = \ln a + b \cdot \ln D$), where $\ln a (= c_1) =$ intercept and $b(= c_2) =$ slope, while $a = e^{\ln a} = e^{\text{intercept}}$. The calculations are shown in Table 4.18.

Table 4.16 Rainfall intensity for each rainfall duration

Hydrological year	1 h	2 h	6 h	12 h	24 h	48 h
1965–1966	2.8	2.4	1.2	0.6	0.4	0.2
1966–1967	13.1	11.7	5.9	3.5	1.7	0.9
1967–1968	1.8	1.2	0.8	0.4	0.2	0.1
1968–1969	8.9	6.4	3.4	2.5	1.3	0.6
1969–1970	1.8	1.4	0.5	0.3	0.2	0.2
1970–1971	25.0	12.5	6.5	3.3	2.5	1.2
1971–1972	9.9	6.2	2.1	1.0	0.5	0.3
1972–1973	8.6	5.0	2.1	1.1	0.6	0.3
1973–1974	11.1	6.5	3.2	2.0	1.4	0.7
1974–1975	8.9	6.4	5.1	3.1	1.5	0.8
1975–1976	27.8	15.6	6.2	3.1	1.8	1.0
1976–1977	8.9	5.2	3.3	1.7	1.6	0.8
1977–1978	9.1	6.9	3.7	3.1	1.6	0.9
1978–1979	14.4	7.4	3.5	3.0	2.1	1.5
1979–1980	23.4	18.1	7.2	3.7	1.9	1.0
1980–1981	13.0	9.4	6.7	5.2	2.8	1.5
1981–1982	6.1	5.5	4.2	3.5	2.1	1.1
1982–1983	16.9	12.5	5.4	3.2	1.6	0.8
1983–1984	11.4	9.4	7.0	4.7	3.2	1.6
1984–1985	5.5	3.9	1.9	1.0	0.5	0.3
1985–1986	30.0	19.5	8.4	6.1	3.3	1.7
1986–1987	12.8	7.1	2.5	1.7	1.0	0.5
1987–1988	18.8	10.8	3.9	3.3	1.9	1.1
1988–1989	5.0	4.7	2.6	1.7	1.6	1.0
1989–1990	5.7	4.5	3.5	2.9	1.8	0.9
1990–1991	12.9	11.1	8.4	6.9	3.8	1.9
1991–1992	8.9	8.2	5.2	2.8	1.4	0.7
1992–1993	11.4	8.4	4.9	3.0	2.0	1.0
1993–1994	7.3	5.6	3.1	1.8	1.1	0.5
1994–1995	14.3	13.2	11.0	6.6	3.3	1.7
1995–1996	13.8	11.3	5.9	3.0	1.6	0.8
1996–1997	30.4	19.5	10.8	5.8	4.4	2.4
1997–1998	20.0	11.8	5.0	4.0	2.3	1.2
Mean	12.7	8.7	4.7	3.0	1.8	0.9
Standard deviation	7.7	4.8	2.6	1.7	1.0	0.5
a	0.167	0.267	0.489	0.747	1.257	2.373
c	9.26	6.58	3.51	2.24	1.33	0.70

Thus, the precipitation curves are given by the following equations:

For $T = 5$ years $\rightarrow i = 19.69 * D^{-0.660}$

For $T = 20$ years $\rightarrow i = 28.88 * D^{-0.661}$

For $T = 50$ years $\rightarrow i = 34.70 * D^{-0.661}$

For $T = 100$ years $\rightarrow i = 39.07 * D^{-0.661}$

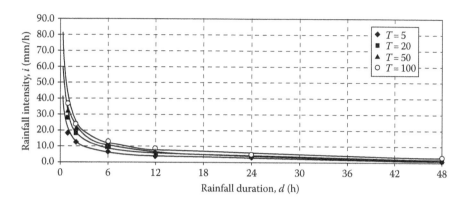

Figure 4.11 Graph 1 – rainfall intensities versus duration.

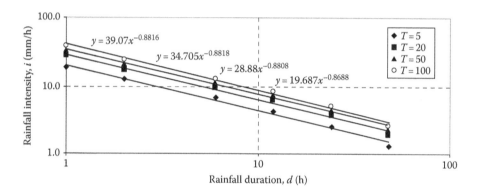

Figure 4.12 Graph 2 – rainfall intensities versus duration on a logarithmic graph.

Table 4.17 Rainfall intensity for each return period and rainfall duration

T	F	$\ln(-\ln(1-1/T))$	1 h	2 h	6 h	12 h	24 h	48 h
5	0.8	−1.50	18.3	12.2	6.6	4.3	2.5	1.3
20	0.95	−2.97	27.1	17.7	9.6	6.2	3.7	1.9
50	0.98	−3.90	32.7	21.2	11.5	7.5	4.4	2.3
100	0.99	−4.60	36.8	23.8	12.9	8.4	5.0	2.6

Table 4.18 Intercept and slope for each return period

T	0.000	0.693	1.792	2.485	3.178	3.871	Slope = b	Intercept	$a = e^{intercept}$
5	2.904	2.502	1.884	1.448	0.927	0.284	−0.660	2.980	19.69
20	3.298	2.874	2.260	1.828	1.308	0.667	−0.661	3.363	28.88
50	3.486	3.054	2.441	2.011	1.490	0.850	−0.661	3.547	34.70
100	3.607	3.170	2.558	2.129	1.608	0.969	−0.661	3.665	39.07

2. The concentration time of the basin is calculated based on the equation of Giandotti:

$$t_c = \frac{4*A^{1/2}+1.5*L}{0.8*\Delta H^{0.5}} = \frac{4*60^{1/2}+1.5*9.65}{0.8*(350-120)^{0.5}} = 3.75 \text{ h}$$

For a hilly area, the runoff coefficient assumes the value $C_1 = 0.24$.
The percentage of impermeable areas is low; thus, $C_2 = 0.05$.
For moderate vegetation conditions, the runoff coefficient is $C_3 = 0.07$.
Due to lack of data for the hydrographic network, an average value is obtained for coefficient $C_4 = 0.08$.
Thus, the composite runoff coefficient of the basin is estimated as follows:

$$C = C_1 + C_2 + C_3 + C_4 = 0.44$$

3. From question 1, for $T = 5$ years, the precipitation curve is $i = 19.69*D^{-0.660}$.
Replacing $d = t_c = 3.75$ h, the result is as follows: $i = 19.69*3.75^{-0.660} \sim 8.2$ mm/h $= 2.3*10^{-6}$ m/s
Therefore, by substituting these values in the rational formula, it follows that

$$Q_{peak} = 0.44*2.3*10^{-6}*60*10^6 = 60.4 \text{ m}^3/\text{s}$$

Similarly, for $T = 100$ years, the precipitation curve is $i = 39.07*D^{-0.661}$.
Replacing $d = t_c = 3.75$ h, we get

$$i = 39.07*3.75^{-0.661} \sim 16.3 \text{ mm/h} = 4.5*10^{-6} \text{ m/s}$$

The runoff coefficient increases by 25% and is equal to 0.55, and by replacing in the rational method, it follows that

$$Q_{peak} = 0.55*4.5*10^{-6}*60*10^6 = 149.5 \text{ m}^3/\text{s}$$

4. The risk is given by the equation $R = 1 - (1 - 1/T)^n$, where n is the useful lifetime of the project. For $n = 50$ years and $T = 5$ years, the risk is $R = 99.999\%$, while for $n = 50$ years and $T = 100$ years, the risk is $R = 39.499\%$.

REFERENCES

Chow, V.T., 1964, *Handbook of Applied Hydrology*, McGraw-Hill New York, NY.

Eadie, W.T., Drijard, D., James, F.E., Roos, M., and Sadoulet, B., 1971, *Statistical Methods in Experimental Physics*. Amsterdam: North-Holland. pp. 269–271. ISBN 0-444-10117-9.

Gumbel, E.J., 1954, *Statistical Theory of Extreme Values and Some Practical Applications: A Series of Lectures*, Vol. 33, Washington: US Government Printing Office.

Gumbel, E.J., 1958, *Statistics of Extremes*, Columbia University Press, New York.

Haan, C, 1985, *Statistical Methods in Hydrology*, The Iowa State University Press, Ames, IA.

Hollander, M. and Wolfe, D.A., 1973, *Nonparametric Statistical Methods*, Wiley and Sons, New York.

Mimikou, M. and Baltas, E., 2012, *Technical Hydrology*. Papasotiriou, Athens, Greece (in Greek).

Singh, V.P., 1992, *Elementary Hydrology*, Prentice Hall, Englewood Cliffs, NJ.

Steel, R.G.D. and Torrie, J.H., 1960, *Principles and Procedures of Statistics, with Special Reference to Biological Sciences*, McGraw-Hill, New York.

Sturges, H.A., 1926, The choice of a class interval, *Journal of the American Statistical Association*, 21(153), 65–66.

Viessman, W. Jr. and Lewis, G.L., 1996, *Introduction to Hydrology*, fourth edition, Harper Collins College Publishers, New York.

Wanielista, M.P., Kersten, R., and Eaglin, R., 1997, *Hydrology*, John Wiley and Sons, New York, 567 pp.

Chapter 5

Groundwater hydrology

5.1 GENERAL

The main topics discussed in this chapter are hydrogeological parameters, the classification of aquifers, the principles of groundwater movement, the hydraulics of water wells (steady and unsteady flow) and the assessment methods of hydrogeological parameters of confined and unconfined aquifers.

Water available in nature and that used by humans can be distinguished into *surface water* and *groundwater*. Surface water is the water found in lakes and rivers, while groundwater is the one stored or moving under the ground. The particular characteristics which distinguish groundwater from surface water resources can be summarized as follows (Latinopoulos, 1986):

1. *Spatial distribution*: Surface water can be found locally (lakes) or following a particular route (rivers), while groundwater occupies much larger areas. As far as the exploitation facilities are concerned, surface water, in most cases, demands more expensive transportation systems, while groundwater can meet the local demand easily with *just* direct pumping.
2. *Temporal variability*: Groundwater presents much slower variability in movement, while in surface waters, the variability is obvious. As a result, surface water reservoirs are usually large and can meet the demands spread over different time periods.
3. *Facilities and operational cost*: Surface water collection projects have a relatively high construction cost (dams, reservoirs, pipelines, etc.), and a low operational cost. On the contrary, the cost of groundwater exploitation facilities (drilling, pumping, etc.) is quite insignificant, but the operational and maintenance costs are important, especially in cases where pumping is from deep aquifers.
4. *Water quality*: This is a very important issue as far as the exploitation and management of water resources are concerned. Groundwater is less exposed to pollution than surface waters. However, restoration/clean-up procedures, in case of pollution, are extremely difficult and expensive.

Finally, it should be mentioned that groundwater aquifers can serve multiple purposes, such as the following (Bear, 1979):

1. *Act as water supply sources*: This is, of course, the most important function. Due to the refilling of inventories by precipitation, underground waters are considered as renewable resources.
2. *Act as reservoir tanks*: In particular, groundwater aquifers, due to their ability to refill their inventories and because of their large area, can store extremely large amounts

of water. The storage capacity of these layers can be greatly enhanced by the artificial recharge technique.

3. *Act as pipelines*: This function can be activated only by human intervention (e.g. by alteration of local hydraulic conditions).
4. *Act as filters*: Using different artificial recharge techniques, surface wastewaters can be filtered in the soil for partial or total purification.
5. *Act as surface water flow regulators*: This function can be accomplished in rivers and in wells by regulating the underground water level (e.g. by pumping) of the aquifers which have a hydraulic connection with surface waters.

These reasons make the importance of underground water in the exploitation and management of water resources quite obvious. In this chapter, basic knowledge and information on this topic is presented, which includes the theory, the methods and the techniques required to address the most common problems, where groundwater flow is dominant. However, the very important problem of groundwater pollution and remediation techniques is not presented.

5.2 SOIL AND AQUIFER PARAMETERS

The *porosity n* of a soil (or a rock) is the property that expresses the volume of voids or pores present in the total volume, expressed as the ratio of the volume of voids U_n to the total soil volume U:

$$n = \frac{U_n}{U} \tag{5.1}$$

Voids ratio e is the ratio of the voids to the total volume of solids U_s, defined by the following equation (Figure 5.1):

$$e = \frac{U_n}{U_s} \tag{5.2}$$

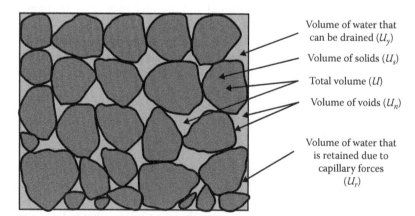

Volume of water that can be drained (U_y)

Volume of solids (U_s)

Total volume (U)

Volume of voids (U_n)

Volume of water that is retained due to capillary forces (U_r)

Figure 5.1 Vertical section of soil.

Therefore, porosity and voids ratio are related by default, according to the following relationship:

$$e = \frac{n}{1-n} \tag{5.3}$$

The *specific yield* or *effective porosity* S_y of a soil or rock refers to the ratio of the volume U_y, which upon saturation can move through the interstices of the medium due to gravity, to the total volume, that is,

$$S_y = \frac{U_y}{U} \tag{5.4}$$

The value of the effective porosity generally is smaller than the overall porosity value. This is due to the effect of capillary forces, which are stronger in cohesive soils; as a result, a portion of the total volume of the water cannot be drained by gravity and remains in the soil. So, after draining, a volume of water U_r remains in the aquifer and is known as specific retention S_r, which can be measured as follows:

$$S_r = \frac{U_r}{U} \tag{5.5}$$

From these relations and the definition of the respective quantities, it is obvious that

$$U_n = U_y + U_r \tag{5.6}$$

$$n = S_y + S_r$$

Directly linked to the effective porosity is the aquifers' property of storage capacity. The exact definition of storage capacity differs in confined and unconfined aquifers (which are described in the next section). In the case of confined aquifers, storage capacity, S, is defined as the volume of water ΔU, removed (or added), from a unit horizontal area A, due to the unit decrease (or increase) $\Delta\varphi$ of the hydraulic head. The following expression describes the storage capacity:

$$S = \frac{\Delta U}{A\Delta\varphi} \tag{5.7}$$

It is a dimensionless parameter. It is obvious that in confined aquifers, the storage capacity depends on the compressibility of water and the elasticity of the rock's solid fracture which encloses it. The higher the value of storage capacity, the greater potential to store or abstract water in this reference volume and, therefore, the greater potential for exploitation of the aquifer.

The storage capacity of an unconfined aquifer is similarly defined. In this case not the hydraulic head but the free water surface is decreasing (or increasing) by Δh, from an abstraction (or replenishment) of a water volume ΔU in a unit horizontal area A. The level of the free surface will drawdown further, even by Δh. Similar to expression 5.7, the storage capacity can be defined as

$$S = \frac{\Delta U}{A\Delta h} \tag{5.8}$$

In confined aquifers, the removal of water is caused by the compressibility of soil grains and the fluid, whereas in unconfined aquifers, the decrease in level reveals removal or water transfer by gravity, from the volume of cavities of a particular area to another. In other words the storage capacity of unconfined aquifers coincides with the effective porosity, and of course, as a magnitude is much higher than the storage capacity of comparable geological formations under pressurized conditions (Aftias, 1992).

5.3 CLASSIFICATION OF AQUIFERS

Water flow in the aquifers is commonly referred to as flow in porous media. The term 'porous media' refers to all soils and rocks that consist of a solid fracture, in the form of a solid granule assembly separated and surrounded by gaps, i.e. pores or cracks. However, when the fractures have large dimensions, flow behaviour changes and it is treated as a special class phenomenon (flow in fractured media).

The basic criterion for the general classification of the aquifers is the position of the maximum water level in soil, as it can be seen in Figure 5.2. Considering a random vertical section of the soil, two different zones can be recognized, where water movement is quite different: (1) aeration zone (or unsaturated zone) and (2) saturation zone (or saturated zone) (Figure 5.3).

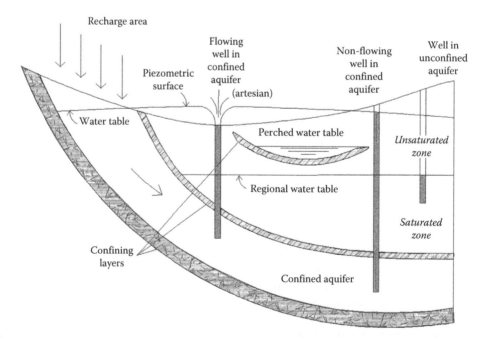

Figure 5.2 Vertical soil structure.

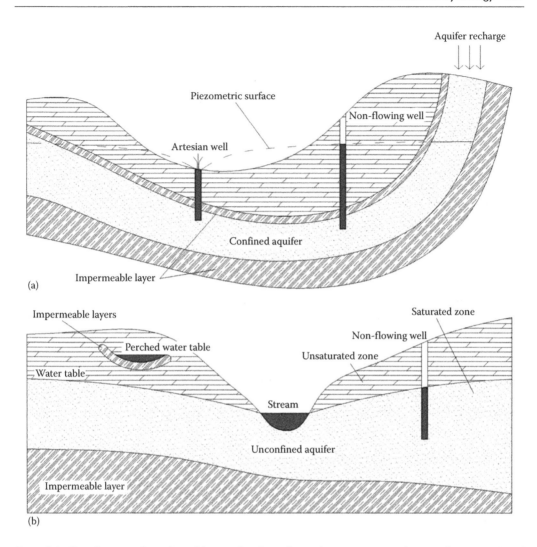

Figure 5.3 Classification of aquifers: (a) a confined aquifer with an artesian and a non-flowing well and (b) an unconfined aquifer showing also the sub-case of a perched water table.

If the upper limit of the zone of saturation exists, this is called phreatic aquifer (or phreatic zone), the synonymous term of water table (of underground water) is used. Because most of the problems related to groundwater resource management refer to water volumes which move or are stored in the saturation zone, the following description is limited to the particular water movement mechanisms and the related phenomena that are observed only in this zone.

Therefore, a conventional classification of the aquifers is made taking into account the geological structure and the local hydraulic conditions as well. So an aquifer is referred to as *confined* or *pressurized*, if it is limited from above and below by impervious geological formations, which are identified as aquitards. The pressurized aquifers have the characteristic that if a well is drilled through them, the water level in the well will be higher than the aquitard and, in some cases, may even reach the surface of the ground. So if the well is properly constructed (concerning filter placement), the water level within this observation well, or piezometer as it is usually called, will indicate the hydraulic head at the specific point. As a result, a *piezometric surface* is defined by the levels of the piezometers drilled at different points along the horizontal area of the specified aquifer.

A confined aquifer is called *artesian*, when the level of its piezometric surface is higher than the soil surface. In this case, if a well is opened, the water will flow out of it freely because of its high pressure, without pumping (artesian well).

The second major category relates to those aquifers where, while the lower limit is an impermeable layer, the upper limit is the free surface of groundwater. The main characteristic of these aquifers, which are called *unconfined* or *phreatic aquifers*, is that they are recharged directly by infiltrating water from the soil surface. Phreatic aquifers can also be classified into the ones restricted by some overlying impermeable layer, whose position is higher than the position of the free surface. This indicates that under conditions of intense water abstraction from a confined aquifer (e.g. intense pumping), it is possible to create favourable conditions for the appearance of a free surface. A subcategory of phreatic aquifers are the perched aquifers (Figure 5.2). In this case, water usually in smaller volumes can be found above the water level of an aquifer due to a curved impermeable layer.

There are also situations where the layer above or below or even both layers, which delineate a confined aquifer, are not totally impervious (semi-permeable). Despite their high resistance to water movement, in the case of large area aquifers, the amount of water which will go into the main aquifer is important. In this case, the aquifer is said to be *pressurized* with leakage (leaky). It is, of course, difficult to assess from a spot investigation whether a layer should be regarded as semi-permeable or not; however, using this method, the practice is to characterize small thickness layers with low infiltration capacity (low permeability) in relation to the main aquifer. Respectively, with the confined aquifer, there are phreatic aquifers with leakage, where certainly the semi-permeable layer forms their lower limit.

5.4 FIELD MEASUREMENTS

The distribution and movement of groundwater can be investigated and analysed either on site or by theoretical methods. But any theoretical study requires some hydraulic data from the field, and the most basic one is the free surface and the hydraulic heads measured in the piezometers and observation wells.

A piezometer consists of a perforated pipe at the lower end, which is positioned vertically by opening up the aquifer under study. Pressure is measured by the piezometer, i.e. the hydraulic head at a particular point of a confined aquifer, exactly at the place where the end is perforated.

In addition to exploratory wells, which aim to investigate the geological and hydrogeological characteristics of the different types of soil and rock formations, there are two other types of wells, with regard to the hydrology of groundwater flow, which are made for observation and exploitation of groundwater, respectively.

An observation well consists of a perforated end, along the whole aquifer thickness, which, in contradiction to the piezometer, is not sealed from the surrounding aquifer. So at least in theory, this well does not cause any obstruction to groundwater flow. It is normal that a pipe of this kind cannot be used to measure the hydraulic head in a confined aquifer, since the resulting value would not be neither the hydraulic head nor the free surface level but their combination.

Finally, the exploitation well is usually constructed for pumping either from an unconfined or from a confined aquifer. It is also used for measuring the water level or the hydraulic head. In the second case, for the confined aquifer, the measurement of the hydraulic head is a mean value along the thickness of the perforated part of the aquifer. The values taken by piezometers and exploitation wells are generally different and are equal only in cases where there is no flow in the particular aquifer.

5.5 MATHEMATICAL PROBLEM OF GROUNDWATER

The water movement in aquifers depends on their hydrodynamic characteristics and local flow conditions. The fundamental law of groundwater hydraulics, which was expressed by the French engineer Darcy in 1856, and bears his name, refers to the following expression with regard to water movement in porous media. Its general form can be written as follows (Hermance, 1999):

$$Q = KAJ \qquad (5.9)$$

where
    Q is the discharge [L³/T]
    K is a porous medium and fluid parameter which is a measure of the porous medium permeability and is called hydraulic conductivity [L/T]
    A is the cross-sectional area of the aquifer through which the flow passes [L²]
    J is the hydraulic gradient of the free or piezometric surface [L/L], which equals $\Delta h/L$ (Figure 5.4)

The hydraulic conductivity is expressed in velocity units and its value depends on the fluid and the porous medium.

The equations of motion (Darcy law) in three dimensions x_i ($i = 1,2,3$) that characterize a general anisotropic and heterogeneous medium of an artesian system can be written as follows:

$$q_i = -K_i \frac{\partial h}{\partial x_i} = nV_i \qquad (5.10)$$

Q Discharge (m³/s)
K Hydraulic conductivity (m/s)
L Distance (m)
Δh Piezometric difference (m)
A Area of reference (m²)

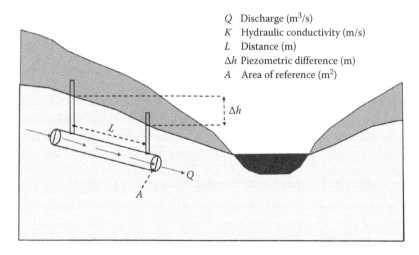

Figure 5.4 Darcy's law.

In these equations, the components of discharge Q in three dimensions have been replaced by the components of the specific discharge or Darcy velocity, $q = Q/A$, which by default equals $q = nV$, where V is the real or intrinsic flow velocity and n is the soil porosity; q has velocity units.

In addition to these parameters, two more parameters of the aquifer are necessary to define the mathematical problem of the groundwater flow: (1) *transmissivity T*, which defines the ability of an aquifer to transfer water and equals to the hydraulic conductivity integrated over the vertical thickness b of the aquifer, and (2) *storage capacity* (or *storativity*) S which, as already mentioned, expresses the amount of water drained per unit area of the aquifer, as a result of a unit change of the hydraulic head.

The flow equation for aquifers, in its general form, is based on the principle of conservation of mass of the fluid (water) in the porous medium (soil). Combining the expression based on this principle with Darcy's law, we end up with the following general differential equation applicable to a heterogeneous and anisotropic medium (Bear, 1979):

$$\frac{\partial}{\partial x_i}\left(K_{ij}\frac{\partial h}{\partial x_j}\right) = S_s \frac{\partial h}{\partial t} \tag{5.11}$$

where
  t is time
  S_s a parameter called specific capacity of the porous medium (usually $S_s = S/b$, where S is the storage capacity and b is the mean thickness of the aquifer)

As already mentioned, most flow problems are solved as horizontal bivariate, using the hypothesis of the hydraulic approach. However, there are special cases where it is required that the solution of the problem, mainly for unconfined aquifers, be on a vertical plane. For this reason, the description of mathematical models which solve the various types of mathematical problems is divided into these two categories.

5.6 GENERAL EXPRESSION OF GROUNDWATER FLOW

This analysis can be generalized easily for 3D flow, considering a differential reference volume $dxdydz$. Following the same pattern of calculations, we end up with the following equation, respective to the continuity equation (Gupta, 1989):

$$\frac{\partial}{\partial x}\left(K_x\frac{\partial h}{\partial x}\right) + \frac{\partial}{\partial y}\left(K_y\frac{\partial h}{\partial y}\right) + \frac{\partial}{\partial z}\left(K_z\frac{\partial h}{\partial z}\right) = S_s\frac{\partial h}{\partial t} + W \tag{5.12}$$

where, besides the 3D approach of the flow domain, the term W is added, which is the volumetric flux per unit volume (L/T), expressing the external inflow or outflow of water (recharge of leakage). This equation generally applies to homogeneous, heterogeneous, isotropic and anisotropic media. For an isotropic (homogeneous or heterogeneous) medium, the aforementioned expression can be written as follows:

$$\frac{\partial}{\partial x}\left(K\frac{\partial h}{\partial x}\right) + \frac{\partial}{\partial y}\left(K\frac{\partial h}{\partial y}\right) + \frac{\partial}{\partial z}\left(K\frac{\partial h}{\partial z}\right) = S_s\frac{\partial h}{\partial t} + W \tag{5.13}$$

and for a homogeneous and isotropic medium, it is written as

$$\frac{\partial^2 h}{\partial x^2} + \frac{\partial^2 h}{\partial y^2} + \frac{\partial^2 h}{\partial z^2} = \frac{S_s}{K}\frac{\partial h}{\partial t} + \frac{W}{K} \tag{5.14}$$

In the case of uniform flow without the recharging term W, in which there is no change in the flow domain with time, the aforementioned equations are much simpler, as the right-hand side equals to zero. So, the equation of uniform flow in a heterogeneous anisotropic medium is written as follows:

$$\frac{\partial}{\partial x}\left(K_x\frac{\partial h}{\partial x}\right) + \frac{\partial}{\partial y}\left(K_y\frac{\partial h}{\partial y}\right) + \frac{\partial}{\partial z}\left(K_z\frac{\partial h}{\partial z}\right) = 0 \tag{5.15}$$

The equation of uniform flow for homogeneous and isotropic medium is written as follows:

$$\frac{\partial^2 h}{\partial x^2} + \frac{\partial^2 h}{\partial y^2} + \frac{\partial^2 h}{\partial z^2} = 0 \tag{5.16}$$

The last expression is known as the Laplace equation.

The governing equation of underground water describes the water movement in both confined and unconfined aquifers. In the case of confined aquifers and without taking into account the term W, the general equation for an anisotropic and heterogeneous medium is modified as follows:

$$\frac{\partial}{\partial x}\left(K_x\frac{\partial h}{\partial x}\right) + \frac{\partial}{\partial y}\left(K_y\frac{\partial h}{\partial y}\right) + \frac{\partial}{\partial z}\left(K_z\frac{\partial h}{\partial z}\right) = S_s\frac{\partial h}{\partial t} \tag{5.17}$$

For an isotropic and homogeneous medium, the equation is as follows:

$$\left(K_x\frac{\partial^2 h}{\partial x^2}\right) + \left(K_y\frac{\partial^2 h}{\partial y^2}\right) + \left(K_z\frac{\partial^2 h}{\partial z^2}\right) = S_s\frac{\partial h}{\partial t} \tag{5.18}$$

In the case of unconfined aquifers, assuming that the flow is horizontal (Dupuit approximation) and without taking into account the term W, the general equation for an anisotropic and heterogeneous medium is modified as follows:

$$\frac{\partial}{\partial x}\left(K_x h\frac{\partial h}{\partial x}\right) + \frac{\partial}{\partial y}\left(K_y h\frac{\partial h}{\partial y}\right) = S_y\frac{\partial h}{\partial t} \tag{5.19}$$

The last expression is known as the Boussinesq expression (Singh, 1992). Linearization of this equation can be made when the change in the drawdown level is relatively small in relation to the depth of the water, where the depth h replaces by the depth b of the aquifer. For a homogeneous aquifer, the equation is modified as follows:

$$\left(K_x \frac{\partial^2 h}{\partial x^2}\right) + \left(K_y \frac{\partial^2 h}{\partial y^2}\right) = \frac{S_y}{b} \frac{\partial h}{\partial t} \tag{5.20}$$

The integration of the mathematical problem is made by adding the initial and boundary conditions. The assignation of a piezometric (or hydraulic) head at the beginning of the phenomenon is defined as an initial condition for an arbitrarily defined time $t = 0$, with the general form $h = f(x,y,0)$, which is a known function at each point x,y of the horizontal flow field.

Three types of boundary conditions, usually used for groundwater flow problems, are as follows: (1) conditions of a known head, $h = f_1(x,y,t)$; (2) conditions of known discharge, $Q'_n = f_2(x,y,t)$, where Q'_n is the per unit length perpendicular to the boundary curve infiltrated discharge; and (3) semi-permeable boundary conditions. The most common application of known condition refers to the boundaries of the aquifer where there is hydraulic contact with surface waters (lakes, rivers, seas), the corresponding condition of known discharge is used for impermeable boundaries (no infiltrated discharge) and the boundary condition of semi-permeable boundary applies to cases of partially bounded riverbed or lake due to the deposition of fine-grained material that reduces the hydraulic communication between the aquifer and the surface water body.

From this discussion, it is clear that the management of confined aquifers where the boundaries are defined geometrically is simpler. For unconfined aquifers, the fact that the upper limit of the flow (free surface boundary) is not fixed geometrically but is determined by the zero pressure condition introduces considerable difficulty. Very often, flow problems in phreatic aquifers are simplified by the use of the so-called Dupuit approximation, according to which the flow can be considered substantially horizontal (assuming the slope of the phreatic horizon is small). The consequences of this hypothesis is that (1) the vertical component of the specific discharge is zero, (2) the horizontal components of the specific discharge are fixed in each vertical line and (3) the hydraulic head in each vertical line is fixed.

5.7 ANALYTICAL SOLUTIONS OF STEADY FLOW

The groundwater flow, which is described by the Laplace equation, can be solved based on the theory of partial differential equations. For conditions of actual flow, the solution must satisfy the boundary conditions with regard to the piezometric heads. The porous medium is assumed as homogeneous and isotropic in all cases.

5.7.1 Confined aquifer

The flow in a well, which fully penetrates in a homogeneous and isotropic aquifer, is radially symmetrical. The radius of the borehole is measured from the centre of the borehole (Figure 5.5). For an isotropic and homogeneous aquifer, the equation describing the groundwater flow is (Freeze and Cherry, 1979)

$$\frac{\partial^2 h}{\partial x^2} + \frac{\partial^2 h}{\partial y^2} = 0 \tag{5.21}$$

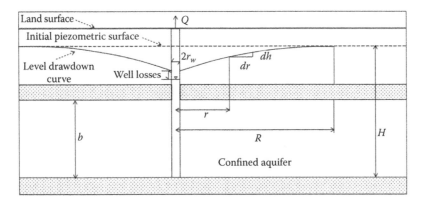

Figure 5.5 Confined aquifer.

By using polar coordinates,

$$x = r\cos\theta$$

$$y = r\sin\theta$$

$$r = (x^2 + y^2)^{1/2} \quad \text{and} \quad \theta = \tan^{-1}\left(\frac{y}{x}\right)$$

The equation of the underground flow becomes:

$$\frac{d^2h}{dr^2} + \frac{1}{r}\frac{dh}{dr} = 0 \tag{5.22}$$

or

$$\frac{1}{r}\frac{d}{dr}\left(r\frac{dh}{dr}\right) = 0 \tag{5.23}$$

This means that the following term is a constant:

$$r\frac{dh}{dr} = C_1 \tag{5.24}$$

where C_1 is a constant.

The discharge can be obtained using Darcy's law, and an area A, which is the lateral surface of a cylinder of radius r and height b, according to the following equation:

$$Q = 2\pi r b K \frac{dh}{dr} \tag{5.25}$$

which, after integration, results in the following equation:

$$h = \frac{Q}{2\pi bK}\ln r + C_2 \tag{5.26}$$

Applying Equation 5.26 at distances R (where the change in the piezometric line is minimal, practically zero, and the head is H) and r, and by subtracting by members, we get

$$h = \frac{Q}{2\pi bK}\ln\frac{r}{R} + H \Rightarrow H - h = \frac{Q}{2\pi bK}\ln\frac{R}{r}[L] \tag{5.27}$$

where
　　H is the hydraulic head at distance R (Figure 5.5) [L]
　　h is the hydraulic head at any distance r [L]
　　Q is the discharge [L^3/T]
　　b is the thickness of the aquifer [L]
　　bK is the transmissivity T [L^2/T]

Equation 5.27 is known as the Thiem equation, which could also be extracted by the integration of Darcy's equation:

$$Q = 2\pi rbK\frac{dh}{dr} \tag{5.28}$$

$$\int_h^H dh = \frac{Q}{2b\pi K}\int_r^R \frac{dr}{r} \tag{5.29}$$

Example 5.1

An aquifer with a thickness of 40 m is overlaid by an impermeable layer of 25 m thickness. After a trial pumping with discharge 0.2 m^3/s for a long time from a borehole with 0.4 m diameter, the water drawdowns at two observation wells located at distances 20 and 80 m were 5 and 2 m, respectively. Estimate the hydraulic conductivity and the drawdown in the pumping borehole.

Solution

If h_1 and h_2 are the hydraulic heads on the two observation wells and r_1 and r_2 are the respective radii, then using Equation 5.27,

$$(H - h_1) - (H - h_2) = h_2 - h_1 = \frac{Q}{2\pi bK}\ln\frac{r_2}{r_1} \Rightarrow -(2 - 5) = \frac{0.2}{2\pi 40K}\ln\frac{80}{20} \Rightarrow K = 0.000368 \text{ m/s}$$

The difference in hydraulic heads is the negative value of the difference of drawdowns for the respective positions.
　　The drawdown s_w in the borehole is calculated accordingly (h_w the hydraulic head in the borehole):

$$h_2 - h_w = \frac{Q}{2\pi bK}\ln\frac{r_2}{r_w} \Rightarrow -(2 - s_w) = \frac{0.2}{2\pi(40)(0.000368)}\ln\frac{80}{0.20} \Rightarrow$$

$$\Rightarrow s_w - 2 = 12.96 \text{ m} \Rightarrow s_w = 14.96 \text{ m}$$

Example 5.2

A pumping well operates with a steady discharge rate of 62.8 m³/h at the centre of a confined aquifer with a shape that can be approximated by a cylinder with a thickness $b = 10$ m and radius $R = 578$ m, under steady flow. The drawdown of the piezometric surface in two observation wells located 50 and 136 m apart from the pumping well is 2 and 1 m, respectively. The pumping stops after 8 h of continuous operation, and in a short span of time a new state of equilibrium is achieved with piezometric surface reduced by 48 cm from the initial surface.

The following variables are to be estimated:

1. The transmissivity (T) of the confined aquifer
2. The drawdown of the piezometric surface 300 m from the well while it is in operation
3. The specific capacity (S_s) of the confined aquifer

Solution

1. For steady flow condition, the Thiem equation (5.27) is valid, so

$$-[s_2(r_2,t) - s_1(r_1,t)] = \frac{Q}{2\pi T} \ln\left(\frac{r_2}{r_1}\right) \Rightarrow 2 - 1 = \frac{62.8}{2\pi T} \ln\left(\frac{136}{50}\right) \Rightarrow T = 10.001 \text{ m}^2/h$$

2. Again from the Thiem equation

$$-[s_3(r_3,t) - s_1(r_1,t)] = \frac{Q}{2\pi T} \ln\left(\frac{r_3}{r_1}\right) \Rightarrow 2 - s_3 = \frac{62.8}{2\pi(10.001)} \ln\left(\frac{300}{50}\right) \Rightarrow s_3(300,t)$$

$$= 0.209 \text{ m}$$

3. After reaching the equilibrium state, we have $\Delta h = 0.48$ m.

   The abstracted volume is

   $$\Delta V = Q \times t = 62.8 \text{ m}^3/h \times 8 \text{ h} \Rightarrow \Delta V = 502.4 \text{ m}^3$$

   The specific capacity of the aquifer is

   $$Ss = \frac{\Delta V}{A \times \Delta h} = \frac{502.4}{\pi \times 578^2 \times 0.48} \Rightarrow Ss = 99.7 \times 10^{-5}$$

   where A is the area of the aquifer.

5.7.2 Unconfined aquifer

The analysis of an unconfined aquifer is made by using the Dupuit approximation, where it is considered that (1) the flow is horizontal and (2) the flow velocity is proportional to the hydraulic slope. A basic difference between the flow of a confined and an unconfined aquifer is that in the unconfined flow case, the aquifer thickness changes and, as a result, the section that provides water to the pumping borehole changes. The equations mentioned in the previous paragraph now become

$$Q = 2\pi r h K \frac{dh}{dr} \tag{5.30}$$

since the discharge does not depend on the thickness of the aquifer b but on the depth of the water h. The last one takes the form

$$hdh = \frac{Q}{2\pi K}\frac{dr}{r} \tag{5.31}$$

Its solution results in

$$h^2 = \frac{Q}{\pi K}\ln r + C_2 \tag{5.32}$$

Applying again the last expression for distances r (with level h) and R (with level H), and by subtracting by members, gives

$$h^2 = \frac{Q}{\pi K}\ln\frac{r}{R} + H^2 \; [L^2] \tag{5.33}$$

or

$$H^2 - h^2 = \frac{Q}{\pi K}\ln\frac{R}{r} \; [L^2] \tag{5.34}$$

This equation could also be extracted by integrating Darcy's equation:

$$Q = 2\pi rhK\frac{dh}{dr} \tag{5.35}$$

$$\int_h^H hdh = \frac{Q}{2bK}\int_r^R \frac{dr}{r} \tag{5.36}$$

All mentioned parameters are presented in Figure 5.6.

Example 5.3

A borehole of 30 cm diameter has its lower end 60 m below the piezometric surface. After 16 h of pumping at a discharge rate of 0.85 m³/s, the piezometric surface is stabilized 15 m below the initial piezometric surface, whereas at an observation well that is located at a distance of 400 m from the borehole, the level drawdown is 2.60 m. Calculate the hydraulic conductivity of the aquifer.

Solution

To calculate the hydraulic conductivity K, the equation that relates the level H at distance R from the borehole and the level in the borehole (5.34) can be used:

$$H^2 - h^2 = \frac{Q}{\pi K}\ln\frac{R}{r} \Rightarrow (60-2.60)^2 - (60-15)^2 = \frac{0.85}{3.14(K)}\ln\frac{400}{(0.30/2)} \Rightarrow$$

$$\Rightarrow K = 1.68\times10^{-3} \; \text{m/s}$$

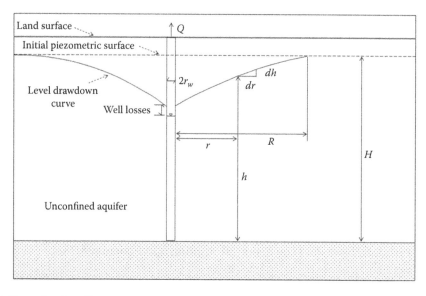

Figure 5.6 Unconfined aquifer.

5.7.3 Semi-confined aquifer

As semi-confined aquifer is called the one that is restricted between a lower impermeable layer and an upper semi-permeable layer. Above the semi-permeable layer, there is an unconfined aquifer. It is assumed that at the beginning, the hydraulic head of the semi-confined aquifer coincides with the level of the unconfined aquifer, as can be seen in Figure 5.7. After pumping from the semi-confined aquifer, a piezometric difference develops between the unconfined and the confined aquifer, resulting in a flow through the semi-permeable layer. The flow is assumed to be horizontal in the confined aquifer and vertical in the

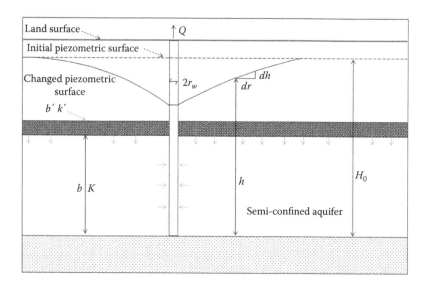

Figure 5.7 Semi-confined aquifer.

semi-permeable layer. In the general groundwater flow equation, a term is added which represents leakage from the unconfined aquifer to the semi-confined one:

$$q' = K' \frac{H_0 - h}{b'} \tag{5.37}$$

where

    q' is the velocity
    K' is the hydraulic conductivity
    b' is the aquifer thickness

Therefore, the equation for the semi-confined aquifer into 1D form is as follows:

$$Kb \frac{\partial^2 h}{\partial r^2} + Kb \frac{1}{r} \frac{\partial h}{\partial r} + K' \frac{H_0 - h}{b'} = 0 \tag{5.38}$$

Using the level drawdown $s = H_0 - h$ and substituting h in Equation 5.38,

$$-\frac{\partial^2 s}{\partial r^2} - \frac{1}{r} \frac{\partial s}{\partial r} + \frac{K'}{Kbb'} s = 0$$

or

$$\frac{\partial^2 s}{\partial r^2} + \frac{1}{r} \frac{\partial s}{\partial r} - \frac{s}{B^2} = 0 \, [\text{L}^{-1}] \tag{5.39}$$

where

$$B^2 = \frac{Kbb'}{K'} = \frac{T}{K'/b'} \tag{5.40}$$

The solution of the last equation is as follows:

$$s = C_1 K_0\left(\frac{r}{B}\right) + C_2 \Gamma_0\left(\frac{r}{B}\right) [\text{L}] \tag{5.41}$$

where

    s is the level drawdown
    $\Gamma_0 (r/B)$ is the first-class modified Bessel function
    $K_0 (r/B)$ is the second-class modified Bessel function
    C_1, C_2 are constants resulting from the boundary conditions

The final equation for a infinite area semi-confined aquifer is as follows:

$$s = \frac{Q}{2\pi T} K_0\left(\frac{r}{B}\right) [\text{L}] \tag{5.42}$$

The $K_0 (r/B)$ values for different r/b are tabulated. For $r/b < 0.05$, Equation 5.42 is written as follows:

$$s = \frac{Q}{2\pi T} \ln\left(1.123 \frac{B}{r}\right) [\text{L}] \tag{5.43}$$

Example 5.4

An aquifer is located between an impermeable layer and a semi-permeable one of 6 m thickness, having a hydraulic conductivity coefficient of 2×10^{-8} m/s. The mean thickness of the aquifer is 150 m and the hydraulic conductivity coefficient is 1.5×10^{-3} m/s. Groundwater is pumped from the aquifer at a discharge rate of 0.20 m³/s through a borehole of 10 in. in diameter. Calculate the drawdown at a distance of 1500 m from the borehole and in the borehole itself.

Solution

At first, the transmissivity of the semi-permeable layer is calculated:

$$T = K \times b = 1.5 \times 10^{-3} \times 150 = 0.225 \text{ m}^2/\text{s}$$

And then the constant B (Equation 5.40)

$$B = \sqrt{\frac{T}{\frac{K'}{b'}}} = \sqrt{\frac{0.225}{\frac{2 \times 10^{-8}}{6}}} = 8216 \text{ m}$$

For the calculation of the level drawdown, first the ratio r/B is estimated:

$$\frac{r}{B} = \frac{0.253/2}{8216} = 1.54 \times 10^{-5} < 0.05$$

Therefore, the following simplified equation can be used:

$$s = \frac{Q}{2\pi T} \ln\left(1.123 \frac{B}{r}\right) = \frac{0.2}{2 \times 3.14 \times 0.225} \ln\left(1.123 \frac{8216}{0.253/2}\right) = 1.58 \text{ m}$$

For the calculation of the level drawdown at the 1500 m distance, the ratio is again estimated:

$$\frac{r}{B} = \frac{1500}{8216} = 0.182 > 0.05$$

In this case the simplified equation is not valid and the level drop is given by the general equation:

$$s = \frac{Q}{2\pi T} K_0\left(\frac{r}{B}\right)$$

with the values of the Bessel function $K_0\left(\frac{r}{B}\right)$ taken from tables.

5.8 THEORY OF IMAGES

The application of the general equation of groundwater flow requires that the aquifer boundaries are infinite. However, all the aquifers are surrounded by either impermeable layers or steady supply boundaries, such as lakes and rivers. In the case where pumping wells are located close to such boundaries, the equations for a infinite aquifer are not applicable; to address these cases, the use of the theory of images is necessary, where boreholes – images – are used in order to create the conditions of a infinite aquifer. In this chapter, two cases are examined, one close to a river and the other close to an impermeable layer.

5.8.1 Well near river

Along the river, there is a constant hydraulic head equal to the level at the surface of the river. Therefore, the cone of the underground water level drawdown approaching the river, should match the river surface. This can be accomplished if a mirror well which enriches the aquifer is situated on the other side of the river, at the same distance from the initial well, as shown in Figure 5.8. The recharge rate of the mirror well will be the same as the pumping borehole, so that the level drawdown, because of the abstraction in the pumping borehole, equals the level rise due to recharge, with the two levels to be assimilated along the river, as shown in Figure 5.9.

Let us assume that the distance from the pumping borehole to the river is α. The distances of a point I with coordinates (x,y) from the pumping and recharge boreholes are

$$r_1 = \sqrt{(a-x)^2 + y^2} \quad \text{and} \quad r_2 = \sqrt{(a+x)^2 + y^2} \tag{5.44}$$

The level drawdown due to the actual borehole is

$$s_1 = \frac{Q}{2\pi bK} \ln \frac{a}{r_1} \tag{5.45}$$

while the mirror borehole creates a level rise equal to

$$s_2 = -\frac{Q}{2\pi bK} \ln \frac{a}{r_2} \tag{5.46}$$

The final level drop to the point (x,y) will be equal to the sum of the level drawdown caused by the actual borehole and the level rise due to the presence of watercourses, i.e.

$$s = s_1 + s_2 = \frac{Q}{4\pi bK} \ln \frac{y^2 + (a+x)^2}{y^2 + (a-x)^2} \; [\text{L}] \tag{5.47}$$

where
  α is the horizontal distance of the river from the borehole
  x,y are the coordinates of the point where the level drop occurs

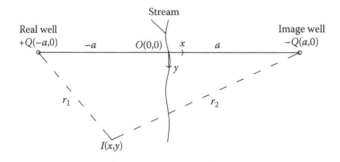

Figure 5.8 Distances of the pumping and enrichment borehole from the river.

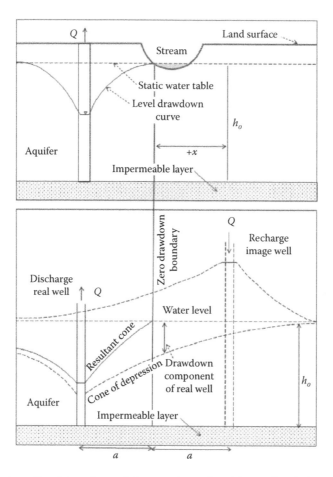

Figure 5.9 Combination of level drawdown of the pumping and enrichment boreholes.

Example 5.5

A 0.30 m diameter borehole is drilled into an aquifer of 40 m thickness with a hydraulic conductivity of 30 m/day. If a drawdown of 2 m is observed in the borehole due to continuous pumping, calculate the following:

1. The pumping rate if the borehole is situated at a distance of 80 m from a river
2. The pumping rate if the borehole is situated at a distance of 4000 m from a river

Solution

1. The drawdown in the borehole is given by the following equation:

$$s = -\frac{Q}{2\pi bK} \ln \frac{y^2 + (a+x)^2}{y^2 + (a-x)^2} \ [L]$$

where x, y are the coordinates of the well, i.e., $x = -80 + 0.15$ and $y = 0$, while a is the distance from the river: $a = 80$ m. By replacing the values, we get

$$2 = -\frac{Q}{2 \times 3.14 \times 40 \times 30} \ln \frac{0 + (80 - 80 + 0.15)^2}{0 + (80 - 0.15)^2} \Rightarrow Q = 1.20052 \ \text{m}^3/\text{day}$$

2. As in case 1, the respective inputs are: $a = 4000$ m, $x = -4000 + 0.15$, $y = 0$, so we have

$$2 = -\frac{Q}{2 \times 3.14 \times 40 \times 30} \ln \frac{0 + (4000 - 4000 + 0.15)^2}{0 + (4000 - 0.15)^2} \Rightarrow Q = 739.47 \text{ m}^3/\text{day}$$

5.8.2 Well near impermeable boundaries

In this case, there is no flow beyond the impermeable layer. If a mirror borehole is placed on the other side of the layer and pumps with the same discharge, then the drawdown will meet in the impermeable boundary, as it can be seen in Figure 5.10. If the radius of influence is R, then the drawdown for each borehole is given by the following equation:

$$s = \frac{Q}{2\pi bK} \ln \frac{R^2}{r_1 r_2} \text{ [L]} \tag{5.48}$$

where
r_1, r_2 are the distances as defined
R is the influence radius

Example 5.6

A borehole of 0.30 m diameter is drilled into an aquifer of 40 m thickness with a hydraulic conductivity of 30 m/day. If a drawdown of 1.5 m is observed in the borehole due to continuous pumping, then calculate the pumping rate if the borehole is situated at a horizontal distance of 100 m from an impermeable layer and the influence radius is 1.0 km.

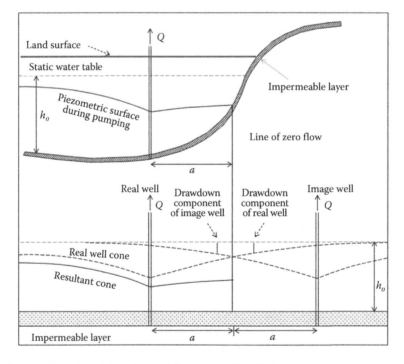

Figure 5.10 Combination of level drawdown of the pumping boreholes.

Solution

Equation 5.48 is applied for $r_1 = 0.15$ m, equal to the borehole radius, and $r_2 = 2 \times 100 = 200$ m, the same as the distance from the mirror borehole:

$$s = \frac{Q}{2\pi bK} \ln \frac{R^2}{r_1 r_2} \Rightarrow 1.5 = \frac{Q}{2 \times 3.14 \times 40 \times 30} \ln \frac{1000^2}{0.15 \times 200} \Rightarrow Q = 1085.4 \text{ m}^3/\text{day}$$

5.9 ANALYTICAL SOLUTIONS OF NON-UNIFORM FLOW

In problems of non-uniform flow, the water level in unconfined aquifers, or the hydraulic head in confined aquifers, is not stable but changes with time. In resolving such problems, the parameter of time is involved, as will be discussed in the following paragraphs.

5.9.1 Well hydraulics

A standard solution of the non-uniform flow equation regards the flow field around the borehole. The equations arising have a direct and extensive application in assessing the characteristics of an aquifer.

Consider the confined horizontal aquifer of similar shape, which has a constant thickness b and infinite extent (in both horizontal dimensions). The aquifer is considered to be of homogeneous and isotropic medium, of transmissivity T and storage capacity S, with no recharge or leakage between the neighbouring layers. At a certain location in the aquifer, a vertical borehole is opened. Water storage in the borehole is considered negligible. At time $t = 0$, the hydraulic head in the aquifer is constant, equal to h_0. At the same moment, water is pumped from the borehole at a constant discharge Q.

Due to the circular symmetry, the flow equation may be written simply using polar coordinates:

$$\frac{\partial^2 h}{\partial r^2} + \frac{1}{r}\frac{\partial h}{\partial r} = \frac{S}{T}\frac{\partial h}{\partial t} \tag{5.49}$$

where the zero recharging term W is omitted. In the case of level drawdown, Equation 5.49 becomes

$$\frac{\partial^2 s}{\partial r^2} + \frac{1}{r}\frac{\partial s}{\partial r} = \frac{S}{T}\frac{\partial s}{\partial t} \tag{5.50}$$

Theis (1935) has shown that the solution of this differential equation for the boundary and initial conditions described earlier is as follows:

$$s = \frac{Q}{4\pi T} W(u) \quad [\text{L}] \tag{5.51}$$

where

$$W(u) = \int_u^\infty \frac{e^{-u}}{u} du \tag{5.52}$$

where

$$u = \frac{r^2 S}{4Tt} \tag{5.53}$$

In these equations
    r is the distance from the centre of the pumping borehole [L]
    t is the time [T]
    S is the storage capacity [dimensionless]
    T is the transmissivity [L²/T]
    s is the hydraulic head drawdown at time t and at a distance r from the borehole [L]

The parameter $W(u)$ is known in mathematical terminology as exponential integral and in technical terminology as the well function. Equation 5.51 is referred to in the literature as the Theis equation. The integral of Equation 5.52 has the following alternative expansion in infinite sequences, used for computational purposes (the first for small values of u and the second for higher values):

$$W(u) = -0.5772 - \ln u + u - \frac{u^2}{2 \cdot 2!} + \frac{u^3}{3 \cdot 3!} - \frac{u^2}{4 \cdot 4!} + \cdots \tag{5.54}$$

$$W(u) \approx \frac{e^{-u}}{u}\left(1 - \frac{1!}{u} + \frac{2!}{u^2} - \frac{3!}{u^3} + \cdots\right) \tag{5.55}$$

Alternatively, the values of the well function can be obtained from Table 5.1 (Chow, 1964). A plot of well function values in the range 10^{-7} to 10^{-1} is presented in Figure 5.11.

Table 5.1 Values of the well function $W(u)$ for various values of parameter u

u	1	2	3	4	5	6	7	8	9
×1	0.219	0.049	0.013	0.0038	0.0011	0.00036	0.00012	0.000038	0.000012
×10⁻¹	1.82	1.22	0.91	0.70	0.56	0.45	0.37	0.31	0.26
×10⁻²	4.04	3.35	2.96	2.68	2.47	2.30	2.15	2.03	1.92
×10⁻³	6.33	5.64	5.23	4.95	4.73	4.54	4.39	4.26	4.14
×10⁻⁴	8.63	7.94	7.53	7.25	7.02	6.84	6.69	6.55	6.44
×10⁻⁵	10.94	10.24	9.84	9.55	9.33	9.14	8.99	8.86	8.74
×10⁻⁶	13.24	12.55	12.14	11.85	11.63	11.45	11.29	11.16	11.04
×10⁻⁷	15.54	14.85	14.44	14.15	13.93	13.75	13.60	13.46	13.34
×10⁻⁸	17.84	17.15	16.74	16.46	16.23	16.05	15.90	15.76	15.65
×10⁻⁹	20.15	19.45	19.05	18.76	18.54	18.35	18.20	18.07	17.95
×10⁻¹⁰	22.45	21.76	21.35	21.06	20.84	20.66	20.50	20.37	20.25
×10⁻¹¹	24.75	24.06	23.65	23.36	23.14	22.96	22.81	22.67	22.55
×10⁻¹²	27.05	26.36	25.96	25.67	25.44	25.26	25.11	24.97	24.86
×10⁻¹³	29.36	28.66	28.26	27.97	27.75	27.56	27.41	27.28	27.16
×10⁻¹⁴	31.66	30.97	30.56	30.27	30.05	29.87	29.71	29.58	29.46
×10⁻¹⁵	33.96	33.27	32.86	32.58	32.35	32.17	32.02	31.88	31.76

Source: Chow, V.T., *Handbook of Applied Hydrology*, McGraw-Hill, New York, 1964.

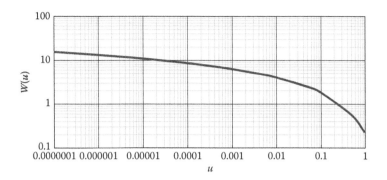

Figure 5.11 Relationship between W(*u*) and *u*.

These equations are used to find the hydraulic properties of the discharge capacity and the storage capacity of an aquifer. The values required to solve the equations we have seen are derived from pumping tests, where a borehole pumps with a constant discharge over a period of several hours to several days. The level drawdown is recorded at specific times in observation boreholes, which are located at different distances from the pumping borehole. An important advantage of the Theis equation is that it can be used for smaller time intervals, i.e. before achieving equilibrium. Several procedures have been proposed for solving these equations. The most essential ones are the Theis, Cooper–Jacob and Chow methods.

Example 5.7

In an aquifer with 35.0 m thickness, hydraulic conductivity $K = 16.0$ m/day and storage capacity $S = 0.00550$, a well is pumped at a rate of 1500 m³/day. Estimate the drawdown at a distance of 30.0 m from the well, after 5 days of continuous pumping.

Solution

The transmissivity T is equal to $T = K \times b = 16 \times 35 = 560$ m²/day
   The parameter u

$$u = \frac{r^2 S}{4Tt} = \frac{(30.0)^2 \times 0.00550}{4 \times 560 \times 5} = 0.00044$$

From Table 5.1 for $u = 0.00044 \Rightarrow W(u) = 7.16$, therefore

$$s = H - h = \frac{Q}{4\pi T} W(u) = \frac{1500}{4 \times \pi \times 560} 7.16 = 1.52 \text{ m}$$

So the drawdown is 1.52 m at a distance of 30.0 m from the well after 5 days of pumping.

Example 5.8

A well is pumped at a rate of 1000 m³/day from an aquifer with $S = 2 \times 10^{-4}$ and $T = 150$ m²/day. Find the drawdown at a distance of 5 m from the well after 2 h of pumping and also at a distance of 500 m after 3 days of pumping.

Solution

With substitution in Equation 5.53 for $r = 5.0$ m and $t = 2$ h (=2/24 days):

$$u = \frac{r^2 S}{4Tt} = \frac{(5.0)^2 \times 2 \times 10^{-4}}{4 \times 150 \times 2/24} = 0.0001$$

From Table 5.1: $W(u) = 8.63$

$$s = \frac{Q}{4\pi T} W(u) = \frac{1000}{4 \times \pi \times 150} 8.63 = 4.58 \text{ m}$$

which is the drawdown at a distance of 5 m, after 2 h.
 Similarly for $r = 500$ and $t = 3$ days,

$$u = \frac{r^2 S}{4Tt} = \frac{(500.0)^2 \times 2 \times 10^{-4}}{4 \times 150 \times 3} = 0.028$$

From Table 5.1: $W(u) = 3.04$. Therefore,

$$s = \frac{Q}{4\pi T} W(u) = \frac{1000}{4 \times \pi \times 150} 3.04 = 1.61 \text{ m}$$

Example 5.9

Using the previous example, estimate the radius of influence after 2 h and 3 days of continuous pumping, respectively.

Solution

We assume the drawdown in the radius of influence is 2 cm, which is practically zero (very small).

$$W(u) = \frac{s 4 \pi T}{Q} = \frac{0.02 \times 4 \times 3.14 \times 150}{1000} = 0.0377$$

From Table 5.1 we get $u = 2.31$
For $t = 2$ h, we get

$$r = \left(\frac{4uTt}{S} \right)^{0.5} = \left(\frac{4 \times 2.31 \times 150 \times 2/24}{2 \times 10^{-4}} \right)^{0.5} = 759.93 \text{ m}$$

Similarly for 3 days,

$$r = \left(\frac{4uTt}{S} \right)^{0.5} = \left(\frac{4 \times 2.31 \times 150 \times 3}{2 \times 10^{-4}} \right)^{0.5} = 4559.61 \text{ m}$$

5.9.2 Processing of drawdown tests

Drawdown and recovery tests are aimed at assessing the basic parameters of the aquifers in the vicinity of a borehole. Pumping tests are one such way of assessing and determining the parameters of controlled experiments rather than historical data of the aquifer. The parameters of the aquifer, which can be assessed, are the transmissivity T and the storage capacity S.

Depending on the nature of the test and the data collected, it is possible to also evaluate other parameters, e.g. the leakage factor λ and information about the boundary conditions.

During testing in a borehole, pumping should be at a constant discharge rate Q_w while changes in drawdown in the borehole being pumped are recorded over time (drawdown test). The same recording is done in one or more standing neighbouring boreholes (interference test) if they have the possibility of level measurement.

The aforementioned parameters are estimated from measured data, with the use of the appropriate equations and regression analysis methods. The choice of the appropriate equation is perhaps the most difficult step in a pumping test, since this equation depends on the boundary conditions of the aquifer, on geology, on leakage between aquifers, etc. Here, assumptions should be made based on previous geological information, since there is no systematic theory for the selection of the appropriate equation.

It is generally assumed that the aquifers behave as a representative aquifer, since the flow in these layers is practically horizontal. It is also assumed that impervious formations between previous ones aren't inserted into the hydraulic calculations.

The drawdown test is an experiment by which characteristics of the aquifer are specified in two phases: the abstraction phase and the recovery phase. The pumping begins with constant discharge Q, so that the level in the borehole begins to drawdown. The level drawdown s is recorded as function of the time t from the start of pumping. When the level stops falling (this happens when the aquifer reaches equilibrium), pumping stops and the level begins to rise again.

Three different methods are used for the assessment of the hydrogeological parameters: the Theis, Cooper–Jacob and Chow methods.

5.9.3 Analysis of confined aquifers

5.9.3.1 Theis method

The equation used in the Theis method is

$$s = \frac{Q}{4\pi T} W(u) \tag{5.56}$$

By taking the logarithm of both sides,

$$\log s = \left[\log \frac{Q}{4\pi T} \right] + \log W(u) \tag{5.57}$$

with

$$u = \frac{r^2 S}{4Tt} \quad \text{or} \quad \frac{t}{r^2} = \frac{S}{4T}\frac{1}{u} \tag{5.58}$$

Taking the logarithm of both sides of the last equation followed by logarithmic analysis gives

$$\log \frac{t}{r^2} = \left[\log \frac{S}{4T} \right] + \log \frac{1}{u} \tag{5.59}$$

where
  s is the level drawdown [m]
  r is the distance from the centre of the borehole [m]
  S is the storage capacity [dimensionless]
  T is the transmissivity [L^2/T]
  Q is the discharge [L^3/T]
  $W(u)$ is the well function

For constant discharge Q, the procedure of the Theis method is based on the following steps:

1. Prepare a log–log graph, with Y-axis the well function $W(u)$ and X-axis the parameter u. The parameters $W(u)$ and u are associated univocally and the corresponding diagram is in the form presented in Figure 5.11.
2. From the pumping data, prepare a log–log graph of the level drawdown with r^2/t quantity. This graph is known as a data graph. The data are acquired from pumping where the discharge remains constant. The level drawdown may be recorded in an observation borehole which is located at a distance r from the pumping borehole at several times. Also, a record can be taken at the same time from a number of boreholes which are spaced at different distances from the initial pumping borehole. This is called drawdown analysis by distance. However, in both circumstances, the term r^2/t can be calculated and plotted in relation to the level drawdown.
3. The data graph is placed to overlap the graph of the curve, and is shifted horizontally or vertically keeping the axes of the two graphs parallel until the data graph coincides with the curve of $W(u)$ and u.
4. An arbitrary point is selected which may not be over the curve. This point is usually chosen so that the coordinates are multiples of 10. Then the values of s and r^2/t of the specific point are recorded.
5. The discharge capacity and the storage capacity are calculated from the following equations:

$$T = \frac{Q}{4\pi s} W(u) \quad [L^2 T^{-1}] \tag{5.60}$$

$$S = 4T \frac{t}{r^2} u \quad [\text{dimensionless}] \tag{5.61}$$

The procedure of the Theis method is presented in Figure 5.12.

5.9.3.2 Cooper–Jacob method

Cooper and Jacob have shown that when the parameter u becomes very small, the function $W(u)$ can be approximated by the first two terms of Equation 5.54. Therefore, according to Cooper and Jacob, the equations become

$$s = \frac{Q}{4\pi T} W(u) \Rightarrow s = \frac{Q}{4\pi T} (-0.5772 - \ln u) \tag{5.62}$$

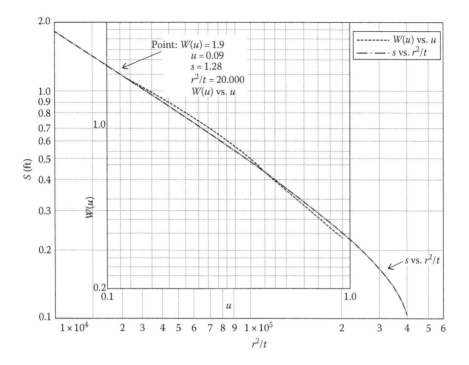

Figure 5.12 Overlapping diagrams and point selection.

The discharge capacity T can be calculated based on the Cooper–Jacob equation:

$$s = \frac{Q}{4\pi T} \ln \frac{2.25 T t}{r^2 S} \tag{5.63}$$

or

$$s = \frac{Q}{4\pi T} \left(\ln\ 0.562 - \ln \frac{r^2 S}{4 T t} \right) \tag{5.64}$$

or

$$s = \frac{Q}{4\pi T} \left(\ln t - \ln \frac{r^2 S}{2.25 T} \right) = \frac{Q}{4\pi T} \ln t - \frac{Q}{4\pi T} \ln \frac{r^2 S}{2.25 T} \tag{5.65}$$

where r is the distance between the point where the level is measured and the point where pumping is located.

In the phase of pumping, the drawdown s is recorded at various time instances. If the pair of values (t,s) are plotted on semi-logarithmic graph, then they are arranged approximately in a straight line. Thus, performing regression between the variables $\ln t$ and s results in a straight line of the following form:

$$s = \alpha \ln t + c \tag{5.66}$$

And from Equation 5.65 it results into

$$\alpha = \frac{Q}{4\pi T} \quad \Leftrightarrow T = \frac{Q}{4\pi\alpha}$$

$$c = -\alpha \, \ln\frac{r^2 S}{2.25T} \quad \Leftrightarrow S = \frac{2.25T}{r^2} e^{-c/\alpha} \tag{5.67}$$

The first of these equations provides directly the discharge capacity. For the storage capacity, the factor of distance r enters into the function. Normally, the level should be measured at some distance r from the borehole, but usually, there is only level measurement in the position of the drilling. In that case, the distance r becomes equal to the borehole radius.

Suppose that at a moment $t = t_1$, pumping from the aquifer stops and the water level starts gradually to return to the initial levels, as shown in Figure 5.13. The discharge capacity for the recovery phase is calculated by the following equation:

$$s = \frac{Q}{4\pi T'} \ln\frac{t}{t - t_1} \tag{5.68}$$

This applies for long time intervals:

$$t - t_1 > \frac{5Sr^2}{T} \tag{5.69}$$

During the return tests, the borehole which initially was pumped with discharge Q stops to pump at time t_1 and the remaining level drawdown s is measured in the borehole or at another observation station (drilling or well) located close to the borehole. According to Hantush (1964), when $t - t_1 > \dfrac{5Sr^2}{T}$, the following relation holds:

$$s(r,t) = \frac{Q'}{4\pi T} \ln\left(\frac{t}{t - t_1}\right) \tag{5.70}$$

That is, the level drawdown is proportional to $\ln(t/[t-t_1])$, so data are arranged in a straight line.

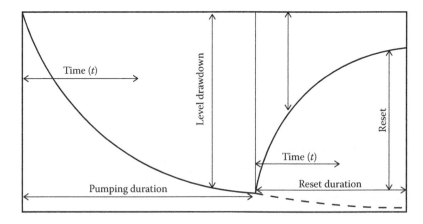

Figure 5.13 Level drawdown diagram during pumping and recovery period.

However, the methodology is again identical. A straight line fits between the variables s and $\ln(t/[t-t_l])$, and the slope of the line is $Q/4\pi T'$, from which T' can be computed. So, this procedure results in two discharge capacities, one for the pumping and one for the return period. As already mentioned, the last equation applies only in the case of long time intervals; therefore, it may be defined by the user that the regression will only consider the high points of the diagram. Also, if bending is presented, more regressions can be made and generate different T, depending on time t.

5.9.3.3 Chow method

The Chow method is a combination of Theis and Cooper–Jacob methods and is defined by the following equations:

$$F_u = \frac{s}{\Delta s/\log(t_2/t_1)}\,[\text{dimensionless}] \tag{5.71}$$

$$\frac{W(u)e^u}{2.3} = \frac{s}{\Delta s/\log(t_2/t_1)}\,[\text{dimensionless}] \tag{5.72}$$

$$F_u = \frac{s}{\Delta s}\,[\text{dimensionless for a logarithmic cycle}] \tag{5.73}$$

The correlation between the magnitudes $F(u), u, W(u)$ is defined by the graph in Figure 5.14.

It has to be mentioned that the earlier equations are applied in the cases when the discharge Q_w is constant during the pumping period and the aquifer is confined with practically very large dimensions. In the case of pumping tests, the discharge value fluctuates at

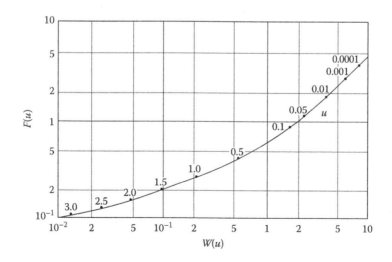

Figure 5.14 Chow diagram governing the relations between the parameters F(u), u and W(u).

the beginning but rapidly stabilizes. Also, in most of the cases, measuring the discharge accurately was not possible, due to lack or to non-continuous hydrometric stations, so the estimation of the pumping discharge was calculated empirically (e.g. using containers of known capacity and clocking of filling time). Also, the limits and the supply of the aquifer were unknown.

Example 5.10

A borehole is constructed for water pumping from a confined aquifer. The pumping rate is 134 m³/h. In an observation well located 70 m away, the following drawdowns were observed, which can be seen in Table 5.2.

Determine the transmissivity of the aquifer and the storage capacity, using the Cooper–Jacob, Theis and Chow methods.

Solution

1. *Cooper–Jacob method*
Using the simple equation of Cooper–Jacob, we correlate linearly the logarithms of time with the level drawdown. The equation gets the following form:

$$s = \frac{Q}{4\pi T} \ln \frac{2.25Tt}{r^2 S} = \frac{Q}{4\pi T} \ln \frac{2.25T}{r^2 S} + \frac{Q}{4\pi T} \ln t = \alpha + b \cdot \ln t$$

where

$$\alpha = \frac{Q}{4\pi T} \ln \frac{2.25T}{r^2 S}$$

$$b = \frac{Q}{4\pi T}$$

Table 5.2 Level drawdown at an observation borehole

Time since the start of pumping (min)	Level drawdown (m)
1	0.18
2	0.34
5	0.52
10	0.65
15	0.73
20	0.80
25	0.84
30	0.89
35	0.92
40	0.96
45	0.97
50	0.99
55	1.01
60	1.03
90	1.11
120	1.18
150	1.23
180	1.26

Data are converted to the same units and the logarithms of time are calculated according to Table 5.3:

Subsequently, the slope and the ordinate of the line which correlates linearly the logarithms of time with the level drawdown is calculated, according to Figure 5.15.

From the linear regression, we derive the transmissivity of the aquifer T, and the storage capacity S, according to the following relations:

$$b = \frac{Q}{4\pi T} = 0.205 \Rightarrow \frac{134/3600}{4\pi \times T} = 0.205 \Rightarrow T = 0.0145\,\text{m}^2/\text{s} = 1252.8\,\text{m}^2/\text{day}$$

Table 5.3 Data measurements of water level drop vs. time

t (min)	t (s)	ln (t)	s (m)
1	60	4.094	0.19
2	120	4.787	0.33
5	300	5.704	0.52
10	600	6.397	0.66
15	900	6.802	0.74
20	1,200	7.090	0.80
25	1,500	7.313	0.85
30	1,800	7.496	0.89
35	2,100	7.650	0.92
40	2,400	7.783	0.95
45	2,700	7.901	0.97
50	3,000	8.006	0.99
55	3,300	8.102	1.01
60	3,600	8.189	1.03
90	5,400	8.594	1.11
120	7,200	8.882	1.17
150	9,000	9.105	1.22
180	10,800	9.287	1.25

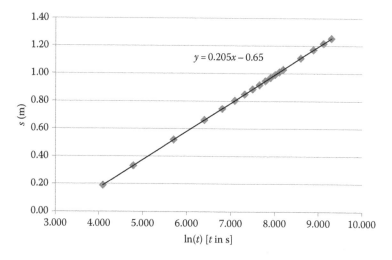

Figure 5.15 Correlation of logarithms of time with the level drawdown.

$$a = \frac{Q}{4\pi T} \ln \frac{2.25T}{r^2 S} = -0.65 \Rightarrow \frac{134/3600}{4\pi \times 0.0145} \ln \frac{2.25 \times 0.0145}{70^2 S} = -0.65 \Rightarrow S = 1.59 \times 10^{-4}$$

2. Theis method

According to the Theis method, the magnitude r^2/t is estimated at the beginning of every time interval, as shown in Table 5.4, and then it is drawn in a double logarithmic diagram in relation to the level drawdown s, as shown in Figure 5.16.

Table 5.4 Conversion of data to appropriate units

t (min)	s (m)	r^2/t (m^2/min)
1	0.19	4900.00
2	0.33	2450.00
5	0.52	980.00
10	0.66	490.00
15	0.74	326.67
20	0.80	245.00
25	0.85	196.00
30	0.89	163.33
35	0.92	140.00
40	0.95	122.50
45	0.97	108.89
50	0.99	98.00
55	1.01	89.09
60	1.03	81.67
90	1.11	54.44
120	1.17	40.83
150	1.22	32.67
180	1.25	27.22

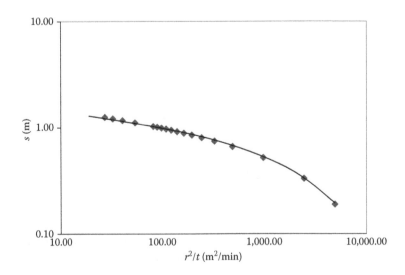

Figure 5.16 Diagram of level drawdown in relation to r^2/t.

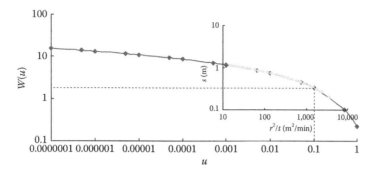

Figure 5.17 Diagram overlapping.

Then, the diagram in Figure 5.16 is placed on the diagram which correlates the well function W with the parameter u and then is shifted horizontally and vertically, until the two curves coincide. This procedure is shown in Figure 5.17 and it requires that the two diagrams have the same logarithmic axes, i.e. the size of each log step of the axis is the same in both diagrams.

As the two curves overlap, a point is selected and the correlation between the parameters in both systems of axes is found. In this case, a point with the following attributes (coordinates) is selected:

1. $u = 0.1$
2. $r^2/t = 9800$ m²/min
3. $W(u) = 2$
4. $s = 0.52$ m

The desired characteristics of the aquifer are derived by replacing these values with Theis' expressions. The transmissivity is derived by the following equation:

$$s = \frac{Q}{4\pi T} W(u) \Rightarrow T = \frac{Q}{4\pi s} W(u) = \frac{134}{4 \times 3.14 \times 0.52} 2 = 1.82\, \text{m}^2/\text{min} = 0.030\, \text{m}^2/\text{s}$$

And the storage capacity is equal to

$$S = \frac{4Tu}{r^2/t} = \frac{4 \times 1.82\ \text{m}^2/\text{min} \times 0.1}{980\ \text{m}^2/\text{min}} \Rightarrow S = 7.4 \times 10^{-4}$$

3. Chow method

Using the Chow method, the following procedure is followed:

a. The level–time drawdown data have already been presented in semi-logarithmic paper in Figure 5.15. In this diagram, a point is selected, such as the one with coordinates $s = 0.52$ m and $t = 5.0$ min. From the same diagram, for a time step on the logarithmic axis, a difference of level drawdown $\Delta s = 0.205$ m arises (same as the slope of the straight line). Therefore

$$F_u = \frac{s}{\Delta s} = \frac{0.52}{0.205} = 2.54$$

b. From Figure 5.14 and for $F_u = 2.54$, one obtains $u = 0.005$ and $W(u) = 4.8$.

c. From Equation 5.60, the transmissivity of the aquifer is calculated:

$$T = \frac{Q}{4\pi s} W(u) = \frac{134/3600}{4 \times 3.14 \times 0.52} 4.8 = 0.0273 \text{ m}^2/\text{s} = 2363.55 \text{ m}^2/\text{day}$$

d. Finally, from Equation 5.61, the storage capacity is

$$S = \frac{4tTu}{r^2} = \frac{4(5 \times 60) \times 0.0273 \times 0.005}{70^2} = 3.34 \times 10^{-5}$$

It should be noticed that the results from the three methods differ significantly.

Example 5.11

A well with 0.50 m diameter is constructed for water pumping from a confined aquifer. The pumping rate is 105 m³/h and the aquifers' thickness is 20 m. The transmissivity T is 47.573 m²/h, while the storage capacity S equals 7.44×10^{-5}.

1. After 6 h of constant pumping, estimate the drawdown in the pumping well and in an observation well that is located 200 m away.
2. In a certain moment between the 6th and the 16th hour (where the daily operation of the pump stops), the system attains a state of equilibrium (steady flow) where the piezometric cone is stabilized. Find the moment when the steady flow begins and the drawdown is 100 and 200 m apart from the well, if the drawdown in the well is 3.35 m.

Solution

1. From the Cooper–Jacob method:
   The drawdown in the well ($r = 0.25$ m) after 6 h is

$$s(0.2, 6) = \frac{Q}{4\pi T} \ln \frac{2.25Tt}{r^2 S} = \frac{105}{4\pi \times 47.57} \ln \frac{2.25 \times 47.57 \times 6}{0.25^2 \times 7.44 \times 10^{-5}} = 3.292 \text{ m}$$

Similarly, the drawdown 200 m from the well ($r = 200$ m) after 6 h is

$$s(200, 6) = \frac{Q}{4\pi T} \ln \frac{2.25Tt}{r^2 S} = \frac{105}{4\pi \times 47.57} \ln \frac{2.25 \times 47.57 \times 6}{200^2 \times 7.44 \times 10^{-5}} = 0.944 \text{ m}$$

2. Let us assume that s_1 is the drawdown at position 1 located 100 m from the well, s_2 is the drawdown at position 2 located 200 m from the well and s_w is the drawdown in the well. When the system enters the state of equilibrium, the following equation becomes valid:

$$s_w - s_1 = \frac{Q}{4\pi T} \ln \frac{r_1}{r_w} = \frac{105}{4\pi \times 47.57} \ln \frac{100}{0.25} = 2.348 \text{ m}$$

$$\Rightarrow 3.35 - s_1 = 2.348 \text{ m} \Rightarrow s_1 = 1.002 \text{ m}$$

Similarly

$$s_w - s_2 = \frac{Q}{4\pi T} \ln \frac{r_1}{r_w} = \frac{105}{4\pi \times 47.57} \ln \frac{200}{0.25} = 2.105 \text{ m}$$

$$\Rightarrow 3.35 - s_2 = 2.105 \text{ m} \Rightarrow s_2 = 1.245 \text{ m}$$

t can be estimated by the Copper–Jacob relationship as follows:

$$t = \frac{r_1^2 S}{2.25T} \exp\left(\frac{s_1 4\pi T}{Q}\right) = \frac{100^2 \times 7.44 \times 10^{-5}}{2.25 \times 47.57} \exp\left(\frac{1.245 \times 4 \times 3.14 \times 45.57}{105}\right) = 8.34 \text{ h}$$

Position 2 gives the same results:

$$t = \frac{r_2^2 S}{2.25T} \exp\left(\frac{s_2 4\pi T}{Q}\right) = \frac{200^2 \times 7.44 \times 10^{-5}}{2.25 \times 47.57} \exp\left(\frac{1.002 \times 4 \times 3.14 \times 45.57}{105}\right) = 8.34 \text{ h}$$

5.9.4 Analysis of unconfined aquifers

The equations described here are not fully applicable to the case of an unconfined aquifer due to the following reasons:

1. The level drawdown of the aquifer
2. The vertical flow being close to the borehole
3. Flow delay due to gravity

In the case when the level drawdown is small in relation to the thickness of the aquifer, the effect of the vertical flow and the level drawdown can be considered as negligible. According to Hantush (1964), the vertical effect is not important if time equals to

$$t > 5b \frac{S_y}{K_z} \text{ [T]} \tag{5.74}$$

where
    t is the time [t]
    b is the thickness of the aquifer [L]
    S_y is the specific yield [dimensionless]
    K_z is the vertical hydraulic conductivity [L/T]

According to Stallman (1971), the flow delay applies for a period of time:

$$t = 10S_y \frac{s}{K_z} \text{ [T]} \tag{5.75}$$

The observed level drawdown are corrected by using the following equation:

$$s' = s - \frac{s^2}{2b} \, [\mathrm{L}]$$

(5.76)

where
  s' is the corrected level drawdown [L]
  s is the observed level drawdown [L]
  b is the thickness of the aquifer [L]

The value of the specific yield can be corrected by using the following equation:

$$S_y = \frac{(b - \bar{s})S'_y}{b} \, [\text{dimensionless}]$$

(5.77)

where
  S_y is the modified specific yield [dimensionless]
  S'_y is the estimated specific yield [dimensionless]
  $\bar{s}$ is the corrected level drawdown at the end of pumping [L]

Example 5.12

A pumping borehole penetrates a phreatic aquifer of thickness 20 m and pumps at a discharge rate of 92 m³/h. The recorded drawdown in time, at an observation borehole located 12 m from the pumping borehole, is given in Table 5.5. The parameters of the aquifer should be specified.

Solution

The data are converted to the appropriate units and the level drawdown is corrected according to the following equation:

$$s' = s - \frac{s^2}{2b}$$

where $b = 20$ m is the thickness of the aquifer. The results are given in Table 5.6.

Table 5.5 Drawdown at an observation borehole in time

Time since the pumping started (min)	Level drawdown (m)
30	0.23
60	0.38
90	0.47
120	0.54
150	0.59
180	0.63
360	0.79
720	0.95
1440	1.11

Table 5.6 Corrected level values

t (min)	t (s)	ln (t)	s (m)	s' (m)
30	1,800	7.50	0.23	0.23
60	3,600	8.19	0.38	0.38
90	5,400	8.59	0.47	0.46
120	7,200	8.88	0.54	0.53
150	9,000	9.10	0.59	0.58
180	10,800	9.29	0.63	0.62
360	21,600	9.98	0.79	0.77
720	43,200	10.67	0.95	0.93
1440	86,400	11.37	1.11	1.08

Then the Cooper–Jacob method is applied, with linear regression for the pairs of values of the logarithms of time and the corrected value of the level drawdown (see also Example 5.10). The results of the regression are presented in the following graph (Figure 5.18).

As usual, the parameters of the aquifer are calculated as follows:

$$b = \frac{Q}{4\pi T} = 0.2202 \Rightarrow T = \frac{92/3600}{4\pi \times 0.2202} \Rightarrow T = 0.0092\,\text{m}^2/\text{s}$$

$$a = \frac{Q}{4\pi T}\ln\frac{2.25T}{r^2 S_y'} = -1.4242 \Rightarrow 0.2202 \times \ln\frac{2.25 \times 0.0092}{12^2 \times S_y'} = -1.4242$$

$$\Rightarrow S_y' = 0.093$$

These values are valid given the next assumptions regarding time:

- $t > 5b\dfrac{S_y}{K_z} = 5 \times 20 \times \dfrac{0.093}{0.0092/20}s = 20217\,\text{s}$ or $5.62\,\text{h}$, in order to neglect vertical flow

- $t > 10S_y\dfrac{s}{K_r} = 10 \times 0.093 \times \dfrac{1.11}{0.0092/20}s = 2244\,\text{s}$ or $0.62\,\text{h}$, in order to neglect the delay of flow

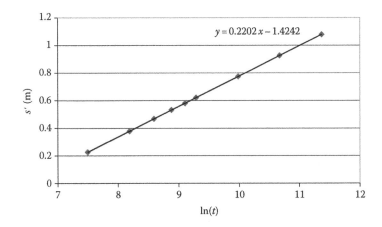

Figure 5.18 Linear correlation of the variables ln (t) and s'.

Overall, the values of S, T apply for time $t > 5.62$ h. Finally, the value of the specific yield is adjusted according to the relation

$$S_y = \frac{b - \bar{s}}{b} S'_y = \frac{20 - 1.08}{20} 0.093 = 0.088$$

where $\bar{s} = 1.08$ m is the corrected value of the level at the end of pumping. In phreatic (or unconfined) aquifers, the specific yield S_y coincides with the storage capacity S.

5.9.5 Anisotropic alluvial deposits

A common situation in alluvial deposits is the layering of materials with different K values. This phenomenon is called anisotropy and it is the rule rather than the exception in alluvial deposits. Anisotropy is caused also when individual particles with a shape different than spherical are deposited with their flat side in a horizontal direction. Moreover, when deposition takes place on a riverbed, flat particles tend to be tilted slightly upward in the direction of the flow. This arrangement is called imbrication and is observed in gravel deposits. Water molecules flowing through imbricated material find lesser resistance in the horizontal rather than in the vertical direction due to imbrication. Also, water travelling through an anisotropic aquifer consisting of separate horizontal sand and gravel layers will find more resistance in the vertical direction where all the water has to move through both sand and gravel layers than the horizontal direction where most of the water moves through just the gravel layers. Consequently, the hydraulic conductivity K_z in the vertical direction will be less than K_x in the horizontal direction. It is not unusual to find K_z values that are only one-fifth or one-tenth of K_x (Bouwer, 1978). The calculation of K_x and K_z values in anisotropic alluvial deposits is explained in the next example.

Example 5.13

Determine the total hydraulic conductivity K_z in a vertical direction and the total hydraulic conductivity K_x in a horizontal direction in an alluvial deposit system which consists of three horizontal layers. The height z_i and the hydraulic conductivity K_i of each layer is given in Figure 5.19.

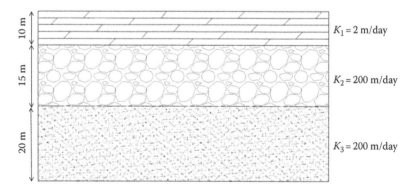

Figure 5.19 The height and hydraulic conductivity of each layer.

Solution

1. If there is horizontal flow through the system, the hydraulic gradient J is the same in each layer (if J were not the same, pressure head differences would exist along the interfaces between the layers, which is impossible in horizontal flow). K_x is the average hydraulic conductivity of the medium in a horizontal direction, Z is the thickness of the entire system and Z_1, Z_2, Z_3 are the thicknesses of each layer, respectively. The flow q_i in each layer per unit thickness of the system can be expressed as follows:

$$q_i = JK_iZ_i \tag{5.78}$$

Summing the q values for each layer to get the total horizontal flow q_x per unit thickness, we get

$$q_x = J(K_1Z_1 + K_2Z_2 + K_3Z_3) = JK_xZ$$

$$\Rightarrow K_x = \frac{K_1Z_1 + K_2Z_2 + K_3Z_3}{Z} = \frac{2 \times 10 + 200 \times 15 + 20 \times 20}{45} = 76.00 \, \text{m/day}$$

2. If there is a vertical flow through the system of Figure 5.19, the flow q per unit horizontal area can be expressed for the top layer as $q = K_1(\Delta H_1/z_1)$, where ΔH_1 is the total head loss in the first layer. Solving the equation for ΔH_1 yields

$$\Delta H_1 = \frac{z_1}{K_1}q \tag{5.79}$$

Since q is the same for all layers, the total head loss ΔH_t can be calculated as the sum of the head losses in each layer:

$$\Delta H_t = \left(\frac{z_1}{K_1} + \frac{z_2}{K_2} + \frac{z_3}{K_3}\right)q = \left(\frac{z}{K_z}\right)q$$

$$\Rightarrow K_z = \frac{z}{\dfrac{z_1}{K_1} + \dfrac{z_2}{K_2} + \dfrac{z_3}{K_3}} = \frac{45}{\dfrac{10}{2} + \dfrac{15}{200} + \dfrac{20}{20}} = \frac{45}{5 + 0.075 + 1} = 7.41 \, \text{m/day}$$

5.10 WELL LOSSES

The total level drawdown on a well consists of two components. One is the level drawdown s_a observed at the outer part of the borehole and the second s_w as the water moves through the filters at the pump. The second level drawdown is known as losses due to borehole shaping and is given by the following equation:

$$s_w = CQ^n \, [\text{L}] \tag{5.80}$$

where

C is the constant, $<0.5 \, \text{min}^2/\text{m}^5$, for a good well design

n is the exponent of the discharge, usually 2

The total losses of a borehole for an equilibrium state (uniform flow conditions) are

$$s_t = \frac{2.3Q}{2\pi T} \log\frac{r_0}{r_w} + CQ^n \; [\text{L}]$$

(5.81)

But for a non-equilibrium state (non-uniform flow conditions), total losses are

$$s_t = \frac{2.3Q}{4\pi T} \log\frac{2.25Tt}{r_w^2 S} + CQ^n \; [\text{L}]$$

(5.82)

The well efficiency is given by the following equation (Figure 5.20):

$$E = \frac{s_a}{s_t} \times 100 \; [\text{dimensionless}]$$

(5.83)

or

$$E = \left(1 - \frac{s_w}{s_t}\right) \times 100 \; [\text{dimensionless}]$$

(5.84)

The specific capacity of a well is the well efficiency for a unit level drawdown:

$$\lambda = \frac{Q}{s_t} \; [\text{L}^2/\text{T}]$$

(5.85)

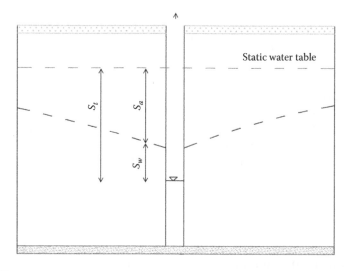

Figure 5.20 Well losses.

Example 5.14

A pumping borehole of diameter 0.4 m penetrates a confined aquifer. The observed drawdown at the borehole after a 12 h pumping with 1.6 m³/min discharge is 7.1 m. The transmissivity and the storage capacity of the aquifer are 560 m²/day and 7×10^{-3}, respectively. (1) The well losses, (2) the well efficiency and (3) the specific capacity of the well must be specified.

Solution

1. The well losses for the non-equilibrium state are calculated as follows:

$$s_t = \frac{2.3Q}{4\pi T} \log \frac{2.25Tt}{r_w^2 S} + s_w$$

$$\Rightarrow s_w = 7.1 - \frac{2.3 \times (1.6/60)}{4 \times 3.14 \times (560/86400)} \log \frac{2.25(560/86400) \times (12 \times 3600)}{0.2^2 \times 7 \times 10^{-3}} \text{ m}$$

$$\Rightarrow s_w = 7.1 - 4.78 = 2.31 \text{ m}$$

2. The well efficiency is the ratio of losses during movement through the soil to total losses:

$$E = \frac{s_a}{s_t} \times 100 = \frac{4.78}{7.10} \times 100 = 67\%$$

3. The specific capacity of the well is calculated as the ratio of discharge to the total losses:

$$\lambda = \frac{Q}{s_t} = \frac{1.6}{7.1} \text{ m}^2/\text{min} = 0.225 \text{ m}^2/\text{min}$$

5.11 AQUIFER RECHARGE

Recharge is called the replenishment of groundwater resources from surface water resources. Rainfall is a natural way of enriching a basin. The artificial recharge of an aquifer is done by a variety of methods when it is necessary to increase groundwater resources. There are five standard techniques of artificial recharge; the first one is further divided into four sub-categories (Mimikou & Baltas, 2012):

1. Surface water recharge
   a. Basin method
      Water is diverted or pumped from streams, filling one or a series of small natural basins to slowly infiltrate and reach the aquifer. The water supply rate is shown to be directly proportional to the difference in level between the water supply and the groundwater level and decreases with time due to filling of soil voids.
   b. Stream–Canal method
      This method enhances the already significant supply of the aquifer from streams, by riverbed modifications and construction of dams and embankments.

c. Trench method
Similar to the stream–canal method, a series of shallow trenches are constructed in a flat bed with the water diverted to them from streams.

d. Flooding method
It is the simplest and most economic method of recharge. It is similar to the basin method, with the exception that the extent of supply is natural with artificial embankments at its ends where the terrain is not conducive to flooding.

2. Supply pits
The excavation of a supply pit is a very effective method of artificial recharge in the case of a permeable soil layer with a shallow depth beneath the impermeable surface. Often the bottom of the pit is covered by gravel.

3. Recharge wells
These are equivalent to the abstraction wells but in reverse operation, since they are supplied with water from the surface. Around these an elevation cone is created which is the image of a depression cone. The formulas of the pumping boreholes are also applied to the supply wells.

4. Induced supply
This is done in a hydraulically connected system of aquifer–watercourse when water is pumped from a nearby well. In this case, the increased pumping from the well creates additional hydraulic gradient in the aquifer from the watercourse and, therefore, better enrichment.

5. Wastewater disposal
Using wastewater that has undergone secondary treatment for aquifer recharge not only enriches the aquifer but also leads to further treatment of wastewater.

These methods are selected on a case-by-case basis.

5.12 SALINIZATION

The mixing of saltwater and freshwater is a common problem of groundwater quality deterioration of coastal aquifers mainly due to the entry of seawater.

5.12.1 Interface of saltwater and freshwater

In Figure 5.21, a section of a coastal aquifer is presented. At each point of the interface, the pressure of the overlying freshwater balances the pressure of the underlying saltwater, and the following equation applies:

$$\rho_s g Z = \rho_f g (Z + h) \tag{5.86}$$

or

$$Z = \frac{\rho_f}{\rho_s - \rho_f} h \ [\text{L}] \tag{5.87}$$

where
ρ_f is the freshwater density $[\text{M/L}^3]$
ρ_s is the saltwater density $[\text{M/L}^3]$
Z is the depth of interface at each point $[\text{L}]$
h is the hydraulic head over the surface of the sea at each point $[\text{L}]$

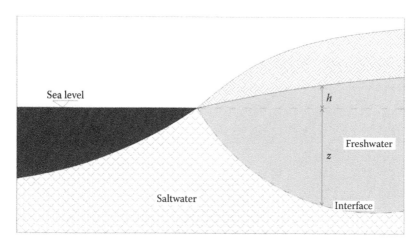

Figure 5.21 Balance of freshwater and saltwater in an unconfined coastal aquifer.

The last equation is called the Ghyben–Herzberg expression and ignores the water movement. For typical values, if density $\rho_s = 1.025$ g/cm^3 and $\rho_f = 1.0$ g/cm^3, the equation gets the following form:

$$Z = 40\,h \tag{5.88}$$

That is, the saltwater is found at a depth equal to 40 times the height of freshwater, as observed in practice, or if the freshwater depth declines by 1 m, the saltwater progresses inland by 40 m.

5.12.2 Saltwater elevation cone

In the case where a layer of saltwater is below the freshwater zone and a well pumps only from the freshwater layer, there is an elevation of the interface of saltwater and freshwater, as can be seen in Figure 5.22. This phenomenon is known as elevation cone.

After its stabilization, the phenomenon may be described by the following equation:

$$Z = \frac{Q\rho_f}{2\pi d K (\rho_s - \rho_f)} \; [\text{L}] \tag{5.89}$$

where
   K is the hydraulic conductivity [L/T]
   d is the depth of the initial saltwater–freshwater interface beneath the borehole bottom [L]

When the elevation becomes critical (e.g. $Z/d = 0.3\text{–}0.5$), saltwater enters the borehole, polluting the supply. Consequently, the maximum discharge for the elevation to stay under the critical limit can be calculated by substituting with $Z = 0.5d$ in the last expression:

$$Q_{\max} = \pi d^2 K \frac{\rho_s - \rho_f}{\rho_f} \; [\text{L}^3/\text{T}] \tag{5.90}$$

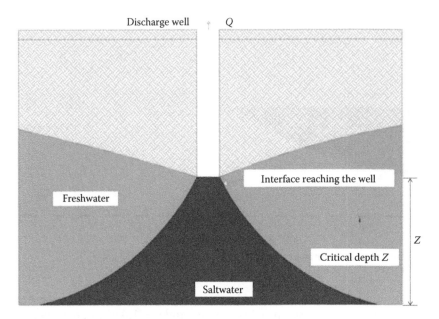

Figure 5.22 Saltwater elevation cone beneath a pumping borehole.

Example 5.15

In a deep aquifer, good-quality water lies at a depth of 125 m under which there is brackish water of specific weight 1.026 g/cm³. A pumping borehole pumps water at a depth of 80 m. Define the maximum pumping rate to avoid pumping of saltwater. The hydraulic conductivity is 800 m/day.

Solution

The depth of the initial saltwater–freshwater interface from the depth of the borehole is $d = 125 - 80 = 45$ m. Consequently, the maximum pumping rate is

$$Q_{max} = \pi d^2 K \frac{\rho_s - \rho_f}{\rho_f} = \pi \times 45^2 \times 1000 \frac{1.026 - 1.000}{1.000} \text{ m}^3/\text{day} = 1645,405 \text{ m}^3/\text{day}$$

REFERENCES

Aftias, E., 1992, *Water Supply*, NTUA, Athens, Greece (in Greek).
Bear, J., 1979, *Groundwater Hydraulics*, McGraw-Hill, New York.
Bouwer, H., 1978, *Groundwater Hydrology*, Vol. 480, McGraw-Hill, New York.
Chow, V.T., 1964, *Handbook of Applied Hydrology*, McGraw-Hill, New York.
Freeze, A.R. and Cherry, J.A., 1979, *Groundwater*, Englewood Cliffs, NJ: Prentice-Hall.
Gupta, R.S., 1989, *Hydrology and Hydraulic Systems*, Vol. 1110, Englewood Cliffs, NJ: Prentice Hall.
Hantush, M.S., 1964, *Hydraulics of Wells*, in *Advances in Hydroscience*, Vol. 1, V.T. Chow (ed.), Academic Press, Inc., New York.
Hermance, J.F., 1999, *A Mathematical Primer on Groundwater Flow*, NJ: Prentice Hall.
Latinopoulos, P., 1986, *Hydraulics of Ground Water*, AUTH, Thessaloniki, Greece (in Greek).
Mimikou, M. and Baltas, E., 2012, *Engineering Hydrology*, Papasotiriou, Athens, Greece (in Greek).
Singh, V.P., 1992, *Elementary Hydrology*, NJ: Prentice Hall.
Stallman, R.W., 1971, Aquifer-test design observation and data analysis, techniques of water resources investigations, Chapter B1, in Book 3, *Applications of Hydraulics*, USGS, Washington, DC.

Chapter 6

Hydrologic design

6.1 INTRODUCTION

Hydrologic design mainly refers to the appropriate sizing of a structure in order to fulfil its specified purpose related to water (e.g. protection from floods) and is considered a fundamental design preceding any other type of design (e.g. hydraulic). In general, hydrologic design is a necessary element in the design of a wide range of engineering works, hydraulic or not (e.g. reservoirs, bridges, roads, buildings, networks). In this chapter, we will focus only on the hydrologic design of hydraulic works that are categorized as (1) reservoirs that store water, (2) flood safety (protection) structures, (3) networks (water supply, irrigation, etc.) and (4) works that improve the water quality (water treatment plants, etc.). More specifically, we will focus on (1) and (2).

The dominant position among hydraulic works is occupied by dams, storing river water in reservoirs and affecting the hydrograph over a period of time. The sizing of reservoirs is an important part of the hydrologic design. Substantial importance is also accorded to the hydrologic sizing of flood safety (protection) structures, such as the spillway in a dam that provides protection from flood overtopping and the diversion of a river at a construction site (during the construction phase). In this chapter, we will discuss in detail the hydrologic design of a reservoir, a spillway, a river's diversion tunnel and other specific design topics. Both deterministic and stochastic approaches in hydrologic design are examined throughout and the differences are also presented.

6.2 SIZING OF RESERVOIRS

6.2.1 General

A reservoir is an artificial lake created behind a dam that stores water inflowing from a drainage basin. The most common example of a reservoir is a river dam (small or large) that is constructed across a river. Also, there are reservoirs that are created underground or in artificial cavities of the earth, where the extraction of water is possible only by mechanical means (i.e. pumping).

Water storage has *only* one common purpose: flow control, both spatially and temporally. In detail, a reservoir changes the distribution, in time, of the water surplus and shortage (quantities above and below a reference level, respectively, i.e. the water demand or the average flow) of the river, at a particular location of the river. For instance, the irrigation needs that are greater during the summer period normally cannot be satisfied by the natural flow of a river, which is low in the summer. So, there is a need of a reservoir construction, in order to retain water during the season of high river flow and supply this water in the summer season, through proper flow management. Usually, the reservoirs do not exclusively serve a single purpose but serve multiple needs, i.e. water for irrigation, hydropower, water supply and recreation. In many cases, these purposes are antagonistic to one another: for example, in the case of energy production and irrigation purposes during the summer season.

Nevertheless, the majority of reservoirs that have been created by big river dams are meant primarily for power generation. In such projects, hydroelectric power production units are installed at the bottom of the dam. High-diameter water pipes (penstocks) draw water from the reservoir and pass it through the unit. The water rotates the turbines of the power generation unit, which transforms the kinetic energy into electric energy. Then the voltage of the electric current is stepped up, so that it can be transported, with small energy losses, across long distances until it reaches distribution centres. The produced electric energy is proportional to the water discharge flowing through the turbines and the height of the waterfall. Single-purpose hydropower plants also provide water for other needs (e.g. irrigation), but this fact does not alter their characterization as single-purpose dams, if these secondary abstractions are not programmed and taken into account into the optimization process of the project, at least during the design phase.

From the hydrologic design point of view of a reservoir, what matters most is the reliability of the design. In order to have a reliable design, the storage volume size must be compatible with the time series characteristics (mainly the statistical) of the reservoir inflows. A tolerable risk must be chosen, for the estimated storage volume to satisfy the production of a defined output (e.g. primary energy) or generally to satisfy the demand.

6.2.2 Sections of volume and reservoir exploitation

The total storage volume of a reservoir comprises the following:

1. The volume from the river bottom until the minimum level of operation, which is defined as the level of water abstraction. This water level must not fall lower than this point, or the reservoir will have operational problems. This volume is called dead volume V_d. In this part of the reservoir, sediments are deposited due to the slowdown of the flow velocity to nearly zero.
2. The volume between minimum and maximum (normal) operational level, which is called active volume V_a. This is the volume used for water subtraction (i.e. for hydroelectric production and other uses) and can be managed in time. The maximum (normal) operation level determines the hydroelectric potential of the project, since the normal energy production is applicable between these minimum and maximum water levels (Figure 6.1).
3. The volume between the maximum operation level and the spillway level which is called flood volume V_f. This volume is used for the containment of the design flood of the spillway. The spillway level is the maximum level that the water is predicted to reach during the design flood (this is the maximum flood that the spillway can route safely). Above this level, a free board is added up to the crest level of the dam. In many hydroelectric projects, where the flow is controlled by flow gates, the maximum flood level does not stay permanently empty waiting for an extremely rare flood event (for which the common return period is 10,000 years or even less frequent), but it can be exploited for energy production, at least partially. So, the maximum operation level and the active volume may not be constant, but they increase seasonally especially when the risk of occurrence of the design flood is extremely small.

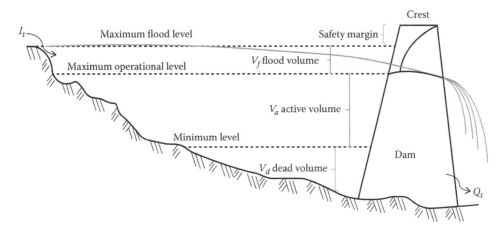

Figure 6.1 Characteristic storage volumes in a reservoir.

The permanent withdrawal Q_t from the reservoir, for various uses can be expressed as a percentage (α) of the mean value of discharge of the river $\bar{I}$ (e.g. the annual mean value of inflows I_t) for a number of N years as

$$a = \frac{\int_0^N Q_t \, dt}{\int_0^N \bar{I} \, dt}, \quad 0 \leq a \leq 1 \tag{6.1}$$

The percentage α is called the efficiency coefficient of the reservoir. If α is relatively low, e.g. 30%, then the reservoir is called low-efficiency reservoir. If α is high, e.g. 95%, then the efficiency of the reservoir is high: a large proportion of water volume is controlled for various uses. The maximum water efficiency level is 100% (ideal reservoir); in that case, the entire river flow can be stored, controlled and used in the reservoir. In the following, the sizing methods for all three volume components will be presented.

6.3 CONVENTIONAL METHOD OF SIZING THE ACTIVE RESERVOIR VOLUME

One of the most common sizing methods of a reservoir is the Rippl method (Rippl, 1883), which is based on the analysis of the fluctuation of the historical inflow volumes of the reservoir. The method is simple, but it has many disadvantages, which will be discussed later here. This method creates many technological problems, especially with regard to the reliability of the design. The problem addressed by the Rippl method is explained in two alternative scenarios (Schultz, 1976):

1. An observed time series of inflows to the reservoir and the time series of abstractions (consumption) are given. The question is to find the necessary volume of the reservoir to satisfy the demand.

2. An observed time series of inflows to the reservoir and an existing (as designed) maximum volume of the reservoir are given. What is asked is the maximum possible abstraction (demand for consumption) from the reservoir.

At this point, only the first scenario is discussed.

6.3.1 Construction of a cumulative inflow curve

The Rippl method uses a simple cumulative curve of the historical volumes of inflows in the reservoir. So if I_T is the time series of inflow discharges into the reservoir (m³/s), the cumulative curve is defined as

$$C_t = \int_0^t I_T \, dT \tag{6.2}$$

where T is the time between $0,\ldots, t$. For practical reasons, the continuous water graph I_T, $T = 0,\ldots, t$, is expressed on a monthly basis and the cumulative curve is

$$C_t = \sum_{i=1}^{t} I_i \Delta t, \quad t = 1,\ldots, N \tag{6.3}$$

where
  i denotes month
  Δt is the period of time (1 month)
  N is the number of the monthly values of the time series ($N = 12 \times n$, n is the number of years of the observed time series)

The time series should be carefully selected to represent a significant drought period.

The Rippl method can be applied to any reservoir independent of the efficiency coefficient α. However, for practical reasons, in order to find the maximum fluctuation of inflows in comparison with the demand curve, we assume that $\alpha = 100\%$ (ideal reservoir) and that the demand Q_t is constant in time and equal to the mean inflow $\bar{I}$:

$$Q_t = \text{constant} = \bar{I} \tag{6.4}$$

Figure 6.2 presents the construction of a cumulative inflows curve ($\alpha = 100\%$), plus the demand curve where $Q_t = \bar{I}$.

6.3.2 Estimation of the active reservoir volume

After the construction of the cumulative curve of inflows and the demand curve, the necessary volume of the reservoir can be estimated by the Rippl method in the next step: from the two points of the cumulative curve that differ the most from the demand curve (maximum surplus and maximum shortage), two parallel lines to the demand curve are drawn that osculate to the cumulative curve. The vertical distance between these two lines is called the maximum range $R_{t,s}$ where t denotes the start point of the inflow curves and s denotes the

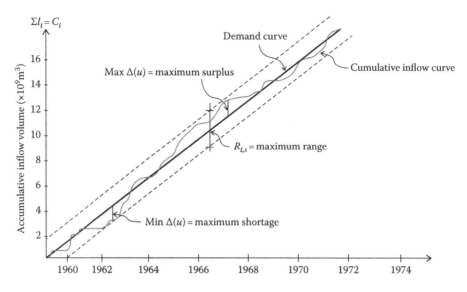

$\Sigma I_t = C_t$

Accumulative inflow volume ($\times 10^9 \mathrm{m}^3$)

Demand curve

Max $\Delta(u)$ = maximum surplus

Cumulative inflow curve

$R_{t,s}$ = maximum range

Min $\Delta(u)$ = maximum shortage

16

14

12

10

8

6

4

2

1960   1962   1964   1966   1968   1970   1972   1974

Figure 6.2 Cumulative curve of inflow volumes in a reservoir.

time period of the data used to estimate the range. The analytical expression of the range, as shown in Figure 6.2, is the following:

$$R_{t,s} = \max_{0 \leq u \leq s} \Delta(u) - \min_{0 \leq u \leq s} \Delta(u) \qquad (6.5)$$

In the aforementioned equation, the range is expressed as the algebraic difference of the maximum surplus and the maximum shortage. The active volume of the reservoir based on this method is $R_{t,s}$. If the reservoir that is designed through this method is full when its operation starts, then theoretically, it will never get empty, and the demand will be always satisfied, as long as the time series of inflows that was used for the design is reproduced during the operation of the reservoir (Chow et al., 1964).

6.3.3 Disadvantages of Rippl's conventional design

During the application of the Rippl method for the sizing of a reservoir, we should not over-see a critical point: in order to verify the assumption that the reservoir with volume $R_{t,s}$ will never empty and will always satisfy the demand, the other assumption must also be considered, namely, the time series of inflows will be reproduced in the future during the operation phase of the project. The occurrence of this event has zero probability. In other words, there is a substantially non-zero probability that the reservoir will not satisfy the demand, and also, there is a non-zero chance that the reservoir will be empty.

Another disadvantage of the method is that the volume $R_{t,s}$ is influenced directly by the duration s of the time series that is used for the estimation. If we assume that the following equation holds:

$$\frac{R_{t,s}}{S_{t,s}} = Ks^H \quad H > 0.5 \qquad (6.6)$$

The volume $R_{t,s}$ is an exponential function of s (with exponent the coefficient H) (Hurst et al., 1965). The departure of the exponent H from 0.5 is the Hurst phenomenon, which shows quantitatively the persistence of the time series I_t, given that $S_{t,s}$ is the standard deviation from time $t + 1$ to $t + s$ and K is a constant. For example, if someone changes the duration s of the time series, $R_{t,s}$ will also change. The dependence of the reservoir volume on the duration of the time series is an important disadvantage, given the fact that long observation periods (something desirable) lead to huge and expensive reservoirs (which is undesirable).

Finally, another disadvantage is the inefficiency of the method to provide, along with the estimated volume, the probability of this volume to be insufficient to meet the demand (since it is obvious that there is a high probability for this to happen).

All disadvantages of this conventional method can be significantly diminished by using a non-conventional method: the cumulative curve of Rippl is applied to a large number of synthetic time series of inflows, which are generated by a stochastic model of simulation of the historical series, as explained in the following. The non-conventional method of reservoir sizing satisfies both necessary conditions that can make the Rippl method technically correct and reliable; e.g. it can provide the capability of determination of failure probability to satisfy the demand and separates the estimation of the reservoir volume from the duration of the observed hydrologic data.

6.3.4 Sequent peak analysis method

Sequent peak analysis method is an alternative conventional method for reservoir sizing. It derives the largest deficit volume of a discharge series with respect to a certain threshold level (usually zero). This method can be applied either for constant or varying demands and it is more suitable for long observation periods. The inflow sequence is assumed to repeat and an algorithm is carried out, over more than one cycle if necessary. The algorithm can be expressed mathematically in the following form:

$$K_t = \begin{cases} K_{t-1} + R_t - Q_t, & K_t > 0 \\ 0, & K_t \leq 0 \end{cases} \tag{6.7}$$

where
Q_t is the inflow
R_t is the outflow, in a certain time step t

At the beginning of the process, we assume that $K_o = 0$.

If the value of K_t is zero, after the first cycle of calculations, then the process ends: the critical value is contained in this cycle. The process also ends after a point wherein the values K_t of each time step are identical to the corresponding ones of the previous cycle. The method is explained in the next example:

Example 6.1: Reservoir sizing with sequent peaks analysis method

Estimate the reservoir size with inflows and demands as seen in Table 6.1, for a period of 6 years. Total inflows are equal to total demand (=24 units).

Solution

Table 6.2 is filled for the first cycle of the algorithm.

As it is observed, during the end of the first cycle, the value of K_t is not zero, so the next cycle is running (Table 6.3).

We perform also a third cycle (Table 6.4):

Table 6.1 Inflows and demands

Year, t	1	2	3	4	5	6
Inflow, Q_t	3	6	8	4	2	1
Demand, R_t	5	4	6	4	3	2

Table 6.2 First cycle of calculations

t	Q_t	R_t	$K(t-1)$	$K_t = K(t-1) + R_t - Q_t$
1	3	5	0	2
2	6	4	2	0
3	8	6	0	0
4	4	4	0	0
5	2	3	0	1
6	1	2	1	2

Table 6.3 Second cycle of calculations

t	Q_t	R_t	$K(t-1)$	$K_t = K(t-1) + R_t - Q_t$
1	3	5	2	4
2	6	4	4	2
3	8	6	2	0
4	4	4	0	0
5	2	3	0	1
6	1	2	1	2

Table 6.4 Third cycle of calculations

t	Q_t	R_t	$K(t-1)$	$K_t = K(t-1) + R_t - Q_t$
1	3	5	2	4
2	6	4	4	2
3	8	6	2	0
4	4	4	0	0
5	2	3	0	1
6	1	2	1	2

By comparison of the second and third cycles, it is obvious that all values of K_t respectively match, meaning the end of the process. The maximum volume of the reservoir is the maximum value of K_t and equals four units (in the first year of the second cycle coinciding with the same year in the third cycle).

6.3.5 Non-conventional stochastic methods of reservoir sizing

The basic notion behind these methods is the observation of the fact that every set of time series of natural inflows in the reservoir is only one case of practically infinite possible scenarios of a variable's value (in this case, the river discharge), which is actually a stochastic variable (Yevjevich, 1967; Mimikou, 1985). So, instead of depending on only one observed time series that gives only a single answer, the reservoir volume in a deterministic way (the probability of the scenario to occur is equal to unit), a number of synthetic time series of inflows can be used, which are statistically equivalent to the historical time series, having the same probability of occurrence. With this method, we get a range of volume for each time series by using the cumulative curve method (Rippl). The resulting volume values have a probability distribution function that can be calculated, which determines a relation between volume and probability of exceedance and actually determines a relation of volume and the risk of design. As far as the selection of the model for the stochastic component is concerned, we may have autoregressive (AR) models using the Markov process, moving average (MA) models, autoregressive moving average (ARMA) models, autoregressive integrated moving average (ARIMA) models, fast fractional Gaussian noise (FFGN) models, etc. (Reddy, 1997).

A short description to these models follows.

- $AR(p)$, autoregressive models

The Markov process considers that the value of an event (i.e. streamflow) at one time is correlated with the value of the event at an earlier period, i.e. a serial or autocorrelation exists in the time series, indicating the order of memory (p) of the process. In a first-order Markov autocorrelation process $AR(1)$, this correlation exists in two successive values of the event. The first-order Markov model, which constitutes the classic approach in synthetic hydrology, states that the value of a variable x in one period is dependent on the value of x in a preceding time period plus a random component. Thus, the synthetic flows for a stream represent a sequence of values, with each value consisting of two parts (Gupta, 1989):

$$x_i = d_i + e_i \tag{6.8}$$

where
   x_i is the flow at ith time
   d_i is the deterministic part of the flow at ith time
   e_i is the random part at ith time

The values x_i are linked to the historical data by ensuring that they belong to the same frequency distribution and possess similar statistical properties (mean, deviation, skewness) as the historical series. The general Markov procedure of data synthesis comprises the following:

1. Determining statistical parameters from the analysis of the historical record
2. Identifying the frequency distribution of the historical data
3. Generating random numbers of the same distribution and statistical characteristics
4. Forming the deterministic part by considering the memory and persistence (influence of previous flows) and combining it with the random part

After the execution of the aforementioned procedure, we have the following equation for flow x_i:

$$x_i = \overline{X} + r_1\left(x_{i-1} - \overline{X}\right) + S\sqrt{\left(1 - r_1^2\right)}t_i \tag{6.9}$$

where (the first two components are the deterministic part while the third is the random part)
   x_i is the streamflow at the ith time
   $\overline{X}$ is the mean recorded flow

r_1 is the lag 1 serial or autocorrelation coefficient

t_i is the random variate from an appropriate distribution with a mean of zero and variance of unity

i is the ith position in series from 1 to N years

- $MA(q)$ moving average models

The other category known as moving average $MA(q)$ models considers a stochastic component (streamflow event) to be a constituent of a number of random variates, with the current variate assigned a weight of unity and random variates generated at antecedent times multiplied by the assigned factors.

The general equations are

$$\overline{Z}_t = a_t - f_1\alpha_{t-1} - \cdots - f_q a_{t-q} \tag{6.10}$$

where $f_1,\ldots f_q$ are weight coefficients. $\overline{Z}_t$ is a linear combination of $q + 1$ white noise variables and we say that it is q-correlated, meaning that $\overline{Z}_t$ and $\overline{Z}_{t+\tau}$ are uncorrelated for all lags $\tau > q$. So q is the order of correlation (Reddy, 1997).

The autovariance equation is

$$\gamma_\kappa = E(Z_t Z_{t-\kappa})$$

$$\Rightarrow \gamma_\kappa = E(\alpha_t - f_1\alpha_{t-1} - \cdots - f_\kappa\alpha_{t-k} - f_{\kappa+1}\alpha_{t-k-1} - \cdots - f_q a_{t-q}) \cdot$$
$$(a_{t-k} - f_1\alpha_{t-k-1} - \cdots - f_{q-k}a_{t-q} - \cdots - f_q a_{t-k-q}) \tag{6.11}$$

And finally

$$\gamma_\kappa = \begin{cases} (-f_k + f_1 f_{k+1} + \cdots + f_{q-k}f_q)\sigma_\alpha^2, & k = 1,2,\ldots,q \\ 0 & k > q \end{cases} \tag{6.12}$$

It is valid that $E(a_t a_t) = \sigma_\alpha^2$ and $E(a_{t-K}a_t) = 0$.

The variance is

$$\gamma_0 = E(Z_t Z_t) = E(\alpha_t - f_1\alpha_{t-1} - \cdots - f_\kappa\alpha_{t-\kappa} - f_{\kappa+1}\alpha_{t-k-1} - \cdots - f_q a_{t-q}) \cdot$$
$$(\alpha_t - f_1\alpha_{t-1} - \cdots - f_\kappa\alpha_{t-\kappa} - f_{\kappa+1}\alpha_{t-k-1} - \cdots - f_q a_{t-q}) \Rightarrow \gamma_0 = \left(1 + f_1^2 + f_2^2 + \cdots + f_q^2\right)\sigma_\alpha^2 \tag{6.13}$$

The autocorrelation function is

$$\rho_k = \frac{\gamma_k}{\gamma_0} = \begin{cases} \dfrac{(-f_k + f_1 f_{k+1} + \cdots + f_q f_{q-k})}{1 + f_1^2 + \cdots + f_q^2}, & k = 1,2,\ldots,q \\ 0 & k > q \end{cases} \tag{6.14}$$

If $q = 1$, we have the $MA(1)$ model:

$$\overline{Z}_t = a_t - f_1\alpha_{t-1} \tag{6.15}$$

$$\gamma_1 = -f_1\sigma_\alpha^2 \tag{6.16}$$

$$\rho_1 = \frac{-f_1}{1 + f_1^2} \tag{6.17}$$

Respectively if $q = 2$, $MA(2)$ is

$$\overline{Z}_t = a_t - f_1\alpha_{t-1} - f_2 a_{t-2} \tag{6.18}$$

$$\gamma_2 = -f_2\sigma_\alpha^2 \tag{6.19}$$

$$\gamma_1 = (-f_1 + f_1 f_2)\sigma_\alpha^2 \tag{6.20}$$

$$\rho_2 = \frac{-f_2}{1 + f_1^2 + f_2^2} \tag{6.21}$$

$$\rho_1 = \frac{-f_1 + f_1 f_2}{1 + f_1^2 + f_2^2} \tag{6.22}$$

- $ARMA(p,q)$ models

$MA(q)$ models are generally inappropriate for direct application to hydrology. An ARMA model, however, combines an autoregressive model $AR(p)$ and a moving average model $MA(q)$ to produce a mixed model known as an $ARMA(p,q)$ model. $ARMA(p,q)$ consists of two polynomials of the order p and q, respectively, as follows (Gupta, 1989):

$$a_i = \varphi_{p,1}a_{i-1} + \varphi_{p,2}a_{i-2} + \cdots + \varphi_{p,p}\, a_{i-p} + n_i - f_{q,1}n_{i-1} - f_{q,2}n_{i-2} - \cdots - f_{q,q}n_{i-q} \tag{6.23}$$

where
    α_i is the ith variable of the sequence (stochastic component) of zero mean and unit variance
    $\varphi_{p,1}\,\varphi_{p,2},\ldots$ are autoregressive parameters or weights
    n_i is the random number at ith time
    $f_{q,1}, f_{q,2},\ldots$ are the random variate weights

Estimation of the parameters φ and f is not a straightforward procedure. In the class of mixed models, the simplest is the $ARMA(1,1)$ model, given by

$$a_i = \varphi_{1,1}\alpha_{\alpha-1} + n_i - f_{1,1}n_{i-1} \tag{6.24}$$

- $ARIMA(p,d,q)$ models

If the value of parameter $\varphi_{1,1}$ is close to its limits –1 and 1, the non-stationary behaviour of the historical hydrologic sequence is indicated. The non-stationarity is accounted for by means of the dth-order difference operator, which represents successive difference of d terms of stochastic variables (i.e. α_i values). This is known as an autoregressive-integrated moving average $ARIMA(p,d,q)$ model (Reddy, 1997).

- FFGN models

The FFGN model is a variation in the family of FGN models with the advantage of smaller computational time required. FFG type I, FFG type II and filtered FGN models are the most characteristic models of this family except FFGN.

An FGN type II model, for example, is generated by (Reddy, 1997)

$$x_t = (h - 0.5) \sum_{u=t-M}^{t-1} (t - u)^{h-1.5} z_u \tag{6.25}$$

where

h is Hurst's coefficient

M is the memory of the process

z_u are independent standard normal variates

The FFGN model has two additive components $x_t = x_t^L + x_t^H$, where x_t^L takes into account the low-frequency effects and x_t^H takes into account the high-frequency effects. The component x_t^L is given by (Reddy, 1997)

$$x_t^L = \sum_{i=1}^{p} W_i X_t (GM/i, B) \tag{6.26}$$

where $X_t(GM/i,B)$ is a standard $AR(1)$ normal process with ρ_1 equal to $(-B)^{-1}$ and with standard normal independent variates as a random component. The parameter B takes a suggested value of 2–4. W_i are weights. The parameter ρ between the range of 15 and 20 is found to be sufficient in practice. The component x_t^H is a separate $AR(1)$ normal process with zero mean.

6.3.6 Synthetic data of inflows and their cumulative curves

The synthetic time series of discharges are generated by a stochastic simulation model such as the ones previously described. The produced group of time series of the physical variables has the same statistical characteristics in the range of a selected order with the historical time series and can have any desired duration (usually must cover the economic life of the project, i.e. 50–100 years). Practically, we can produce an infinite number of time series of this type from the simulation, by changing the stochastic component (the white noise) of the simulation (Monte Carlo technique). Every single produced time series has at least theoretically the same probability of occurrence in the future as the historical time series.

In the case of the runoff process, depending on the ground water storage and permeability characteristics of the basin, a considerable amount of rainfall may become groundwater and it takes a longer time for this groundwater to appear as runoff in the stream. This is especially true in the case of large basins. So rainfall of a previous year may contribute to the runoff of the next year. This storage effect, variously known as the memory of the process or the persistence, is truly reflected in the first-order AR model as the value of the variable at any time that is partly dependent on the previous variable. Therefore, the $AR(1)$ model is a common choice to describe annual flow series. If the memory extends for two time periods, perhaps $AR(2)$ may have to be tried, something very rare.

The historical characteristics that are embedded in a synthetic time series depend on the type of simulation used. For instance, the Markov type of simulations keeps the first and second order of statistical moments and the characteristics of autoregression of the historical time series. A Markov chain is a stochastic process with the Markov property (the memoryless property of a stochastic process). The term 'Markov chain' refers to the sequence of random variables that a process moves through, with the Markov property defining serial dependence only between adjacent periods (as in a 'chain'). It can thus be used for describing systems like some types of stream runoffs that follow a chain of linked events, where what happens next depends only on the current state of the system. On the contrary, other types of simulations, like FFGN models, keep, in addition, the statistical persistence.

At this point, given the fact that the produced group of time series has a usually long duration, in order to cover the lifetime of the project, it must be underlined that the designer does not perform a hydrological prognosis (which is actually a prediction of the future). The simulated synthetic time series represent only statistically equivalent time series with the observed one. After the production of m synthetic time series (e.g. 100 monthly time series of 100 years duration), which have been generated by a stochastic simulation model, synthetic time series are plotted for every synthetic time series ($i = 1,2,..., m$). The procedure then is done according to Equation 6.3 or 6.4 for the total duration of the cumulative curve.

6.3.7 Design risk: Estimation of volume based on acceptable risk

From every cumulative curve of a synthetic time series, the maximum range, and consequently a storage volume V_i, is calculated by the Rippl method assuming that there is full efficiency ($a = 100\%$) and a constant discharge equal to the average inflow. Volumes $V_i = 1,2,..., m$, have a cumulative curve of probability distribution function $Fv_i(v)$ that can easily be found. This distribution correlates the volumes of storage with the probability of exceedance $j(\%)$ according to the following relationship:

$$F_{vi}(v) = 1 - \frac{j}{100} \tag{6.27}$$

The probability $j(\%)$ is defined as the risk of design with volume $V_i = v$ and represents the probability that the specified volume v does not meet the demand. By defining in this way an acceptable risk of design, the volume $V_i = v$ that corresponds to the analytical or graphical expression of the distribution $Fv_i(v)$ can be instantly calculated. These calculations are shown in a graphical way in Figure 6.3 for a sum of $m = 100$ volumes of V_i between the values of 3 and 10×10^9 m³.

It is obvious that if someone chooses a relatively small risk $j = 20\%$, that is, $100 \times 0.20 = 20$ years, the economic life of the project, the volume will not meet the demand, and the volume must be 7.65×10^9 m³. On the other hand, if $j = 90\%$, then the volume is 3.20×10^9 m³.

It was found (Schultz, 1976) that if non-conventional sizing based on a historical time series of n years is repeated using only a part of the historical time series (i.e. from 15 years, only 10 is used), the resulting $Fv_i(v)$ is practically the same. This presumes that the stochastic method is independent of the duration of the historical time series, in contradiction to the totally dependent conventional method. This fact is an important improvement in the area of hydrologic design of a reservoir.

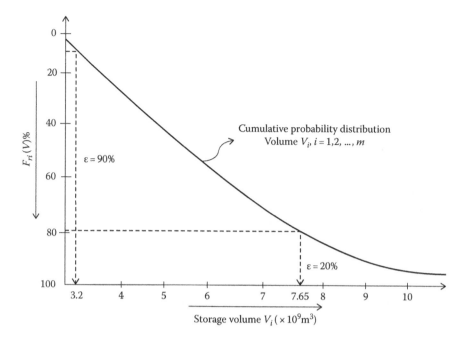

Figure 6.3 Determination of design risk in the storage volume of a reservoir.

6.3.8 Advantages of the non-conventional sizing method

From the previous analysis, it is obvious that the use of the non-conventional, stochastic method of reservoir sizing can now solve the problems that exist in the conventional method. The basic assumption of the conventional method, which was the impossible reproduction of the historical time series in the future, is no longer needed and it is replaced by a more logical assumption that a number of statistically equivalent time series can occur, intending to cover all possible scenarios that nature can produce. Additionally, instead of deterministic design with only one answer for the required volume, the non-conventional method creates a volume-risk distribution curve, as seen in Figure 6.3.

An immediate result is the ability to choose the acceptable risk, or vice versa, with the volume given (that is, in a number of cases, dictated by technical or economical restrictions) to know the particular risk of this volume. In both cases, the qualitative result is the reliability enhancement of the design. Finally, the independence of the stochastic method from the time duration of the observed time series of the discharge solves the problem of dependence of risk (which must be selected by technical, economic or safety criteria) on the duration of observations (which is irrelevant to the decision criteria and, definitely, is dependent on the algorithm used).

6.3.9 Sensitivity of the reservoir design on the mechanism of synthetic discharge generation

A point that needs extra attention, regarding the stochastic method of sizing, is that the anticipated statistical equivalence of the synthetic time series with the historical ones must be compatible with the needs and the special requirements of the design.

It is known that every stochastic simulation model, depending on the model (infrastructure) capabilities, keeps in the synthetic time series a group of historical statistical features.

Also there are some special requirements in the reservoir design to keep some historical statistical features. In order to standardize these requirements, we must conduct a sensitivity analysis, relative to the design of the inflow generation mechanism.

It has been found (Wallis and Matalas, 1972) that the storage capability of a reservoir is directly related to the generation mechanism of the inflows, in other words, to the chosen stochastic simulation model. More particularly, it was found that the minimum storage capacity V_m of a reservoir that must meet a specified demand is related to the efficiency coefficient α, the value of the autocorrelation coefficient of first-order r_1 and the Hurst coefficient H of the inflows.

For relatively low-efficiency coefficients ($\alpha < 0.80$), V_m depends mainly on the autocorrelation coefficient r_1. For this reason, all stochastic models that sustain this statistical characteristic, i.e. Markov models and models with moving averages, can be used. For high-efficiency coefficients ($\alpha \geq 0.80$), V_m depends mostly on the Hurst coefficient H, which expresses the persistence of the time series. In these cases, the former models that do not retain the persistence of the historical time series are improper and different models must be used, such as FFGN that retain (apart from the first and second order of moment and autocorrelation) the statistical feature of persistence.

Example 6.2: Estimation of storage volume using Rippl and stochastic $AR(1)$ methods

At a river site, where a dam is planed to be constructed, we have the monthly inflows I_t for a period of 20 years, given in Table 6.5. The lifetime of the project is 50 years. The time series of outflows from the reservoir Q_t is assumed to be constant and equal to the annual mean value of inflows.

The following are asked

1. The sizing of the reservoir using the conventional Rippl method for 10 and 20 years of time series data
2. The sizing of the reservoir with the non-conventional method for risk = 20% (without keeping the statistical persistence) for the whole time series of 20 years
3. The sizing of the reservoir with the non-conventional method for risk = 20% (without keeping the statistical persistence) using only the first 10 years of the time series
4. The value of risk that corresponds to the volume that was estimated with the conventional method, with the use of the results of the non-conventional method
5. Comparison of the results of the non-conventional method between the case of 20-year time series and the case of 10-year time series

Remarks

- The time series of inflows is not constant and as a first step needs to be stabilized.
- For the non-conventional method of sizing, the model $AR(1)$ will be implemented.
- At least 40 synthetic time series of 50-year length will be constructed.
- The random part of the time series of inflows is assumed to follow the Gaussian distribution.

Solution

1. In order to calculate the storage volume with the conventional Rippl method, we create the cumulative curve of inflow volume for the time series of 20 years. The range that is created by the maximum surplus and maximum shortage is the needed storage volume, as it is seen in Figure 6.4. In this example, the maximum surplus for a time series of 20 years is 0.72×10^9 m^3, and the maximum shortage is 0.47×10^9 m^3. In total, the storage volume is 1.19×10^9 m^3.

Table 6.5 Monthly inflows to the reservoir (m³/s)

Years/months	October	November	December	January	February	March	April	May	June	July	August	September
1	6.40	15.7	54.3	29.5	20.7	22.0	18.4	14.9	7.80	4.10	2.80	0.10
2	10.2	28.5	33.4	21.6	34.6	63.8	41.8	32.8	18.3	5.10	3.20	9.70
3	13.8	86.5	96.1	64.8	112.0	45.2	48.9	32.2	12.5	10.3	4.00	1.70
4	10.4	8.10	53.9	23.8	25.1	35.3	27.1	25.0	14.8	6.40	2.90	3.90
5	9.50	25.9	50.8	30.1	28.6	30.6	37.7	31.9	16.9	8.60	2.90	0.20
6	2.70	37.1	56.1	65.0	37.4	48.7	34.6	35.1	16.8	5.90	2.30	5.80
7	7.60	49.8	58.5	50.1	9.30	32.5	33.8	26.7	7.50	5.50	4.00	4.80
8	4.50	5.30	35.1	66.9	59.3	39.8	40.8	31.4	21.4	4.10	2.50	2.40
9	4.70	12.90	41.3	36.6	51.2	49.5	42.3	30.6	12.8	5.00	3.10	3.30
10	1.40	11.2	69.1	62.2	39.2	49.0	45.9	17.8	10.4	5.40	2.30	1.80
11	6.50	14.3	33.4	59.9	28.6	60.4	52.8	25.4	7.8	2.90	2.40	3.10
12	4.60	24.8	27.4	22.1	27.4	44.1	50.8	31.3	8.9	6.00	3.20	5.00
13	33.5	22.1	10.7	20.5	43.2	34.2	38.6	32.0	11.6	4.7	2.70	2.80
14	12.3	19.6	53.9	23.4	48.1	30.7	48.5	51.0	12.4	4.40	2.40	5.10
15	31.5	50.1	16.7	9.30	10.7	32.0	25.9	20.1	7.70	4.60	3.50	2.80
16	9.60	21.4	27.4	8.30	14.4	22.7	25.0	23.2	7.20	3.60	2.40	2.00
17	7.00	30.5	53.9	31.2	26.7	11.5	17.0	12.0	10.3	3.50	2.40	1.30
18	2.47	7.85	24.8	23.8	61.1	49.4	60.1	37.8	11.4	4.57	3.75	5.71
19	4.87	9.94	45.5	84.9	76.7	29.9	75.7	40.4	9.68	6.13	2.95	2.71
20	3.87	34.7	29.1	49.6	27.5	55.5	41.7	56.8	23.4	7.32	2.86	2.41

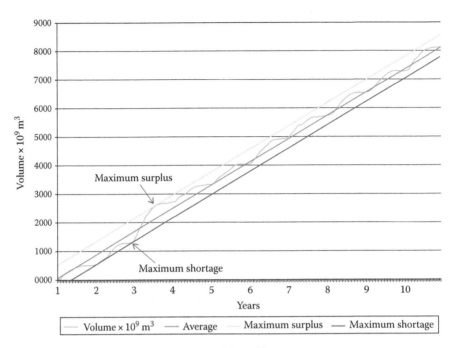

Figure 6.4 Cumulative inflows using the Rippl method for a 20-year time series.

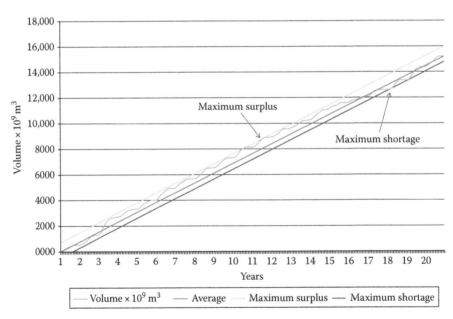

Figure 6.5 Cumulative inflows with the Rippl method for a 10-year time series.

Respectively for only the first 10 years, we plot Figure 6.5.

In this example, the maximum surplus for a time series of 10 years is 0.44×10^9 m³, and the maximum shortage is 0.34×10^9 m³. In total, the storage volume is 0.78×10^9 m³.

2. In order to calculate the active volume of the reservoir with a given risk using the non-conventional method, we must initially produce synthetic time series

statistically equivalent with the actual time series of inflows to the reservoir, and then, proceed with the estimation of the volume of a given risk.
These are the steps that we must follow:

- Production of 40 synthetic time series of 50-year duration.
- Estimation of the maximum range and subsequently of the storage volume for every one of these time series (converted to volume units), using the Rippl method
- Sorting of these volumes in descending order
- Correlation for each one of these volumes of the probability $P = m/(N + 1)$ where m is the order and N is their number
- Plotting of a cumulative curve of probability distribution of volumes, where the volume is plotted in the x-axis and the respective probability of surpass (risk) $J = P\%$ is plotted in the in y-axis
- Selection from the diagram of the respective value for $R = 20\%$

Procedure of synthetic time series of inflows
The steps that we must follow to produce synthetic time series of inflows are the following:

a. Creation of a data file of actual values of inflows (m³/s).
b. Estimation of the statistical characteristics of the time series of inflows like the mean, the variance, the standard deviation, the skewness coefficient and the kurtosis coefficient. If N is the number of the monthly values of the time series, then these coefficients are calculated by the following equations:

$$\text{Mean: } M = \sum_{i=1}^{N} \frac{X_i}{N} \tag{6.28}$$

$$\text{Variance: } VAR = \frac{1}{N} \sum_{i=1}^{N} (X_i - M)^2 \tag{6.29}$$

$$\text{Standard deviation: } SD = VAR^{1/2} \tag{6.30}$$

$$\text{Skewness coefficient: } SK = \frac{1}{N} \sum_{i=1}^{N} (X_i - M)^3 SD^{-3} \tag{6.31}$$

$$\text{Kurtosis coefficient: } KUR = \frac{1}{N} \sum_{i=1}^{N} (X_i - M)^4 SD^{-4} \tag{6.32}$$

c. Estimation of the autocorrelation coefficients of the time series for half of its length and retaining the autocorrelation coefficients of the first and second order. The autocorrelation coefficient of the time series, with itself shifted for every time step (coefficient of T order), is

$$r_t = \frac{\sum_{t=1}^{N-T} \left(X_t - \overline{X}_t \right) \times \left(X_{t+T} - \overline{X}_{t+T} \right)}{\sqrt{\sum_{T=1}^{N-T} \left(X_t - \overline{X}_t \right) \times \sum_{t=1}^{N-T} \left(X_{t+T} - \overline{X}_{t+T} \right)^2}} \tag{6.33}$$

where

T is the time step of the shift
$\overline{X}_t$ is the average of the time series in time t
$\overline{X}_{t+T}$ is the average of the time series in time $t + T$

d. Stabilization of the time series of the monthly outflows. The stabilization is possible through the abstraction from each value of the average value and the division of the result with the standard deviation of the respective month. In the following, we can see the relationship that is used for the stabilization:

$$X1(I,J) = (X(I,J)-M(J)/SD(J) \tag{6.34}$$

where

$X1(I,J)$ is the stabilized value of the discharge
$X(I,J)$ is the monthly outflow in m^3/s
J is the month
I is the year
$M(J)$ is the average value of month J
$SD(J)$ is the standard deviation of month J

e. Estimation of the statistical characteristics of the stabilized time series.
f. Estimation of the autocorrelation coefficients of the stabilized time series for half of its length and retaining the autocorrelation coefficients of the first and second order.
g. Application of the model $AR(1)$ for the estimation of the $F1 = R(1)$ coefficient where $R(1)$ is the coefficient of autocorrelation of the first order of the stabilized time series and of the theoretical deviation C1 given by the following relationship:

$$C1 = C \times (1 - F1^2)^{0.5} \tag{6.35}$$

where C^2 is the variance of the stabilized time series.

h. Production using a computer of random normal numbers with average 0 and standard deviation 1. Calculate the values of synthetic stabilized time series of outflows using the following relationships:

$$X(1) = C1 \times NR(I) \quad \text{and} \quad X(I) = C1 \times NR(I) + F1 \times X(I - 1) \tag{6.36}$$

where

C1 and F1 are the theoretical deviation and the coefficient of the model $AR(1)$
$NR(I)$ is the random number

i. Estimation of the statistical characteristics of the synthetic time series. The values of average and variance must coincide with the respective values of the actual time series of inflows.
j. Estimation of the autocorrelation coefficient of the first order of the synthetic time series which must coincide with the respective value of the actual time series.
k. Production of 40 synthetic time series, of 50-year length in the same procedure. In this example, after the application of the procedure mentioned earlier, the following results are produced (Tables 6.6 through 6.8):

Finally, we plot the next figure and select the value of volume for risk 20%, which is 2.2×10^9 m^3.

Table 6.6 Statistical characteristics for 20 years
for a non-stabilized time series

Mean	24.27
Variance	442.00
Standard deviation	21.02
Skew coefficient	1.05
Kurtosis coefficient	4.05
$R(1)$	0.63
$R(2)$	0.33

Table 6.7 Statistical characteristics for 20 years
for the stabilized time series

Mean	1.71E−8
Variance	0.95
Standard deviation	0.98
Skewness coefficient	0.91
Kurtosis coefficient	3.88
$R(1)$	0.42
$R(2)$	0.18

Table 6.8 Statistical characteristics for 20 years
for the model $AR(1)$

F1	0.42
C1	0.89
Mean	25.23
Variance	455.731
Standard deviation	21.35
Skewness coefficient	0.76
Kurtosis coefficient	2.62
$R(1)$	0.66
$R(2)$	0.35

3. If the process is repeated using only the data of the first 10 years, it gives (Tables 6.9 through 6.11).
   From the aforementioned diagram, the volume value for risk 20% is selected, which is 1.85×10^9 m^3.
4. In the case of the use of a 20-year time series, the volume of 1.19×10^9 m^3, derived from the Rippl method, corresponds to a risk of 100% (Figure 6.6), while with the use of 10 years of data and volume 0.78×10^9 m^3, the risk is also = 100% (Figure 6.7).
5. If 20 years of data set is used, the final volume for risk of 20% is 2.2×10^9 m^3, while in the other case (10 years of data) with the same risk, the volume is 1.85×10^9 m^3. The difference is 16% and justifies the fact that the length of the time series is unimportant considering the use of stochastic models, like $AR(1)$, for the estimation of the storage reservoir volume.

Table 6.9 Statistical characteristics for 10 years for a non-stabilized time series

Mean	25.81
Variance	500.19
Standard deviation	22.37
Skewness coefficient	1.09
Kurtosis coefficient	4.39
$R(1)$	0.65
$R(2)$	0.34

Table 6.10 Statistical characteristics for 10 years for the stabilized time series

Mean	1.52E−07
Variance	0.91
Standard deviation	0.95
Skewness coefficient	0.44
Kurtosis coefficient	2.98
$R(1)$	0.46
$R(2)$	0.27

Table 6.11 Statistical characteristics for 10 years for the model $AR(1)$

FI	0.46
CI	0.85
Mean	25.82
Variance	489.69
Standard deviation	22.13
Skewness coefficient	0.89
Kurtosis coefficient	3.23
$R(1)$	0.65
$R(2)$	0.31

6.3.10 Non-conventional method of sizing including the persistence in inflows

In the cases of design with high-efficiency coefficient, more than 80%, the retention of persistence of the historical time series of reservoir inflows is achieved by proper models, like the FFGN model (Reddy, 1997).

Persistence is a long-term statistical feature of the discharge time series; according to its definition, periods of high discharges tend to follow other periods of high discharges, while the same phenomenon tends to exist also in periods of low discharges.

Persistence characterizes the fluctuation of discharges. Hurst (1951), during his study on long-term reservoirs, defined that the maximum range of the cumulative fluctuations from the average time series of inflows of s years is equal to the reservoir storage capacity, which is needed to satisfy a yearly demand equal to the average inflow of s years.

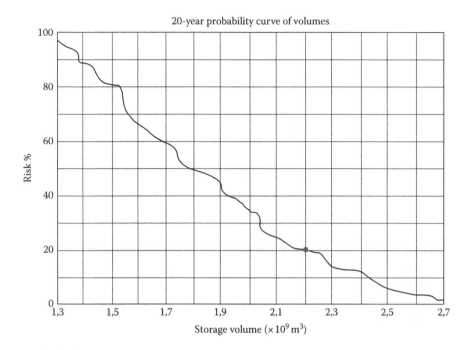

Figure 6.6 Probability curve of volumes for 20 years.

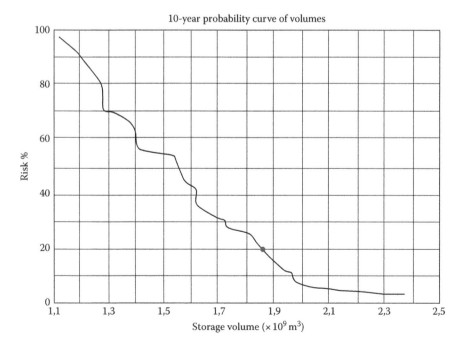

Figure 6.7 Probability curve of volumes for 10 years.

Hurst found (Equation 6.6) that the range divided by the standard deviation of the inflows varies exponentially with the duration of the time series with an exponent H ($H > 0.50$) which expresses the persistence of the time series. In other words, for a specified duration of historical data, the sizing of a reservoir with the use of synthetic time series that are produced by models not taking into account statistical persistence, will lead to smaller reservoir volume in relation to models with statistical persistence, for a given value of risk $j\%$. The use of models with persistence leads to increase in design risk for a given volume in comparison with models without persistence. This results in the use of models with statistical persistence that gives more conservative and reliable reservoir design in high exploitation systems (Mimikou, 1985).

6.3.11 Sizing of a reservoir with a spillway with the use of the method of maximum shortage

The determination of the reservoir volume with the method of the maximum range of the inflow cumulative curves from the demand curve is not suitable in many actual cases of reservoirs with overflow structures. Gomide (1975) introduced the term 'maximum cumulative shortage' or just 'shortage', which is defined in the following text.

Assume that the reservoir is initially full and $V_0 = 0$ corresponds to the full storage capacity. In this case, overflow will occur when the water balance equation for the reservoir $V_t = V_{t-1} + I_t - Q_t$ (I_t the inflow, Q_t the outflow) is positive (evapotranspiration and other minor parts of the equation are omitted for simplicity reasons). Assuming $V_t = 0$ and for time equal to $t + 1$, the storage changes according to the balance equation. In Figure 6.8, it is obvious that the needed storage capacity is equal to $D_{t,s}$ expressed analytically by the following equations (Bayazit, 1982):

$$V_t = \begin{cases} V_{t-1} + I_t - Q_t, & \text{if } V_{t-1} + I_t - Q_t \geq 0 \\ 0, & \text{else} \end{cases}$$
(6.37)

$$D_{t,s} = -\min(0, V_1, V_2, \ldots, V_s)$$
(6.38)

The statistical variables of shortage $D_{t,s}$ and range $R_{t,s}$ (which are both stochastic variables if estimated for a number of synthetic time series) are completely different even if the reservoir

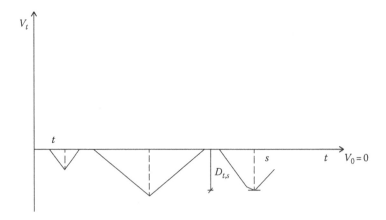

Figure 6.8 Graphical definition of shortage $D_{t,s}$.

does not overflow. The average of $D_{t,s}$ is smaller than the average $R_{t,s}$ but the reverse occurs in case of variances of the two variables.

The asymptotic equations of Fellow (1951) for individual inflows in the case of complete management of the reservoir are as follows:

$$E(R_{t,s}) = 1.596s^{1/2}, \quad \text{Var}(R_{t,s}) = 0.226s \tag{6.39}$$

Gomide (1975) has derived similar equations:

$$E(R_{t,s}) = 1.253s^{1/2}, \quad \text{Var}(R_{t,s}) = 0.261s \tag{6.40}$$

where
  s is the duration of the time series
  E denotes the expected value

In the examined literature (Gomide, 1975; Pegram, 1980), significant effort has been put on the statistical behaviour of the shortage, especially in the case of complete management. Figures have been published for the mean value and the variance of the $D_{t,s}$ distribution.

Nevertheless, there are a few studies devoted to the partial management of the reservoir (the case of $E(I_t) > E(Q_t)$). The reason for this is that it is difficult to deduce analytical equations when the inflows are linearly correlated (Bayazit, 1982). In Figure 6.9, we can see the diagram for the cumulative distribution function $F(D_{t,s})$ of $D_{t,s}$, for a group of autocorrelation coefficients r_1, of first order of inflows, where $D_{t,s}$ have been estimated by synthetic time series of Markov models $AR(1)$, for a 50-year lifetime of a project.

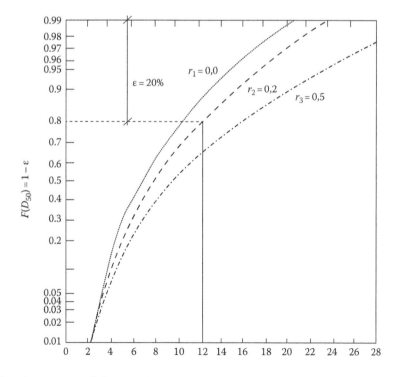

Figure 6.9 Cumulative curve of shortage $D_{t,s}$ probabilities, for total exploitation and various values of r_1. (From Bayazit, M., *J. Hydrol.*, 58, 1, 1982.)

The sizing of the reservoir that is based on the shortage $D_{t,s}$ follows the same procedure as the non-conventional method described for the range $R_{t,s}$. Initially, many synthetic time series of inflows have been deduced by a proper stochastic simulation model (Markov model in the case of persistence rejecting and FFGN when keeping of persistence is targeted). For every time series, we estimate the shortage $D_{t,s}$ and then we construct the cumulative probability distribution $F(D_{t,s})$, which, according to Figure 6.9, depends on the range of autocorrelation of the time series of inflows. Figure 6.9 covers a wide variety of cases; for annual discharges, r_1 is almost in every case smaller than 0.5 and the lifetime of projects is usually 50 years. Finally, by choosing an acceptable risk $j\%$, we can estimate the respective volume equal to the shortage $D_{t,s}$ that corresponds to this risk from the distribution. For example, in Figure 6.9 for $j = 20\%$ and $r_1 = 0.20$, we have $D_{50} = V_{50} = 12 \times 10^9$ m^3.

6.4 SIZING OF A RESERVOIR IN A RIVER SITE WITHOUT MEASUREMENTS

Very often dams and reservoirs are to be constructed in remote areas without any nearby flow measurement station, with the nearest one located at a large distance along the same river or, even worse, in another river. The remoteness of the position sometimes prohibits the construction of a measurement station even after the decision for the project construction is taken. In this case, the hydrologic design must start from a prior level. The time series of discharges at the position of the dam is calculated by transferring the hydrologic information from nearby stations. This transfer is feasible through rainfall–runoff models that are calibrated at existing measurement locations. Then, given the assumption that they are also valid for the dam position, they are applied, using as input data the rainfall on the dam position (acquired by rain gauges which are distributed nationwide).

Nevertheless, this transfer might have many problems, such as the possibility that the equation of rainfall–runoff is not valid in the second position, or the distance between the two positions is significant, so that the calibrated parameters of the model differ. One method that is often used in hydrologic information transfer, especially in relatively extended but hydrologically homogenous areas, is the development of regional rainfall–runoff models.

6.4.1 Regional hydrologic models

Regional hydrologic models are empirical relations developed (usually with the method of single or multiple regression analysis) between a dependent variable (i.e. discharge), which lacks the observations at a specific outlet of a river basin, and several independent physical variables of the basin (i.e. rainfall, basin drainage area, river length, slope), which can be estimated from archives and maps. The model parameters (proportionally to the correlation level) express the geographical variation of the examined dependent variable within a hydrologically homogenous area. The latter is an area where the hydro-meteorological conditions are very similar.

A regional model can be calibrated with data from sites where measurements of the dependent variable (i.e. discharge) exist. These sites should be distributed in the hydrologically homogenous area. This way, the transfer of the hydrologic data (i.e. discharges) to ungauged sites within the same area occurs, in accordance to the laws of geographical variation of the parameters of the empirical equations developed. Many regional models have been developed in many cases, for instance, a regional model to predict sediment transport at positions without measurements (Mimikou, 1982), for the estimation of average monthly inflows, (Mimikou and Rao, 1983), for the estimation of annual losses and the unit hydrograph (Mimikou, 1984), for the estimation of the design flood (Mimikou, 1984) and for the estimation of the discharge duration curves (Mimikou and Kaemaki, 1985). A monthly inflow regional model is described for sizing of a reservoir at sites without discharge measurements.

6.4.2 Regional model for the derivation of monthly inflows

A simple regional model of rainfall–runoff has been developed (Mimikou and Rao, 1983), which is valid for both linear and nonlinear basins. The model is expressed by the following equation:

$$Q_{p,t} = \sum_{i=0}^{k} a_{t,i} \left(\frac{P_{p,t} + P_{p,t-1} + \cdots + P_{p,t-n}}{n+1} \right)^i \quad i = 0,1,\ldots,k \tag{6.41}$$

where

$Q_{p,t}$ is the monthly discharge (m³/s) on year p ($p = 1,2,\ldots, N$) and month t ($t = 1,2,\ldots, 12$)
$P_{p,t}$ is the respective monthly average surface rainfall on the area of interest
k and n are parameters of the model
$a_{t,i}$ are the model's coefficients

k and n determine the structure of the model and, more specifically, the order and the memory, respectively. The order k expresses the monthly rainfall–runoff relation and is equal to the order of the relation (linear or nonlinear). To estimate k, one must study the monthly rainfall–runoff relationship of the area of interest. In the case of a linear basin, k is equal to 1, while in a nonlinear basin, it is polynomial:

$$Q_{p,t} = a_{t,0} + a_{t,1}P_{p,t} + a_{t,2}P_{p,t}^2 + \cdots + a_{t,k}P_{p,t}^k \tag{6.42}$$

k is equal to the order of this polynomial equation. In the case when the variance of the points ($Q_{p,t},P_{p,t}$) in the monthly correlation for a particular basin is high, then the model needs to use the memory parameter n. Then, the correlation is restarted with discharge $Q_{p,t}$ and the average $P_{p,t}$ during months $t, t - 1,\ldots, t - n$, where n is the memory of the correlation of rainfall–runoff. This phenomenon is actually present during the period when runoff is coming from snow melting, which precipitated during the previous n months. With the use of parameter n, we can correct the correlation for these months (spring period). We can see that k and n are characteristics of the basin for every month. Parameter k, nevertheless, is more logical to be considered a constant characteristic of the basin and to express the linearity or not of the basin.

The model was calibrated in two basins, with sizes 217 and 640 km². From the monthly correlations, one basin was found linear ($k = 1$) and the other nonlinear ($k = 2$). From the same relationships when the runoff coefficient $C_t = Q_t/P_t$ was found for a particular month to be greater than 1 (April, May, June), the use of n was necessary ($n = 2$). After that, all coefficients $a_{t,i}$, $i = 0,1,\ldots, k$, were calculated with the method of least squares. Regional relationships that correlate these coefficients with the monthly runoff coefficients under the same diagram were developed. It was found that the coefficients form a line for $i = 0$, an $-S$ curve for $i = 1$ and a parabolic curve for $i = 2$, as shown in Figure 6.10.

The monthly runoff coefficients c_t for all basins were plotted on the same figure, divided by the basin area (km²), along with the values of monthly rainfall (divided also by the

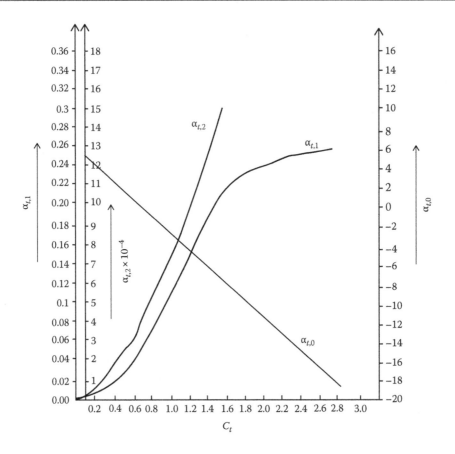

Figure 6.10 Graph of regional variance of coefficients $a_{t,i}$.

basin area). Six different curves were found that vary throughout the year as shown in Figure 6.11. The model works at locations without measurements through the following: by using the data of the average monthly rainfall divided by the area of the basin, we estimate the runoff coefficient C_t, for every month, and then we multiply by the area of the basin at the point of interest, as shown in Figure 6.11.

Then, using Figure 6.10, we estimate $a_{t,i}$, $i = 0,1,\ldots, k$. At this point, k must be estimated indirectly. The following is the proposed solution for the estimation of k: for basins smaller than 217 km^2, $k = 1$; for bigger than 640 km^2, $k = 2$; and for intermediate basin areas, the parameter k depends on the designer's judgement. The n parameter of memory is equal to zero for months with $C_t < 1$, while for $C_t > 1$, $n = k$. With all parameters estimated, we finally use Equation 6.25 and estimate the discharge $Q_{p,t}$, for every monthly rainfall data $P_{p,t}$.

6.5 SIZING OF THE DEAD RESERVOIR VOLUME

6.5.1 General

In order to design the dead volume of the reservoir, we must estimate the volume of sediments transported into the reservoir. All reservoirs that are constructed on natural river flows are subject to sedimentation. The sediments that enter the reservoir with the river flow, due to the reduction of the flow velocity, deposit at the bottom: the coarser-grain sediment

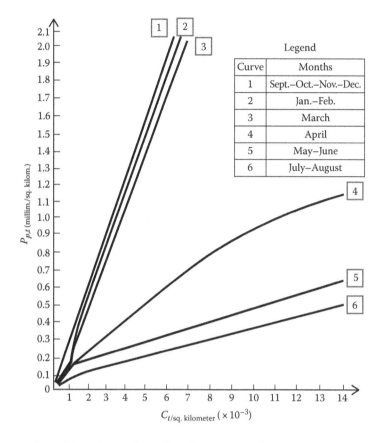

Figure 6.11 Curves of monthly variance of runoff coefficient to rainfall.

first followed by the finer ones. Against the backdrop of sedimentation, the basic problems for the design are

1. The estimation of the annual rate of sedimentation and the total volume of deposited sediment throughout the project lifetime
2. The estimation of the time period that the sediments will start to affect the basic functions of the project

In the case when this is estimated to happen during the economic lifetime of the project, an additional volume should be provided to accommodate deposited sediments, and design changes should be made, like the position of the water outlet in the dam.

6.5.2 Sedimentation in a reservoir: Definition of dead storage

The life cycle of solids, starting with the generation of sediments through the erosion process, and the transport through the water bodies, is affected by climatic, topographical and other factors (see also Chapter 8). The amount of sediments that reach a reservoir can be estimated by various methods. After that the estimation of the proper volume (dead volume) of the reservoir is the next task. The dead volume is usually defined by various technical characteristics of the project, like the point of water abstraction. The estimated dead volume should be larger than the estimated sedimentation. Definitely the latter is a lower limit of the dead volume.

Table 6.12 Reservoir operation types

Reservoir operational characteristics	Type
Sediments are always or almost always sunk to the bottom of the reservoir	I
Medium- to large-level fluctuations	II
Empty reservoirs (i.e. flood protection)	III
Reservoirs with sediments (bottom sediment only)	IV

Some technical terminologies regarding the sedimentation of a reservoir are as follows:

Storage capacity: It is defined as the quotient of the amount of transported solids that deposit to the total transported solids. It depends on the flow velocity and the falling velocity of the particles in the reservoir. The particles' falling velocity depends on their size, the viscosity of the water, its chemical composition, etc.

Sediment indicator: It is defined as the quotient of the period of retention to the average velocity throughout the reservoir.

Specific weight of the deposited sediments: The average inflow of solids (ton/day) must be converted to an equivalent of volume with the use of an estimated specific weight of solids. Many factors (Lara and Pemberton, 1963) affect the specific weight of solids:

1. The operation of the reservoir
2. The texture and size of the particles
3. The sedimentation rate
4. The currents and the flora in the reservoir

Among these factors, the most important is the first (1). According to the operation-mode, reservoirs are categorized into four main types, as shown in Table 6.12.

6.5.3 Estimation of dead volume

After the evaluation of the reservoir type by the designer according to Table 6.12, based on the operation characteristics, the initial unit weight of the sediments is estimated using the following equation:

$$\gamma = W_c P_c + W_m P_m + W_s P_s \tag{6.43}$$

where
γ is the initial unit weight in lb/ft^3
P_c, P_m, P_s are the percentages of clay (size < 0.004 mm), silt (0.004 mm < size < 0.0625 mm) and sand (>0.0625 until 2.0 mm), while the coefficients of clay, silt and sand, W_c, W_m, W_s, respectively, are estimated by the reservoir type using Table 6.13

The initial unit weight γ increases in time and becomes W after T years:

$$W = \gamma + K \log_{10} T \tag{6.44}$$

Table 6.13 Estimation of coefficients W_c, W_m, W_s

Reservoir type	Clay W_c	Silt W_m	Sand W_s
I	26	70	97
II	35	71	97
III	40	72	97
IV	60	73	97

K is a constant that depends on the gradation of the sediments and the reservoir type:

$$K = F_c P_c + F_m P_m + F_s P_s \qquad (6.45)$$

The coefficients F_c, F_m, F_s are given in Table 6.14.

Equation 6.44 is valid for T years. Actually every additional year, extra compaction takes place and we end up in increased compaction in the lifetime. Miller (1953) has presented an analytical expression for the approximation of the integration of the average unit weight W after T years of operation:

$$W_T = \gamma + 0.4343K\left[\frac{T}{T-1}(\ln T) - 1\right] \qquad (6.46)$$

For example, in a reservoir of type I, with the following proportions of sediments: 40% clay, 30% silt and 30% sand, we have $W_c = 26$, $W_m = 70$, $W_s = 97$ and the initial unit weight is 60.5 lb/ft³. The value of K from Equation 6.45 and Table 6.14 is $K = 8.11$. The unit weight is shown in Table 6.15.

The last column of Table 6.15 defines the volume of the sediments for 10, 20, 50 and 100 years in the reservoir using the relation $S_T = (G/W_T) \times T$, where G is the average solids

Table 6.14 Estimation of K

Reservoir type	Clay	K Silt	Sand
I	0	5.7	16.0
II	0	1.8	8.4
III	0	0	0
IV[a]	–	–	–

[a] Type IV only has a bottom sediment.

Table 6.15 Average unit weight after T years

T (years)	W_T (lb/ft³)	W_T (ton/m³)	Volume of sediments S_T (10^6 m³)
10	65.99	1.055	11
20	68.08	1.088	22
50	71.04	1.136	53
100	73.36	1.173	102

transfer into the reservoir. For the previous example, $G = 1.2 \times 10^6$ m³/year. In other words, if the lifetime of the project is 50 years, then the dead volume must be 53×10^6 m³ or bigger if other restrictions are applied, like the selection of the abstraction point.

6.5.4 Estimation of active volume loss due to reservoir sedimentation

After the estimation of the total volume of sediments that deposit at the bottom of the reservoir, especially when the dead volume is determined by it, one must determine the distribution of sediments in time. This distribution leads to the estimation of possible loss of active volume, in case the surface of the sediments is extremely rough with formations that enter the active volume (above the minimum level) and reduce the storage capacity of the reservoir. It is also possible that deposited sediments create technical problems like the covering of the reservoir outlet. The method that is used for the estimation of the sediment distribution is the empirical method of reduced surface (Borland and Miller, 1960; Lara, 1962). A study in 30 reservoirs in the United States showed that a predefined relationship exists among the reservoir shape and the percentage of the deposited material at different depths. The shape of the reservoir is defined by the relationship of depth and storage capacity by the classification in Table 6.16, where m is the reverse of the slope of the curve relating storage to depth (horizontal axis is depth and vertical axis is storage capacity) plotted in a double logarithmic paper. Analytically, it can be estimated as the slope in the relationship: $\log V = a + m \log H$, where H is the depth and V is the storage.

Figure 6.12 presents the empirical relation (%) of the depth of the reservoir and the volume of deposited sediment and also the shapes of the reservoirs of types I and IV which are the most common. Sometimes the type of the reservoir which was classified based on Table 6.12 can be changed. For example, if the reservoir belongs to type III (Table 6.16) with regard to the shape (i.e. between hills) but the level is lowered frequently, or the sediment is mostly clay, we should classify it as type IV (Table 6.12), because a large percentage of the sediment is deposited near the bank, a characteristic of type IV in Table 6.12.

The basic equation that the method uses is

$$S = \int_{0}^{Y_o} A\,dy + \int_{Y_o}^{Hm} Ka\,dy \tag{6.47}$$

where
S is the total volume of sediments that are deposited in the reservoir
0 is the initial level of the dam
Y_0 is the zero level in the dam after the inflow of sediments
A is the area of the reservoir
dy is the differential depth
H_m is the total depth of the reservoir at the normal level

Table 6.16 Types of reservoir shapes

Reservoir type	Classification	m
I	Lake	3.5–4.5
II	Flooding plane	2.5–3.5
III	Between hills	1.5–2.5
IV	Canyon	1.0–1.5

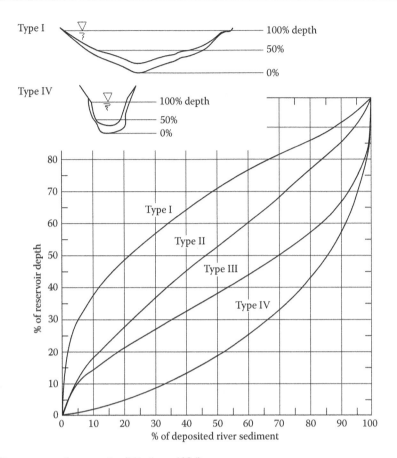

Figure 6.12 Shape types of reservoirs (Mimikou, 1994).

K is the proportionality constant for coefficient transformation of relative areas of
sediments to absolute areas for a given reservoir

A is the coefficient of relative surface of sediments

Equation 6.47, after integration, becomes

$$\frac{1-U_o}{a_o} = \frac{S - V_o}{H_m A_o} \tag{6.48}$$

where

U_o is the relative volume of the reservoir at the new zero level

a_o is the coefficient of relative surface of the reservoir at the new zero level

V_o is the total volume of the reservoir at the new zero level

H_m is the initial total depth of the reservoir (from the maximum normal level)

A_o is the total area of the reservoir at the new zero level

We define

$$h_p = \frac{1-U_p}{a_p} \tag{6.49}$$

and

$$h'_p = \frac{S - V_{PH}}{H_m A_{PH}} \qquad (6.50)$$

where
    P is the relative depth, meaning a percentage of reservoir depth that is measured from the river bank
    V_{PH} is the total volume of the reservoir at depth PH
    A_{PH} is the total area of the reservoir at depth PH

From Equations 6.49 and 6.50, h_p is equal to h'_p, when we have the new zero level Y_o.

After evaluation of data collected in the United States in a large number of reservoirs of every type, dimensionless curves were developed: for the storage of sediments which are presented in Figure 6.12; for the relative surface of the reservoir a, based on the relative depth P, in Figure 6.13; and for the relation of h_p based on the relative depth P, in Figure 6.14.

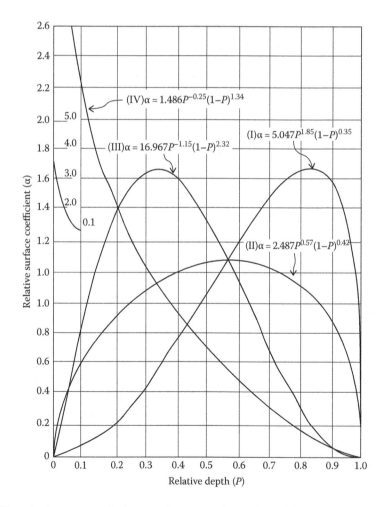

Figure 6.13 Dimensionless curves of relative surface α to relative depth P (Mimikou, 1994).

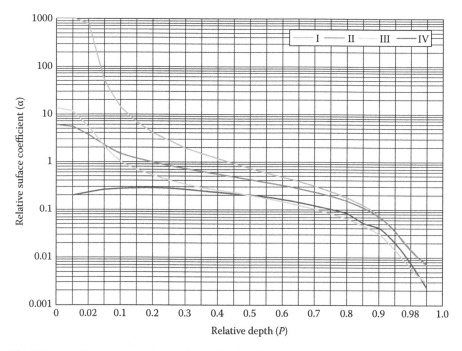

Figure 6.14 Dimensionless curves of h_p to relative depth P (Mimikou, 1994).

Example 6.3: Estimation of dead reservoir volume

The methodology for the distribution of sediments at different depths in a reservoir for instance, in Sections 6.5.2 and 6.5.3, for a 50-year lifetime project, with $S_T = 53 \times 10^6$ m^3 volume of sediments, is the following:

- We determine the initial data on the project lifetime: $S_T = 53 \times 10^6$ m^3, the maximum depth of the reservoir H_m at the normal level in our example is 100 m, the lower normal operational is 400 m and the level of the bottom is 300 m, above mean sea level. Also the depth–volume and depth–surface relations of the reservoir are given in Table 6.17.

Table 6.17 Relations between reservoir depth and storage volume, and reservoir depth and surface

Level (m)	Depth H (m)	Surface (km²)	Volume (10⁶ m³)
300	0	0.000	0.0
310	10	0.025	0.23
320	20	0.110	0.93
330	30	0.230	2.63
340	40	0.380	5.68
350	50	1.050	16.00
360	60	2.180	26.00
370	70	3.050	60.00
380	80	5.350	115.00
390	90	10.000	215.00
400	100	17.000	350.00
410	110	23.200	530.00

Table 6.18 Calculations of the h'_p relationship

Level (m)	Relative depth P	V_{PH} (10^6 m³)	$S_T - V_{PH}$ (10^6 m³)	$H_m A_{PH}$	h'_p
320	0.2	0.93	52.07	11	4.730
330	0.3	2.63	50.37	23	2.190
340	0.4	5.68	47.32	38	1.245
350	0.5	16.00	37.00	105	0.352
360	0.6	26.00	27.00	218	0.124

The logarithmic relation between depth H and storage volume V with the data of Table 6.17 for $H > 50$ m is as follows (with high correlation coefficient):

$$\log V = -6.75 + 4.64 \log H$$

The correlation has a slope $m = 4.64$ that is near the upper limit of m for type I reservoirs (Table 6.16). If we use all data from Table 6.17 (not only for $H > 50$), the average slope is $m \approx 3.4$, implying reservoir between types I and II. Finally, we decide to classify it as type I based on the general characteristics. We estimate the function h'_p from Equation 6.50, for this particular reservoir, as shown in Table 6.18. The volume V_{PH} and the areas A_{PH} from Table 6.17 are used.

The actual function h'_p of the reservoir is plotted in the diagram (h, h_p) of Figure 6.14. The section of (h, h'_p) with the theoretical (h, h_p) curve, for the type of the reservoir (here is type I), gives the new (relative) zero level (where $h = h'_p$) – $P_0 = 0.473$ – and the new zero level is $P_0 H_m = 47.3$ m, and $y_0 = 300 + 47.3 = 347.0$ m

Then we have the following estimation of the sediment distribution presented in Table 6.19:

- Columns 1, 2 and 3 of Table 6.19 are used from Table 6.17.
- Column 4 gives the relative depth P for every level.
- Column 5 gives the relative surface of P with the use of Figure 6.17.
- Column 6 (sediment area) is derived when we multiply column 5 with the constant $K = A_{P_0} H_m / a_0 = 0.50$, where $A_{P_0} H_m$ is the reservoir surface at the new zero level, and a_0 is the respective coefficient of relative surface.
- Column 7 (sediment volume) has been derived from column 6 as $(A_i + A_{i+1})/2 \times (H_i - H_{i-1})$.
- Column 8 is the cumulative sum of column 7. If the final sum is different from S_T, then we correct K with a multiplying factor $K_1 = K(S/S_1)$. In this example, $S_1 = 51.7 < S_T$, so $S_1 = 0.50 \times (53/51.7) = 0.512$. The process has been repeated from column 6. The corrected values are in parentheses.
- Column 9 gives the absolute cumulative sediment volume. It starts from value S and is reduced in every level by abstracting the respective amount of column 8.
- Column 10 gives the revised area of the reservoir as a difference of columns 2 and 6.
- Column 11 gives the revised values of the active volume as a difference of columns 3 and 9. We notice a reduction of the active volume (in 50 years of the reservoir lifetime) from 100% to 15% when moving from the lower to the higher levels.

Table 6.19 Estimation of sediment distribution in a reservoir

Level (m)	Area (km²)	Volume V (10⁶ m³)	Relative depth P	Relative surface a	Sediment area (km²)	Sediment volume (10⁶ m³)	Cumulative sediment volume (10⁶ m³)	Absolute cumulative sediment volume (10⁶ m³)	Revised area of reservoir A' (km²)	Revised volume of reservoir (10⁶ m³)
(1)	(2)	(3)	(4)	(5)	(6)	(7)	(8)	(9)	(10)	(11)
400	17.00	350.0	1.0	0.0	(0.00) 0.00	(4.6) 4.50	4.50 (4.60)	53.0	17.00	297.0
390	10.00	215.0	0.9	1.8	(0.92) 0.90	(9.35) 9.15	13.65 (13.95)	48.4	9.08	166.6
380	5.35	115.0	0.8	1.86	(0.95) 0.93	(9.10) 8.875	22.52 (23.05)	39.05	4.4	76.0
370	3.05	60.0	0.7	1.69	(0.87) 0.845	(8.00) 7.775	30.30 (37.50)	29.95	2.18	30.0
360	2.18	26.00	0.6	1.42	(0.73) 0.71	(6.45) 6.300	36.60 (37.50)	21.95	1.45	4.05
350	1.05	16.00	0.5	1.10	(0.56) 0.55	(5.36) 5.250	41.85 (42.86)	15.50	0.49	0.5
347.3	0.5 (0.512)	10.14	0.473	1.00	(0.512) 0.50	(4.46) 4.375	46.22 (47.32)	10.14	0.0	0.0
340	0.380	5.68	0.4	0.75	(0.38) 0.375	(3.05) 3.000	49.28 (50.37)	5.68	0.0	0.0
330	0.230	2.63	0.3	0.45	(0.23) 0.225	(1.70) 1.675	50.90 (52.07)	2.63	0.0	0.0
320	0.110	0.93	0.2	0.22	(0.11) 0.11	(0.70) 0.675	51.57 (52.77)	0.93	0.0	0.0
310	0.025	0.23	0.1	0.05	(0.03) 0.025	(0.15) 0.125	51.70 < 53 (52.9 ≈ 53)	0.23	0.0	0.0
300	0.000	0.00	0.0	0.00	(0.00) 0.00		0.00	0.0	0.0	0.0

6.6 SIZING OF THE RESERVOIR'S FLOOD VOLUME

The third component of the volume of a reservoir, as discussed previously, is the flood volume. The sizing of the flood volume is a process that is strongly related to the hydrologic design of the spillway. The reason is that the spillway must protect the dam from its design flood which when routed through the reservoir will increase the maximum operation level. Thus, the reservoir flood volume is determined. Design methods differ according to the adopted definition of the design flood as it will be described in the following. Example 6.4 deals with the estimation of the design flood of a spillway in combination with a case study of flood routing inside a reservoir.

Moreover, the reservoir storage can be expressed as a function of water surface elevation and of the surface area of the reservoir (storage–water level–surface curve), which is derived based only on topographic maps. Thus, storage can be estimated in every time step based on the inflow hydrograph by using appropriate routing methods. In general, the flood routing method (Puls Method) and the river routing (Muskingum Method) are described in Chapter 7.

6.7 HYDROLOGIC DESIGN OF FLOOD SAFETY (PROTECTION) STRUCTURES

The hydrologic design of flood protection structures is concerned with the estimation of the design flood of the structure. This is the flood up to which the structure protects downstream life and properties. The design flood is produced by the occurrence of the design storm over the basin.

The design flood is the flood that is expected to occur in the project site area at an exceedance frequency equal to $1/T$, where T is the return period. Risk J is the probability of occurrence of the design flood with frequency equal to $1/T$ (or greater), at least once in the selected time of n years period (i.e. the economic life time of the structure). The risk J, assuming n years of project lifetime, is equal to

$$J = 1 - \left(1 - \frac{1}{T}\right)^n \tag{6.51}$$

The risk should be technically and economically acceptable, in relation to the intended purpose and the safety level that this construction will provide. For instance, a spillway of a dam must provide safety so that the stored water does not overtop the crest of the dam throughout the economic lifetime of the project, while a diversion of a river at a dam construction site must be designed to provide flood protection only during the construction period of the project.

It is more proper to define J as the 'hydrologic risk', since the total risk of a project depends on many other factors such as the structural and the seismic risk. For example, in the case of an earth dam, the hydrologic risk (for failure by overtopping) is approximately 25% of the total risk of the project. The selection of J defines the design flood, with frequency $1/T$, of the structure. The criteria for selection of J are usually both economic and technical. It is obvious that as J decreases resulting in more safety to the structure, the bigger and more expensive the project becomes. For this reason, in projects of paramount importance, a techno-economic analysis is carried out to optimize the relation of construction cost to the cost of the consequences from partial or total failure of the project.

For the estimation of the design flood in the basin of interest, the estimation of the total depth h of the design storm is required. This can be estimated using the idf curves described in Chapter 5, which relate depth, duration and frequency of storm (h,t,T) or alternatively, intensity, duration and frequency of storm (i,t,T). These curves are developed for a particular basin from frequency analysis of annual maximum rainfall depths. As input data are used, the duration of the design storm and its frequency $1/T$, and described next are the design criteria of the project.

Thereafter, a rainfall–runoff model is used, and the design flood hydrograph (produced by the design storm), showing the peak of the flood, is estimated. The estimation of the design flood, with the use of the unit hydrograph, is simple. In the example that is presented next, the estimation method of a design flood applying a unit hydrograph is explained in detail.

In the case of nonlinear basins, where the application of unit hydrograph generates errors in the hydrologic design, the unit hydrograph should not be used for the estimation of the design flood. In this case, a method of nonlinear estimation of flood peaks must be implemented, as explained in the following.

Finally, in the case where the design takes place at a site without discharge measurements, the estimation method of the design flood must follow different approaches. In this case, a regional model transferring hydrologic data from other places where measurements are available is needed, as described previously in the chapter. This could be a flood regional model from a nearby location in the same river, taking into consideration the water contribution from intermediate basins (between the measurement site and the point of interest). Alternatively, it could be a flood regional model from other nearby locations of the major area with hydrologically homogeneous characteristics around the design site (Mimikou and Rao, 1983).

Safety structures are levees, rainwater networks, spillways, river diversion tunnels, etc. In this chapter, we will focus on the hydrologic design of spillways and river diversion tunnels. The procedure followed for the estimation of the design flood is almost identical for all flood protection works, but the design criteria (return period T, duration and time distribution of the design storm) vary in accordance with the particular importance of the structure under study.

6.7.1 Spillway hydrologic design

A spillway is the main safety feature for a dam built in the route of a river. It provides safety against overtopping when this can happen, i.e. an earth dam, and generally, in all hydraulic projects, the spillway controls the outflow of a flood runoff and the containment of the flood inside the reservoir, for the protection of all structures downstream of the dam, like a power station, or the foundation of the dam.

Spillways, especially on earth dams, are expensive structures and in many cases are considered the crucial factor for the economic feasibility of this type of water resources projects. In extreme cases, the cost of the spillway can exceed even the cost of the dam itself, and in normal cases, a spillway costs 40%–80% of the total cost of the dam. Nevertheless, despite this enormous burden on the total cost of the project, until now, all attempts to surpass its necessity, i.e., the construction of dams resistant to total overtopping, or improvements in design and construction, are at early stages. A basic obstacle to the achievement of economic design seems to be the lack of contemporary design criteria, based on up-to-date knowledge and experience that complies with the design criteria and safety factors, adopted by the society and used in other fields and technological areas like highway safety. This persistence of the design criteria of a spillway at very conservative levels is contrary to the new perspectives and possibilities in hydrologic design given by the improvements of the methods of flood analysis, an area where important steps have been made during the last years.

The types of spillways generally applicable to dams are the following (Nicolaou, 1971; Moutafis, 2003):

1. With free flow from inside of the dam. It consists of a crest and a channel with side-walls across the external side of the dam, ending in an energy stilling basin or a jump structure
2. An open channel of free flow that is sited at the edges of the dam. It also ends in an energy dissipation basin
3. A tunnel through the dam sloping in the first stage and horizontal layer that ends also in an energy dissipation basin
4. A side canal along the crest of the dam that ends up in an inclined open channel or tunnel
5. A funnel-shaped spillway with a tunnel in the abutments of the dam
6. A bell-mouth spillway inside the dam
7. A free fall inside the dam

A spillway is a structure of major importance especially with regard to earth dams. Types (2)–(5) are suitable for earth dams, because the failure of a spillway in earth dams is considered as a total failure of the project, resulting in flooding of the downstream area. For other dam types, like arch or gravity dams, a spillway is considered more like a structure that keeps the flood inside the reservoir, discharging the floodwater in a way that prevents possible damages downstream.

The regulation of discharges from a spillway is achieved by the addition of floodgates for water level control (above the crest of the spillway), inside the reservoir, especially when we want to exploit a part of the flood volume for useful purposes. In this case, we add floodgates on the crest that open and close partially or completely, depending on the regulation schedule. In this case, the spillway is called a controlled spillway. On the contrary, the spillways without floodgate installation, from the crest level and above, release any flood volume, waiting for the design flood. V_F is the flood volume estimated by the designer for the containment of the flood.

From the point of view of the hydrologic design of a spillway, the design flood must be estimated. This is an extremely rare flood (of frequency 1/10,000 or less) with a respectively low risk. Finally, it is worth mentioning that a spillway, apart from a safety construction, in many cases has also other uses of ecological importance: an example is the opening of the floodgates of 'Bonnet Carre' spillway in the Mississippi river; the water diverted to Pontchartrain lake with the purpose of the lake enrichment with water and nutritious elements (Lane et al., 2001).

6.7.2 Spillway design flood

The design flood of the spillway is an extremely rare flood from which the spillway protects the dam. The more rare the flood is (or the smaller the frequency or the bigger the return period are), the bigger the volume of the flood, the more costly the spillway and the bigger the required flood volume of the dam are. As a matter of fact, the design frequency is directly associated with the cost of the safety construction. A decrease in the selected frequency (or increase of the return period) of design leads to an increase in the project's safety and also an increase in the project budget and vice versa. Nevertheless, the selection of the design frequency of a spillway is based mainly on safety criteria.

6.7.3 Criteria of spillway design

As spillway design criteria, we consider the hydrologic parameters, which are decided and selected by the designer on the basis of technical and economical parameters of the project. Hydrologic design criteria of a spillway are the frequency or the return period of the design flood and the duration and distribution of the design storm.

Frequency or return period of the design flood: As mentioned before, cost is a function of the return period and safety. Even if an optimal solution can be found after a feasibility study among safety and budgeting of the project, the design frequency is selected mainly by safety restrictions. In many cases, mostly in the United States, the probable maximum flood (PMF) at the site of the project is the design flood of a spillway. Theoretically, its frequency cannot be exceeded. Recently, a criticism has been raised against this practice by many designers, in the United States, for being very conservative and unrealistic. Various statistical studies have proven that a frequency of occurrence with a size class of 10^{-4}, leading to practically zero risk, during the economic life time of the project is acceptable for the spillway design (Benson, 1973).

The duration and the distribution of design storm: The design flood, at the site of the dam, is usually estimated by the unit hydrograph method using the design storm over the basin. The design storm is estimated from the idf curves of the basin based on the frequency and duration of the storm. The frequency is equal to the frequency of the design flood (this match does not apply for higher design frequencies as it will be mentioned later), and its duration is also a design criterion selected on the basis of technical and economic-specific features of the project. Small durations give small flood volumes and vice versa. So, if the project has the containment of the flood through the reservoir as an important objective, then high frequencies must be selected. In the case when the peak of the flood is important, then we must focus on all hydrologic variables relative to the peak (i.e. the time distribution of storm), and through a trial and error procedure, we must find the duration of the storm which maximizes the peak.

For the duration of the design storm, there are also other suggestions, such as the one proposed by the Institute of Hydrology of England in 1978. The duration is estimated through climatic and morphologic features of the basin. The empirical formula proposed is

$$D = (1 + SAAR/1000) * T_p \qquad (6.52)$$

where
    D represents the duration of the design storm (h)
    $SAAR$ is the average annual precipitation over the basin (mm)
    T_p is the time to the peak from the start of the unit hydrograph at the project location (h)

The last factor is known (Mimikou, 1984) to be correlated to the morphological features of the basin (area, stream length, etc.). In the case of the design of a spillway, where we have to estimate the additional outflow of the flood (after the containment), it is suggested that the duration D must be increased by a time lag T_r in T_p:

$$D = (1 + SAAR/1000) * (T_p + T_r) \qquad (6.53)$$

The time distribution of the storm could be selected from the four standard types and their respective group of curves (level of confidence), according to the respective quarter (first, second, third or fourth) when the storm depth reaches its peak, as discussed in Chapter 2 (Huff, 1970). The selection of the first quarter storm fits better the needs of a storm sewage network, whereas the fourth quarter storm is more appropriate for the spillway of a large dam.

6.7.4 Probable maximum flood (PMF)

The PMF, in the design of a dam spillway, is usually estimated by the probable maximum precipitation (PMP) which is used in this case as the design storm. The definition of PMP as per the National Weather Service (AMS, 1959), mentioned in the international literature (Shaw, 1994), is the theoretically maximum depth of storm for a given duration that is physically possible for a given time spot in the year. PMF is extracted by using PMP of a known duration, using the unit hydrograph at the point of interest.

The basic advantage for the use of PMF is the provision of maximum possible safety. Nevertheless, there is a lack of rationality considering hydrologic risk to be actually zero, and issues arise with impacts on the design, as far as the definition and the conditions of application of PMP are concerned. The definition of PMP is quite abstract and can lead to errors, like the assumption that rainfall events can be used in basins with different climatic and morphological features, the unjustified estimation of the maximum volume of storm (precipitated water) at the point of interest, and the omission of use in the PMP of antecedent humidity conditions of the basin.

For the reasons mentioned, the use of PMF has been abandoned because it leads to over-conservative estimations and has been replaced by a flood with frequency 10^{-4}. Thus, the probable maximum flood of the project has a frequency of 10^{-4}.

6.7.5 Estimation of the design storm and flood

The estimation of the total depth h of the design storm of a spillway for the basin in the position of the dam is possible through the idf curves that have been developed for the basin, through frequency analysis of the extreme storms. As entry data, we use the duration of the storm and frequency $1/T$, which are, as mentioned before, the design criteria of the spillway. Total depth h (mm) of the storm must be distributed in time according to one of the four types of distributions.

One usual distribution adopted for peak maximization is the second quarter distribution with 50% level of confidence (see Example 6.4). If the method used for the estimation of the design flood is based on the unit hydrograph, then the time distribution relies on the base of the unit hydrograph (i.e. 2 h). The unit hydrograph is multiplied, as mentioned, by the depth of the excess rainfall for every time step of the rainfall event and finally summed up. To these values, the basic flow of the stream is added and the total design flood of the spillway is calculated. Additionally, when the unit hydrograph has a relatively long duration (i.e. longer than 6 h), it should be converted into one of smaller duration according to a known method, e.g. the S-curve (Linsley et al., 1975; Mimikou and Baltas, 2012). The procedure is described in the following example.

Based on the fact of the extreme rareness of this phenomenon, rain losses (infiltration, percolation) are considered very small (equal or less than 1 mm/h on the average). The time distribution of the rain losses is homogeneous during the rainfall, or (more correctly) follows an exponential curve like Horton's (1933), which better simulates the mechanism of

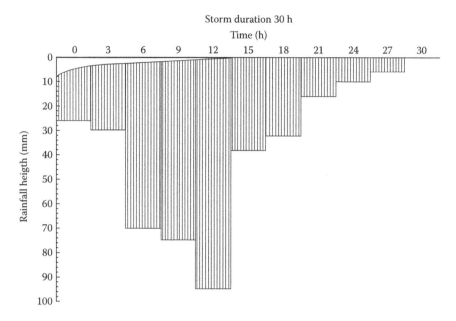

Figure 6.15 Design storm of a dam spillway.

infiltration in the ground. By abstracting the rain losses, a net design rainfall of 2 h distribution is acquired. This is a design rainfall of frequency 10^{-4}, with a second quarter distribution for 50% level of confidence (Figure 6.15).

6.7.6 Derivation of flood hydrograph in specific cases

The theory of unit hydrograph and the application of it, for the estimation of flood hydrographs, have been analysed in Chapter 3. The unit hydrograph usually is not derived by a single flood event but by using a group of recorded storm events that give an average result, as long as they do not differ substantially from each other. The way of deriving a unit hydrograph when the net rainfall is known is relatively easy, based on the theory.

The derivation of unit hydrographs at sites without measurements is accomplished by several methods such as Snyder, SCS, etc. (Mimikou, 1994). An important problem when applying these empirical methods is the estimation of the parameters of the relationships of peak discharge and time duration between the centre of the storm volume and the flood peak. The use of parameters applicable to other areas is prohibited, because it has been proven that they are extremely site specific.

We must also mention that linear methods, like the unit hydrograph, are inappropriate for nonlinear basins, where the runoff volumes are not directly proportional to storm volumes. Researchers (Rogers and Zia, 1982; Mimikou, 1983c) have proven that there is an analytical criterion in order to classify basins as linear or nonlinear. This is the slope of a group of curves of standard peak discharges distribution, which are plotted in a double logarithmic paper, correlating the peak discharges Q_p and the respective storm volumes V. It has been proven (Mimikou, 1983c) that the slope m of the curve

$$\log Q_p = b + m \log V \tag{6.54}$$

is a necessary and sufficient criterion of linearity test. If the basin is linear, then the slope is 1. If m is less than 1, the basin is nonlinear and the value depicts the degree of nonlinearity of the basin (i.e. close to 1, it is relatively nonlinear, while near zero, it is extremely nonlinear). The same research has also showed that the percentage of nonlinear basins increases with the increase of the area size of the basins examined.

Also, it was found that the use of linear methods, like the unit hydrograph, in nonlinear water basins, results in significant mistakes in design, either of underestimation or overestimation of the design flood in the order of 60%. A question raised concerns the proper method to be applied if the basin is found to be nonlinear, a test that is obligatory before the hydrologic design. A simple method is proposed (Mimikou, 1983) for the prediction of peak discharges of selected recurrence periods, applicable to linear and nonlinear basins. The volume of direct flood discharge V_n can easily be estimated for a given return period T from idf curves or any other method, by subtracting from the rainfall volume discharge, of the same period T and known duration t, a properly estimated percentage of losses for the particular basin (based on previous historical data). Additionally, based on historical data, an analysis of the relations of total and net rainfall for hydrographs that have been derived from storm events of the same duration with the one selected for the hydrologic design can be conducted. In this way, the average percentage of increase of the direct runoff due to basic runoff is estimated. This increase is added to the previously estimated net volume V_n of flood discharge to give the total flood volume V. After that, we can estimate the peak discharge Q_p of a given frequency from Equation 6.54.

Example 6.4: Estimation of the design flood of a spillway and flood routing through the reservoir using the storage indication method (see also Section 6.2.4.5)

In a small basin covering an area of A km², a small reservoir is about to be constructed for irrigation purposes. The estimated lifetime of the project is 50 years and the risk 5‰. Idf curves on the area are estimated using Table 6.20, with the maximum values of storm depth for the respective duration of the storm (12, 24, 36, 48 and 72 h), for a time period of 18 years.

Table 6.20 Estimation of storm curves

Year	12 h	24 h	36 h	48 h	72 h
1	58.7	89.8	–	–	–
2	60.4	67.1	–	77.9	82.9
3	48.5	51.2	64.4	–	–
4	46.8	52.9	54.7	62.7	79.8
5	35.4	52.9	71	76.4	91.8
6	61.3	65.7	101.6	112.1	113.5
7	22.2	43.1	45.1	56.1	63.7
8	27.3	45.9	58.1	73	73.8
9	33.8	38.2	44.5	50.4	53.6
10	36.7	38.6	43.1	50.9	68.2
11	33	38	42.5	44.1	50.6
12	51.8	56.1	75.2	84.7	126.3
13	39.7	47.7	63.8	70.8	92.6
14	60.2	67.4	69.7	80.6	89.2
15	45.5	57.7	60.2	–	93.2
16	36.5	45.2	51.9	74.9	90.3
17	52.9	75	78.7	82.9	–
18	48.7	60.1	90.4	102.6	118.6

Table 6.21 Design storm

% total time	0	37	75	90	100
% total depth	0	34	69	95	100

The duration of the design storm is equal to $t = 4$ h. The design storm follows the distribution given in Table 6.21.

For the estimation of the total hourly storm, intermediate values are needed, which will be obtained through linear regression. The 2 h unit hydrograph of the basin has been constructed and is presented in Table 6.22.

Estimate the design flood of the spillway. Total losses are considered constant for the whole duration of the storm and equal to 3 mm/h.

Solution

In Table 6.23, the depth of the storm is shown in mm for durations of 12, 24, 36, 48 and 72 h.

The equation of the idf curve is

$$i = \frac{K^* T^a}{(t + b)^m}$$

For each storm duration, we have $a = 1.282/S_x$ and $c = x - 0.45S_x$ (Table 6.24).

From equation $X = c - \ln\{\ln (T) - \ln (T - 1)\}/a$, where $X = i$ (intensity of storm), and for return periods $T = 10, 20, 50, 100, 1,000, 10,000$ years, we estimate the maximum intensity of storm for the given storm durations (Table 6.25).

Table 6.22 Unit hydrograph of 2 h of the basin

Time (h)	Discharge (m³/s)
1	0
2	8.1
3	51.0
4	90.0
5	122.6
6	143.1
7	124.3
8	106.3
9	91.7
10	77.1
11	65.1
12	54.9
13	44.6
14	36.9
15	28.3
16	22.3
17	12.3
18	12.0
19	8.6
20	4.3
21	2.1
22	0

Table 6.23 Storm depth (mm) for durations of 12, 24, 36, 48 and 72 h

Year	12 h	24 h	36 h	48 h	72 h
1	58.7	89.8	–	–	–
2	60.4	67.1	–	77.9	82.9
3	48.5	51.2	64.4	–	–
4	46.8	52.9	54.7	62.7	79.8
5	35.4	52.9	71.0	76.4	91.8
6	61.3	65.7	101.6	112.1	113.5
7	22.2	43.1	45.1	56.1	63.7
8	27.3	45.9	58.1	73.0	73.8
9	33.8	38.2	44.5	50.4	53.6
10	36.7	38.6	43.1	50.9	68.2
11	33.0	38.0	42.5	44.1	50.6
12	51.8	56.1	75.2	84.7	126.3
13	39.7	47.7	63.8	70.8	92.6
14	60.2	67.4	69.7	80.6	89.2
15	45.5	57.7	60.2	–	93.2
16	36.5	45.2	51.9	74.9	90.3
17	52.9	75.0	78.7	82.9	–
18	48.7	60.1	90.4	102.6	118.6

Table 6.24 Estimation of mean value, S_x, a and c

	12 h	24 h	36 h	48 h	72 h
Mean value	3.70	2.30	1.76	1.53	1.19
S_x	0.99	0.58	0.48	0.39	0.31
a	1.29	2.21	2.68	3.27	4.15
c	3.25	2.04	1.55	1.35	1.05

Table 6.25 Calculation of maximum intensity of storm

T (years)	Duration of rain (h)				
	12 h	24 h	36 h	48 h	72 h
10	5.00	3.06	2.39	2.04	1.60
20	5.56	3.38	2.65	2.26	1.77
50	6.28	3.80	3.00	2.55	1.99
100	6.82	4.12	3.26	2.76	2.16
1,000	8.61	5.16	4.12	3.47	2.72
10,000	10.39	6.21	4.98	4.17	3.27

Table 6.26 Calculation of y = log i and x = −log t

T (years)	(−) log t				
	−1.08	−1.38	−1.56	−1.68	−1.86
10	0.6987	0.4850	0.3777	0.3096	0.2029
20	0.7447	0.5290	0.4240	0.3541	0.2476
50	0.7978	0.5800	0.4774	0.4057	0.2995
100	0.8337	0.6147	0.5135	0.4407	0.3346
1,000	0.9348	0.7128	0.6152	0.5396	0.4339
10,000	1.0166	0.7926	0.6974	0.6200	0.5146

Calculation of m and A_T

T	10	20	50	100	1,000	10,000
m	0.634	0.635	0.636	0.637	0.638	0.640
A_T	1.373	1.419	1.473	1.510	1.612	1.695

Average m = 0.637.

For the calculation of slope m and the constant term A_T of the line,

$$\log i = A_T - m \times \log (t + b),$$

with b = 0, that is, $\log i = A_T + m \times (-\log t)$, $y = \log i$ and $X = -\log t$, as shown in Table 6.26.

In the same way, slope a and the constant term log K of the line $AT = \log K + a \times \log T$, with $y = AT$ and $X = \log T$, are calculated (Table 6.27).

In conclusion, the idf curve equation is

$$i = \frac{19.233 \times T^{0.106}}{t^{0.64}}$$

From this equation and for t = 4 h, the intensity of the storm is calculated. By multiplying with t, we have the value of h of storm. For the spillway T = 10,000 years, i = 21.11 mm/h and h = 84 mm.

For the distribution of the total depth of storm, we use Table 6.28. We are interested in the active rainfall; so from the values of depths of rain, we abstract the rainfall losses as φ = 3 mm/h.

For the estimation of the design flood, we assume that the basin is linear, so the unit hydrograph can be applied (Table 6.29).

Design flood

Q_i are evaluated based on the principle of proportionality

$$Q_i = h_i/h \times Q,$$

Table 6.27 Calculation of a and K

T (years)	log T	A_T	a	log K	K
10	1.00	1.373	0.106	1.284	19.233
20	1.30	1.419			
50	1.70	1.473			
100	2.00	1.510			
1,000	3.00	1.612			
10,000	4.00	1.695			

Table 6.28 Distribution of the total storm depth

t (h)	% total time	% total depth	% not cumulative	Total h (mm)	Active h (mm)
0	0	0	0	0	0
1	25	22.97	22.97	19.39	16.39
2	50	46.00	23.03	19.44	16.44
3	75	69.00	23.00	19.42	16.42
4	100	100.00	31.00	26.17	23.17

where

Q is the ordinates of the unit hydrograph
h_i is the active rainfall depth of every hour
Q_i is the discharge for every hour (Table 6.30)

Flood routing though the reservoir
In order to calculate the flood routing through the reservoir, we need an extra curve in addition to the inflow discharge versus time curve of the design storm. This curve is the $((2S/\Delta t) + O)$curve versus Q. The quantity $((2S/\Delta t) + O)$ is called storage indication and it is calculated by solving the mass balance equation of the reservoir: $I - Q = \Delta S/\Delta t$, where I is the inflow to the reservoir, Q is the discharge from the reservoir and $\Delta S/\Delta t$ is the change of storage inside the reservoir.

Using mathematical transformations, we have this solution:

$$\frac{I_1 + I_2}{2} - \frac{Q_1 + Q_2}{2} = \frac{S_1 - S_2}{2} \Rightarrow (I_1 + I_2) + \left(\frac{2S_1}{\Delta t} - O_1 \right) = \left(\frac{2S_2}{\Delta t} + O_2 \right) \tag{6.55}$$

where 1 and 2 are indicators for two consecutive time steps.

The curve of $((2S/\Delta t) + O)$ versus Q is nonlinear and can be estimated by the characteristics of the spillway and the reservoir. A relationship that is used to estimate the discharge from a spillway is

$$Q = cLH^{3/2} \tag{6.56}$$

where

Q is the discharge
L is the length of a rectangular spillway
H is the relative height above the spillway's crest
c is a parameter, usually equal to 3

In this example, for simple reasons, the curve of $((2S/\Delta t) + O)$ versus Q has been estimated and is given in Table 6.31:

The estimation of flood routing through the reservoir is shown analytically in Table 6.32.

- Column 2 is the sum of two consecutive values of column 1.
- For $n = 1$ (first time step) in line 1, we have $I(n) + I(n + 1) = 2S(n + 1)/\Delta t + Q(n + 1)$. Through Table 6.31 with linear interpolation, we estimate values of column 5 using column 4. Column 6 is derived by substitution of the value of Q in $2S(n + 1)/\Delta t + Q(n + 1)$, in column 4.
- Column 3 in line 1 has zero value, but in line 2, we estimate it using columns 5 and 6 from the previous line 1.
- Column 4 is the sum of columns 2 and 3 in every line (Figure 6.16).

Table 6.29 Base duration of the unit hydrograph D = 22 h

t (h)	Q (m³/s)	Q_displ	Q_displ	Q_displ	Q_displ	Q_displ	Q_displ	Q_displ	Q_displ	Q_displ	Q_displ	S 2h	S 2h	S 2h	S_displ	S − S_displ	UH 1 h
1	0.00											0.00	0.00	0.00		0.00	0.00
2	8.10											8.10	8.10	8.10	0.00	8.10	16.20
3	51.00	0.00										51.00	51.00	51.00	8.10	42.90	85.80
4	90.00	8.10										98.10	98.10	98.10	51.00	47.10	94.20
5	122.60	51.00	0.00									173.60	173.60	173.60	98.10	75.50	151.00
6	143.10	90.00	8.10									241.20	241.20	241.20	173.60	67.60	135.20
7	124.30	122.60	51.00	0.00								297.90	297.90	297.90	241.20	56.70	113.40
8	106.30	143.10	90.00	8.10								347.50	347.50	347.50	297.90	49.60	99.20
9	91.70	124.30	122.60	51.00	0.00							389.60	389.60	389.60	347.50	42.10	84.20
10	77.10	106.30	143.10	90.00	8.10							424.60	424.60	424.60	389.60	35.00	70.00
11	65.10	91.70	124.30	122.60	51.00	0.00						454.70	454.70	454.70	424.60	30.10	60.20
12	54.90	77.10	106.30	143.10	90.00	8.10						479.50	479.50	479.50	454.70	24.80	49.60
13	44.60	65.10	91.70	124.30	122.60	51.00	0.00					499.30	499.30	499.30	479.50	19.80	39.60
14	36.90	54.90	77.10	106.30	143.10	90.00	8.10					516.40	516.40	516.40	499.30	17.10	34.20
15	28.30	44.60	65.10	91.70	124.30	122.60	51.00	0.00				527.60	527.60	527.60	516.40	11.20	22.40
16	22.30	36.90	54.90	77.10	106.30	143.10	90.00	8.10				538.70	538.70	538.70	527.60	11.10	22.20
17	16.30	28.30	44.60	65.10	91.70	124.30	122.60	51.00	0.00			543.90	543.90	543.90	538.70	5.20	10.40
18	12.00	22.30	36.90	54.90	77.10	106.30	143.10	90.00	8.10			550.70	550.70	550.70	543.90	6.80	13.60
19	8.60	16.30	28.30	44.60	65.10	91.70	124.30	122.60	51.00	0.00		552.50	552.50	552.50	550.70	1.80	3.60
20	4.30	12.00	22.30	36.90	54.90	77.10	106.30	143.10	90.00	8.10		555.00	555.00	555.00	552.50	2.50	5.00
21	2.10	8.60	16.30	28.30	44.60	65.10	91.70	124.30	122.60	51.00	0.00	554.60	555.00	555.00	555.00	0.00	0.00
22	0.00	4.30	12.00	22.30	36.90	54.90	77.10	106.30	143.10	90.00	8.10	555.00	555.00	555.00	555.00		
									124.30	106.30	91.70	554.60	555.00				
										143.10	124.30	546.90					
											124.30	503.60					
												456.90					
												381.00					

(Continued)

Table 6.29 (Continued) Base duration of the unit hydrograph $D = 22$ h

t (h)	Q (m³/s)	$Q_{displ.}$	$Q_{displ.}$	$Q_{displ.}$	$Q_{displ.}$	$Q_{displ.}$	$Q_{displ.}$	$Q_{displ.}$	$Q_{displ.}$	$Q_{displ.}$	$Q_{displ.}$	S 2h	S 2h	$S_{displ.}$	$S - S_{displ.}$	UH 1 h
	0.00				4.30	12.00	22.30	36.90	54.90	77.10	106.30	313.80				
					2.10	8.60	16.30	28.30	44.60	65.10	91.70	256.70				
					0.00	4.30	12.00	22.30	36.90	54.90	77.10	207.50				
						2.10	8.60	16.30	28.30	44.60	65.10	165.00				
						0.00	4.30	12.00	22.30	36.90	54.90	130.40				
							2.10	8.60	16.30	28.30	44.60	99.90				
							0.00	4.30	12.00	22.30	36.90	75.50				
								2.10	8.60	16.30	28.30	55.30				
								0.00	4.30	12.00	22.30	38.60				
									2.10	8.60	16.30	27.00				
									0.00	4.30	12.00	16.30				
										2.10	8.60	10.70				
										0.00	4.30	4.30				
											2.10	2.10				
											0.00	0.00				

Table 6.30 Estimation of the design flood

t (h)	UH l h	Q_1	Q_2	Q_3	Q_4	Q (discharge)
1	0.00	0.00				0.00
2	16.20	26.55	0.00			26.55
3	85.80	140.63	26.63	0.00		167.26
4	94.20	154.39	141.06	26.60	0.00	322.05
5	151.00	247.49	154.86	140.88	37.54	580.77
6	135.20	221.59	248.24	154.68	198.80	823.31
7	113.40	185.86	222.27	247.94	218.26	874.33
8	99.20	162.59	186.43	222.00	349.87	920.88
9	84.20	138.00	163.08	186.20	313.26	800.55
10	70.00	114.73	138.42	162.89	262.75	678.79
11	60.20	98.67	115.08	138.26	229.85	581.85
12	49.60	81.29	98.97	114.94	195.09	490.29
13	39.60	64.90	81.54	98.85	162.19	407.49
14	34.20	56.05	65.10	81.44	139.48	342.08
15	22.40	36.71	56.22	65.02	114.92	272.88
16	22.20	36.39	36.83	56.16	91.75	221.12
17	10.40	17.05	36.50	36.78	79.24	169.56
18	13.60	22.29	17.10	36.45	51.90	127.74
19	3.60	5.90	22.36	17.08	51.44	96.77
20	5.00	8.20	5.92	22.33	24.10	60.54
21	0.00	0.00	8.22	5.91	31.51	45.64
22			0.00	8.21	8.34	16.55
23				0.00	11.59	11.59
24					0.00	0.00

Table 6.31 $\left(\dfrac{2S}{\Delta t} + O\right)$ versus Q curve

Q (m³/s)	0	100	200	300	400	500
$\left(\dfrac{2S}{\Delta t} + O\right)$ (m³/s)	0	1,000	3,000	5,000	7,000	10,000

6.8 HYDROLOGIC DESIGN OF A RIVER DIVERSION

6.8.1 Introduction

Many times, during the construction period of hydraulic projects, especially of considerable size, the completion of the project is impossible in one dry season (summer). For this reason, a diversion of the river is necessary around the construction site. The main purpose of the diversion is to allow the construction to take place unhindered in a dry environment but also to protect the project and its resources (personnel, machinery, etc.) from a flood.

The main technical phase of a diversion usually consists of the construction of an open channel, or closed tunnel (depending on the topographical features) where the river water will inflow at its upstream end and outflow far downstream from the project site. The design the diversion channel is placed near one riverbank at the construction site.

Table 6.32 Estimation of the flood routing using storage indication method

t (h)	I (n)	I(n) + I(n + 1)	2Sn/Δt − Qn	2S(n + 1)/Δt + Q(n + 1)	Q(n + 1)	S(n + 1)
(0)	(1)	(2)	(3)	(4)	(5)	(6)
1	0	26.55	0	26.22	2.66	11.78
2	26.55	193.81	20.9	214.71	21.4	96.66
3	167.26	489.31	171.91	661.22	34.2	313.51
4	322.05	902.82	592.82	1495.64	85.7	704.97
5	580.77	1404.08	1324.24	2728.32	144.5	1291.91
6	823.31	1697.64	2439.32	4136.96	256.8	1940.08
7	874.33	1795.21	3623.36	5418.57	320.9	2548.84
8	920.88	1721.43	4776.77	6498.20	370.1	3064.05
9	800.55	1479.34	5758	7237.34	404.7	3416.32
10	678.79	1260.64	6427.94	7688.58	419.3	3634.64
11	581.85	1072.14	6849.98	7922.12	427.5	3747.31
12	490.29	897.78	7067.12	7964.90	428.9	3768.00
13	407.49	749.57	7107.1	7856.67	425.4	3715.64
14	342.08	614.96	7005.87	7620.83	417.5	3601.67
15	272.88	494.00	6785.83	7279.83	406.1	3436.87
16	221.12	390.68	6467.63	6858.31	388.1	3235.11
17	169.56	297.30	6082.11	6379.41	384.1	2997.66
18	127.74	224.51	5611.21	5835.72	337.2	2749.26
19	96.77	157.31	5161.32	5318.63	311.1	2503.77
20	60.54	106.18	4696.43	4802.61	285.3	2258.66
21	45.64	62.19	4232.01	4294.20	259.9	2017.15
22	16.55	28.14	3774.4	3802.54	240	1781.27
23	11.59	11.59	3322.54	3334.13	216.7	1558.72
24	0	0.00	2900.73	2900.73	172.1	1364.32

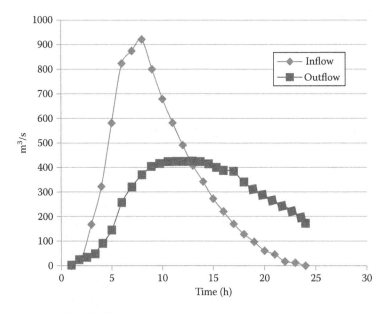

Figure 6.16 Flood routing through the reservoir.

Special importance must be given to the simplicity of the design, in order to cut down the cost, since the diversion channel is an auxiliary temporary construction that stops operating after the completion of the project. However, in many cases, especially in dams, the diversion channel may have a supplementary use after the construction of the dam, e.g. serve as the spillway. Nevertheless, the simplicity of the structure must not endanger the safety of the construction site and the whole project. Another auxiliary technical work that usually supports a diversion is a small embankment (pre-dam) constructed upstream of the main project. Inside the reservoir of the pre-dam, river water accumulates, and at a second phase, it is discharged through the diversion channel. The hydrologic design of the diversion consists of the evaluation of the design flood which has a frequency of occurrence greater than the spillway due to the limited time of the required protection (1–2 years for a typical dam). The frequency of occurrence ranges from 1/10 to 1/100 years, depending on the tolerable risk accepted by the designer and the cost of the diversion in each case. For example, the risk J of a flood equal to or greater than the frequency $1/T = 1/10$ at least once during the 2 years of construction is $J = \{1 - (1 - 1/T)^2\} = \{1 - (1 - 1/10)^2\} = 19\%$. If the diversion design flood had frequency 1/100, then J would be significantly smaller: $J = \{1 - (1 - 1/100)^2\} = 1.99\%$, resulting in an increase in the diversion's construction cost.

6.8.2 Design criteria of a diversion

The design criteria of a river diversion are exactly the same as that of a spillway but with quantitative differences. These criteria are the frequency, the return period of the design flood and the duration of the design flood, as described in the following.

The frequency and the return period of the design flood. Due to the relatively small duration of the demanded protection time of the project by the diversion, the frequency of the design flood is chosen much higher than the one of a spillway. A common choice is between 10 and 100 years for less or more conservative designs, respectively. The corresponding range of risk J for a usual construction period of 2 years is 19% and 1.99%, respectively. The final choice of the design frequency in this range is based on the designer's experience, the hydrologic response of the river (i.e. often floods need greater protection and a smaller design frequency) and the tolerable diversion cost. At this point, it is worth mentioning that, for the range of frequencies for which a diversion is designed, there is a mismatch between the frequencies of the design flood and the design storm (British Institute of Hydrology, 1978). In particular, the pairs of frequencies of the design storm and the design flood, resulting from the storm are shown in Table 6.33.

This means that a storm with smaller frequency is required for the derivation of a flood peak of a known frequency. This mismatch is getting smaller as the return period increases, and diminishes in the range of 500–1000 years.

The design storm duration and distribution. These parameters are chosen based on technical and hydraulic characteristics of the specified project. In the case of dams, usually a smaller duration is chosen compared to the design storm of the spillway. Rarely, there are cases wherein the same duration is chosen for the diversion and the spillway. The distribution is chosen according to the specific geographical and climatic data of the area of design.

Table 6.33 Pairs of frequency of the design storm and the frequency of the design flood

Return period of the flood peak	2.33	5	10	20	30	50	100	250	500	1000
Return period of the design storm	2	8	17	35	50	81	140	300	520	1000

6.8.3 Estimation of the design flood of diversion

For the estimation of the design flood of the diversion, first of all, the estimation of the corresponding design storm over the river basin at the position of the project is required. The depth h of the design storm has been calculated with a known frequency $1/T$ or a return period T. For example, from Table 6.33, a flood with frequency 1/50 must have a frequency of the design storm 1/81 and a return period of 81 years; for the duration t, the idf curves (h, t, T) that have been constructed for the basin are used. Then, we apply one of the four used distributions. Definitely the percentage of losses for the case of diversion will be higher than the one estimated for the design storm of a spillway, due to the fact that it is, in the first case, a much more frequent phenomenon. In the same way, a conversion of the design storm to an appropriate unit hydrograph takes place, at the project position, so the design flood of the diversion can be calculated.

For the case of nonlinear basins, but also for those where there are no discharge measurements available at the construction site, models for the estimation of the design flood of the diversion are applied, like the ones applied to the design flood of spillways.

Example 6.5: Estimation of the design flood of a river diversion

It has been estimated by statistical analysis that the idf curves of a particular basin have the following analytical expression:

$$i = \frac{19.66 T^{0.115}}{t^{0.63}} \tag{6.57}$$

At the outlet of this basin where there are gauging stations of discharge and water level of the river, a dam is about to be constructed, so it is essential to estimate the design flood of the diversion. Based on the technical, hydraulic, economic and the other characteristics of the project, we select a design frequency with a diversion value 1/50. Also we choose for the duration of the design storm the value 48 h. From the idf curves, we come up with the following intensity and storm depth for the design storm (Table 6.34):

Notice that frequency 1/81 for the design storm of diversion corresponds to the desired frequency 1/50 of the design flood.

After that, the storm depth has been distributed in 48 h with 2 h time step, based on the distribution of second quarter, with confidence level 50%. In the next step, we deduct the losses equal to 35% of the total rainfall, respectively, for the design storm of the diversion. The distributed design storm and the losses are shown in Figure 6.17. Net 48 h storm is implemented next on a 2 h unit hydrograph that was derived from station data on the basin outlet, and is shown in Figure 6.17. The flood that is produced for the diversion is shown in Figure 6.17. In the total flood, the baseflow has been added, as shown in Figure 6.17.

Table 6.34 Intensity and storm depth

	Frequency	Intensity i (mm/h)	Storm depth h (mm)
Design storm of the diversion	1/81	2.84	136.32

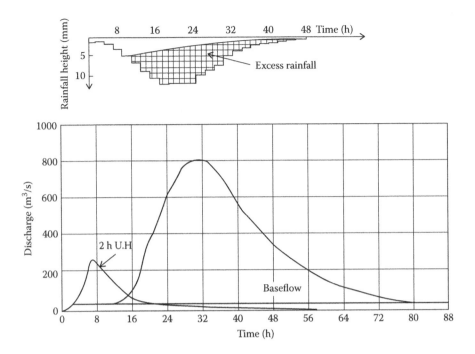

Figure 6.17 Design flood of a river diversion.

6.9 HYDROLOGIC DESIGN OF OTHER WATER STRUCTURE–SPECIFIC ISSUES

6.9.1 General

So far, from the hydrologic design point of view, the interest was particularly in reservoirs and their components. However, there are also other water resources projects, like irrigation and water supply networks and wastewater treatment plants (WWTPs). The social benefits of these projects are of great importance during the effort of exploitation of a country's water resources, especially in developing countries. The necessity of construction of projects of this type throughout a country, rural and urban, creates an apparent need for optimum design and operation, so as to cover effectively the social needs for water, with the guaranteed quantity and without cost overruns.

In the effort for optimum design and operation of hydraulic development projects, the maximization of hydrologic information for each system that is designed has a significant part. This is accomplished by certain techniques like models and statistical analyses that enhance the quality and quantity of historical data (like discharges, rainfall depths) measured at the point of interest. The enhancement of these data is usually based on a proper process of optimizing recorded files in a format which allows us to take directly more information for the hydrologic design of a project. For instance, in an irrigation system, the basic goal of the hydrologic design is to forecast the water quantities, properly adjusted in time and space. If we assume that the water comes from a multiple purpose reservoir, the importance of existence of rainfall information in the area is obvious for the system design, in order to optimize the abstractions from the reservoir, which are naturally supplemented by the rainwater in this area. This procedure has as direct result on the optimum programming of abstractions of water and the saving of increased abstractions from the reservoir. This is translated from a hydrologic point of view to the ability of reliable prediction of rainfall over an area in a way that everyone can estimate directly, at every moment, the rainwater of a given duration that will precipitate, with a tolerable risk of failure of that prediction. In the following sections, we deal with this problem, but also with other ones relevant to the optimization of hydrologic information during the design phase of various hydrologic projects.

6.9.2 Construction of operational depth–duration–frequency curves of rainfall for irrigation needs

We present here the prediction of rainfall water over an agricultural area for irrigational needs through the construction of monthly depth–duration–frequency rainfall curves over the basin (Gabriel and Neumann, 1962; Mimikou, 1983).

Due to the complexity of this issue, first of all the theory of the rain fall depth model over a period on n days is analysed.

Let us assume that

$$n_j = \begin{cases} 1 & \text{if } j \text{ is a rainy day} \\ 0 & \text{if } j \text{ is a dry day} \end{cases}$$

The number of rainy days is

$$N_n = \sum_{j=1}^{n} n_j, \quad N_n = 1, 2, \ldots, n \tag{6.58}$$

We set x_v, $v = 1, 2, \ldots, n$, the daily depth of rainfall, in the v rainy day. The total depth of rainfall in a period of n days, $S(n)$, is calculated with the help of the next relationship:

$$S(n) = \sum_{v=0}^{N_n} x_v, \quad n = 1, 2, \ldots, n \tag{6.59}$$

The function of $S(n)$ is

$$F_n(x) = P[S(n) \le x] = P\left[\sum_{k=0}^{N_n} x_k \le x\right] = P\left[\sum_{k=0}^{N_n} x_k \le x, \quad \cup \quad \{N_n = v, v = 0, \ldots, n\}\right] \tag{6.60}$$

By using the part of the sample area for which $S(n) \le x$, we get

$$F_n(x) = P[S(n) \le x] = \sum_{v=0}^{n} P\left[\sum_{k=0}^{v} x_k \le x, N_n = v\right]$$

$$= P[N_n = 0] + \sum_{v=1}^{n} P\left[\sum_{k=1}^{v} x_k \le x, N_n = v\right] \tag{6.61}$$

If we assume that the information of the number of rainy days does not necessarily mean that the quantity of rain is also known, we transform the aforementioned relationship into the following:

$$P[S(n) \leq x] = P[N_n = 0] + \sum_{v=1}^{n} P\left[\sum_{k=1}^{v} x_k \leq x\right] \cdot P[N_n = v] \tag{6.62}$$

$P[N_n = v]$ is calculated through a Markov chain relationship (Mimikou, 1994), while the other factor $P\left(\sum_{k=1}^{v} x_k\right)$ can be calculated by a frequency analysis (usually log-normal distribution) of the total rainfall amount for the specific period of n days, for the years that data exist (Gabriel and Neumann, 1962). Thus, we can correlate the rainfall depth $S(n)$ with the probability $P[S(n) \leq x]$ for various periods of n days, where n is not the rain duration of course but the period that concerns an irrigation project for instance. This means that for each month of analysis during the irrigation period, we have the relationship between the rainfall depth $S(n)$ and total depth in n days with the probability that this depth is less or equal to a certain value x. The results of such research can be presented in the form of curves that are useful for the planning and management of water systems, as described in Example 6.5.

If we want to construct monthly curves of rainfall, we use as raw data the average daily areal rainfall depths of the basin for the months that are important for irrigation, i.e. May, June, July and August. For every month and every duration n, we plot the relationships of frequency and rainfall depth. If we define as risk j the probability of the irrigation water prediction through these curves that do not satisfy the given needs (i.e. the irrigation of a given area and a crop that is dependent on this water), then we have

$$J = P[S(n) \leq x] \tag{6.63}$$

So the idf curves of depth–duration and frequency can provide directly also the risk of each prediction of irrigation water for a given duration of n days in every month of irrigation. Through these curves, we can estimate the following:

1. The rainfall depth of the area of concern for a certain duration of n days of an irrigation period and an estimated accepted risk j
2. The risk j for a certain water depth x during a certain period of n days of an irrigation period (supplementary to the water abstractions from the reservoir)

It has been proven (Mimikou, 1983) that if the idf curves (Figure 6.19) are plotted on a double-axis logarithmic paper for every frequency $P[S(n) \leq x] = j$ apart (with the duration in horizontal axis and the rainfall depth in vertical axis), there are straight lines (Figure 6.20), parallel to each other, following the next analytical relationship:

$$X = k\{P[S(n) \leq x]\}^{\lambda} n^{m} \tag{6.64}$$

where k, λ, m are constants depending on climatic factors (see Figure 6.20). The stochastic curves of depth–duration and frequency of Equation 6.62 are similar to the idf curves (h, t, T) that have been presented earlier. Nevertheless, they are totally different, since in the idf curves, t is the duration of the rainfall, but here, n is a period of days independent of single rainfall events. Also, the rainfall depths in regular idf curves are dealt like extreme events (maximum) throughout a year, but from the stochastic point of view, they are dealt like a constant physical process. This results in a stochastic prediction of depth–duration–frequency (or risk) of rain to be operationally better than the classical idf curves, since someone can adjust the duration n of the curves to any desired duration appropriate for the design but also to the operational needs of the water system. This way, we can avoid the individual analysis of extreme rainfalls of a limited duration (usually 2–3 days at most) that is mandatory when using classical idf curves.

6.9.3 Discharge prediction in a water recipient of outflows from a wastewater treatment plant

The prediction of discharges of a river that accepts sewage of an urban or industrial area, after biological treatment, offers a crucial information. The reason is the potential of lowering the outflow quality of the WWTP, which reduces the energy consumption, during high discharges of the river, due to the self-cleaning ability of the recipient.

6.9.3.1 Conditions and requirements of the method

Let us assume a site A in a river with industrial and domestic sewage outflow from a nearby area, after the treatment in a WWTP. A flow measurement station is located at site B (basin outlet) and rainfall-measuring gauges are also installed in the basin, as shown in Figure 6.18.

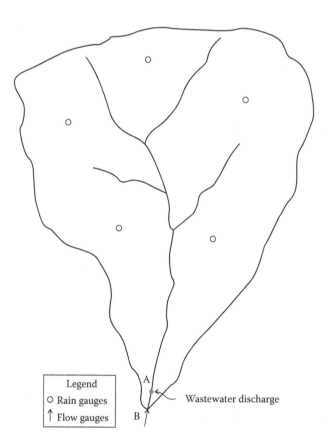

Figure 6.18 General layout of the basin.

The method concerns n days of the year when the discharge at site A and the nearby site B are high enough. In this case, water indicators (dissolved oxygen, concentration of ammonium nitrogen, etc.) that define the quality of water treatment (% BOD removal etc.) are high enough that allow the moderation of the treatment plant operation, lowering the levels of the desired efficiency, due to the enhanced self-cleaning ability of the river. So, initially an estimate of a threshold river discharge was made, which, when exceeded, the plant can operate with a lower efficiency. For example, in a treatment plant that can provide second and third degree of treatment, it is possible these days to operate only the second degree of treatment, with substantial economic benefits in energy saving. An indicative factor for the selection of this crucial discharge is the duration (in days) with greater or equal actual discharges in an average year. This is obtained from the curve of daily discharges at site B. This duration must be long enough so that the reduction of full capacity of the plant operation is meaningful from a technical and economical point of view. Usually, we properly combine the two indicative elements (criteria), which are the substantial moderation of the plant operation and the duration of average annual time that exceeds the crucial discharge, for the optimum selection of it.

The requirements of the prediction model at site B are mainly regarding its operational features. It must have the ability of real-time implementation, with direct warning of discharges greater or equal to the selected crucial discharge. Also it must be equipped with relatively long prediction time, in order to have the ability to activate the system for the exploitation of a sudden rise of the river discharge. This last requirement complicates the solution of the problem, given the fact that in hydrologic simulation models, the increase of the prediction time is against the credibility of the prediction itself. This problem is confronted partially by the use of an automated correction mechanism (updating mechanism), which during the prediction time corrects the final prediction with updated data inserted in the model in real time.

6.9.4 Models of discharge predictions of water recipient of sewage

The demand for long-term predictions restricts substantially the number of real-time prediction models that can be selected. So models like the one presented previously, when the prediction time is limited to the duration of the routing of the river, are inappropriate for this case. The selection must be among models of one variable (e.g. the river discharge) or with two or more variables. In this case, a combination of two or more models is required in order to obtain the entry variables in the initial model (i.e. a rainfall prediction model).

The single-variable models of the river discharge predict the future values using mainly the correlation of the successive measurements of the variable, using the memory characteristic of the variable's time series. Based on this, it is obvious that the most proper single-variable models for discharge predictions in seasonal (most commonly daily) basis is the stochastic models of the $ARMA$ (p, q) category. Basic advantages of these models are the fast process of calibration, their simplicity and the requirement of minimum field data (only the time series of discharges in site B of the previous figure is needed). A basic disadvantage is the increase of the average square error of the prediction, when the time of the prediction increases. It has been proven (Quimbo, 1967) that the most appropriate stochastic model for simulation, and therefore prediction, of daily discharges is the self-correlation model of second-order $AR(2)$.

In the category of models with two variables (rainfall–runoff), someone can use deterministic and stochastic models as well. For better functional capability in real time, and also for automation (of input and output data), 'black-box' models are superior. Calibration of these models is based on the respective time series of discharges at site B and rainfall (through the daily data of the basin) from the rainfall gauges in Figure 6.18. For the hydrologic use of these models, we must follow an empirical (and not automated) process of calibration, in order to take into account additional basin information like snowfall and precedent moisture. In order to use both models for prediction in real time, we must combine parallel single-variable models (stochastic) of rainfall prediction, which take the role of input data. So, with a stochastic model, i.e. $AR(2)$, we predict L time units ahead of the value of rain (average daily rainfall) in time $(t + L)$, and then we compute the discharge at site B, and consequently at the site of interest A at time $(t + L)$. The models of rainfall–runoff have increased credibility compared to single-variable models, because they better describe the response of the basin. Nevertheless, more effort for calibration and implementation is needed.

An even better description of the basin's balance mechanism is provided by holistic models of the basin that incorporate, except rainfall, storage and evaporation. These models must be connected in parallel with three other stochastic single-variable models for prediction of L time units ahead of rainfall values of storage and evaporation of the basin. These values are used as input data in the model for discharge prediction at site B for time $(t + L)$.

With criteria mainly the field data, which are provided for the basin and also the computing capacity, and the existing means of hydrologic data transfer (system automation) for the implementation of the model, we can select one of the aforementioned models for the prediction of the discharge at site B (Figure 6.18). If, for example, only discharge data or rainfall and discharge data are available, there is a limitation to single-variable models or two-variable models, respectively. Also an assumed deficiency of computational means and scarcity of the hydrologic data transfer system could prohibit credible implementation of a holistic model even though the field data exist (for its calibration).

After the selection, calibration and implementation of the prediction model in real time (connection with the system of direct data transfer of entry data), we must assure that a credible notification system towards the treatment plant operators exists. Each one of the models that have been described must provide the prediction in real time t of the discharge L time units ahead (i.e. 10 days). Exactly at time t, the system of direct notification of the plant must automatically be activated (with a sound or an electronic message on a pc screen), in order to predict which discharge will exceed the crucial value during time $(t + L)$. Then, proper adjustments must be done in the plant's operation. In the case when the model has a correcting mechanism, and new real-time data enter the pc in time $(t + 1)$, $(t + 2)$,..., $(t + L - 1)$, an extra change, i.e. the cancellation of the prediction that the discharge will exceed the crucial value in time $(t + L)$, must also be transmitted automatically to the plant.

Example 6.6: Optimization of irrigation uptakes from a reservoir

A mountainous basin of a river, with an area of 2744 km^2, is to be irrigated by a nearby multipurpose reservoir. Since there is scarcity in the uptakes from the reservoir, a prediction model has to be developed, with acceptable risk in the anticipated amount of rainwater over the basin, during the irrigation period (May until August), in order to optimize the irrigation programming, from the reservoir, for water saving purposes (Ross, 1970). For this reason, a stochastic model has been used. The raw input data give the average daily rainfalls of the basin (from 18 rain gauges using the Thiessen aggregation method) for a period of 20 years. In this analysis, rainy day is defined as the day with at least 0.25 mm of rainfall occurring; otherwise the day is dry (Chin, 1977).

The simulation of the daily events of rain for 20 years for the months May, June, July and August has been performed for every monthly sub-period of the irrigation period.

Thus, we have computed for every month the initial probabilities, the transition probabilities and the probability to cope with ($v = 1,2,..., n$) rainy days in any period of n days ($n = 1,2,..., 30$) in 1 month: $P[N_n = v]$. After that, the proper probability distribution has been defined as the cumulative amount of rainfall,

$$\sum_{k=1}^{v} X_k \qquad (6.65)$$

for $v = 1, 2, \ldots, 30$, with selection between a bi-parametrical gamma, tri-parametrical gamma and log-normal distribution, the last one being better according to the x^2 test criterion. So the probabilities have been calculated:

$$P\left[\sum_{k=1}^{v} X_k \leq x\right] \tag{6.66}$$

for every v and x values between 5 and 100 mm (with 5 mm step). This combined probability acquires the total probability distribution, for the amount of rain $S(n)$, in a period of n days with risk j equal to the probability. In this way, for every month of this period, the depth–duration–risk curves $j\,(x, y, j)$ have been plotted, as seen in Figure 6.19. After this step, the lines have been plotted in double-axis logarithmic paper in Figure 6.20, based on the previous equation, while the parameters k, λ, m, for May, June, July and August are shown in Table 6.35.

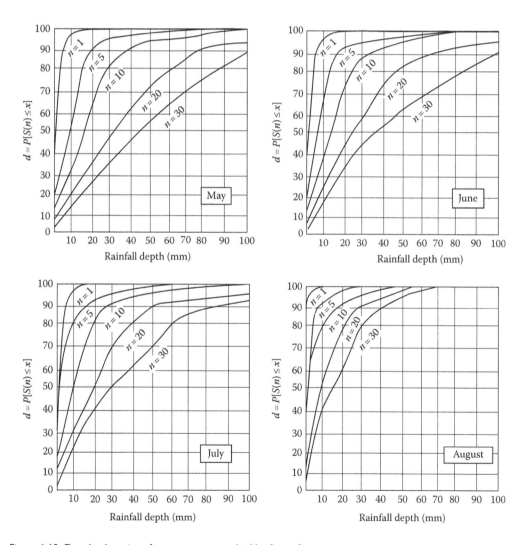

Figure 6.19 Depth–duration–frequency curves (risk) of rain for irrigation purposes.

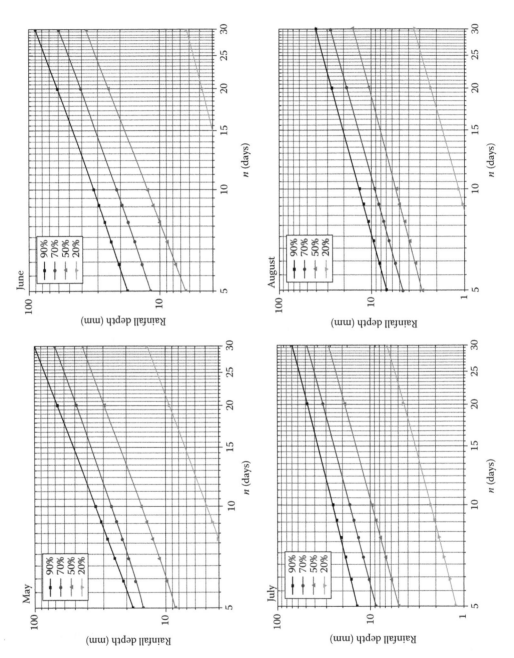

Figure 6.20 Logarithmic curves of depth–duration–frequency (risk) of rain for irrigation purposes.

Table 6.35 Parameters of the stochastic curves of frequency of rain

Parameters	Months			
	May	*June*	*July*	*August*
k	1.08×10^{-2}	2.8×10^{-3}	1.9×10^{-3}	0.83×10^{-3}
λ	1.33	1.60	1.62	1.65
m	1.02	0.93	0.96	1.00

From Figures 6.19 and 6.20, the expected rainfall depth over the basin is estimated for a given period of n days with standard risk *j*, i.e. for a period of 20 days, in May, with risk 20% (failure probability), the expected rainfall depth is 10 mm. Also the risk can be found for a certain period, i.e. for 20 mm of rain in 20 days in May, risk is 36%.

6.9.5 Duration–discharge curves and their use in the estimation of the hydroenergy potential

The duration–discharge curve in a river position is a very useful tool for the hydrologic design of hydroenergy installations and other hydraulic works like flood protection, irrigation networks and works for sustaining water quality. Especially in the development of hydroelectric projects, it is very useful during the preliminary phase of the design for the estimation of the hydroelectric potential of the position, mainly in projects with a small reservoir or without one. For projects with large reservoirs, these curves are not useful enough given the fact that they do not provide the information of a time series of discharge of that certain position.

The curve displays the discharge Q in relation to the percentage of time D that a discharge is equal or surpasses a certain value. In the *x*-axis, there is the percentage of time and in the *y*-axis the discharge Q. We can have curves on a daily, monthly, etc. basis. In Figure 6.21, a duration–discharge curve is displayed.

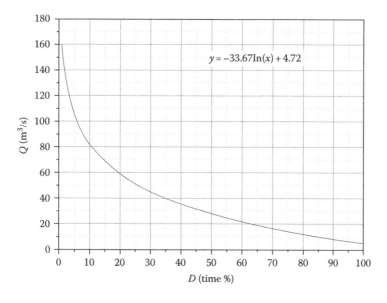

$$y = -33.67\ln(x) + 4.72$$

Figure 6.21 Duration–discharge curve.

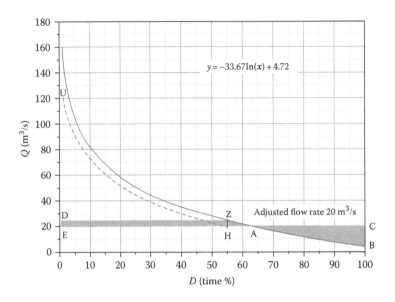

Figure 6.22 Duration–discharge curve for adjusted flow rate at 20 m³/s.

The duration–discharge curve provides the necessary data for sizing the required storage in order to maintain a specific discharge value. For example, if the natural flow is 8 m³/s for the 100% of time and the desired discharge is 20 m³/s for the 100% of time, then the following steps are applied (Figure 6.22):

1. The parallel from point E (value of 20 m³/s) to the horizontal axis is drawn. The intersection of the parallel with the curve is point A.
2. The section AEDZ is drawn on the left side of the curve, in such a way that the areas of sections AEDZ and ABC on the other side of the curve, are equal.
3. The new curve is created with downward parallel transposition of the previous curve from point Z to point H and has the shape of UHAC.

REFERENCES

Bayazit, M., 1982, Ideal reservoir capacity as a function of yield and risk, *Journal of Hydrology*, 58, 1–9.

Benson, M.A., 1973, Thoughts on the design of design floods, in *Floods and Droughts: Proceedings of the Second International Symposium in Hydrology*, Water Resources Publications, Fort Collins, CO, pp. 27–33.

Borland, W.M. and Miller, C.R., 1960, Distribution of sediment in large reservoirs, *Trans. Am. Soc. Civ. Engs.*, 125(1), 166–180.

Chin, E.H., 1977, Modeling daily precipitation occurrence process with Markov chain, *Water Resources Research*, 13(6), 949–956.

Chow, V.T., Maidment, D.R. and Larry, W., 1964, *Handbook of Applied Hydrology*, McGraw-Hill Co., New York.

Fellow, W., 1951, The asymptotic distribution of the range of sums of independent random variables, *The Annals of Mathematical Statistics*, 22, 427–452.

Gabriel, K.R. and Neumann, J., 1962, A Markov chain model for daily rainfall occurrences at Tel Aviv, *Quarterly Journal of the Royal Meteorological Society*, 88, 90–95.

Gomide, F.L.S., 1975, *Range and Deficit Analysis Using Markov Chains*, Hydrology papers, Colorado State University, Fort Collins, CO.

Gupta, R.S., 1989, *Hydrology and Hydraulic Systems*, Prentice Hall, Englewood Cliffs, New Jersey.

Horton, R.E., 1933, Determination of infiltration capacity for large drainage basins, *Eos, Transactions, American Geophysical Union*, 18, 371–385.

Huff, F.A., 1970, Spatial distribution of rainfall rates, *Water Resour. Res.*, 6(1), 254–260, DOI: http://dx.doi.org/10.1029/WR006i001p0025410.1029/WR006i001p00254.

Hurst, H.E., 1951, Long term storage capacity of reservoirs, *Transactions of the American Society of Civil Engineers*, 116, 770–779.

Hurst, H.E., Black, R.P. and Simaika, Y.M., 1965, *Long-Term Storage: An Experimental Study*, Constable, London, U.K., 145pp.

Institute of Hydrology, 1978, *Methods of Flood Estimation: A Guide to the Flood Studies Report*, Wallingford, U.K.

Lane, R., Day, J.W., Kemp, G.P. and Demcheck, D.K., 2001, The 1994 experimental opening of the Bonnet Carre spillway to divert Mississippi river water into lake Pontchartrain, Louisiana, *Ecological Engineering*, 17(4), 411–422.

Lara, J.M. and Pemberton, E.L., 1963, Initial unit weight of deposited sediments, in *Proceedings of the Federal Interagency Sedimentation Conference*, USDA-ARS 970, pp. 818–845, Washington, DC.

Lara, S.M., 1962, Revision of the Procedure to Compute Sediment Distribution in Large Reservoirs, Bureau of Reclamation, Denver, Colo., USA.

Linsley, R.K., Kohler, M.A. and Paulhus, J.L.H., 1975, *Hydrology for Engineers*, McGraw Hill, New York.

Miller, C.R., 1953, Determination of the Unit Weight of Sediment for Use in Sediment Volume Computations. Bureau of Reclamation, Denver, CO.

Mimikou, M., 1982, An investigation of suspended sediment rating curves in western and northwestern Greece, *Hydrological Sciences Journal*, 27(3/9), 369–383.

Mimikou, M., 1983a, Monthly rainfall frequency curves during irrigation periods, *International Journal of Water Resources Development*, 1(4), 311–321.

Mimikou, M., 1983b, Daily precipitation occurrences modelling with Markov chain of seasonal order, *Hydrological Sciences Journal*, 28(2), 221–232.

Mimikou, M., 1983c, A study of drainage basin linearity and non-linearity, *Journal of Hydrology*, 64, 113–134.

Mimikou, M., 1984a, Envelope curves for extreme flood events in northwestern and western Greece, *Journal of Hydrology*, 67, 55–66.

Mimikou, M., 1984b, Floodflow forecasting during dam construction, *International Water Power & Dam Construction*, 5, 15–17.

Mimikou, M., 1984c, Regional relationships between basin size and runoff characteristics, *Hydrological Sciences Journal*, 29(1), 63–73.

Mimikou, M., 1985, *Stochastic Hydrology*. Edited by National Technical University of Athens.

Mimikou, M., 1994, *Water Resources Technology*, 2nd edn., Papasotiriou, Athens, Greece, p. 225 (in Greek).

Mimikou, M. and Baltas, E., 2012, *Engineering Hydrology*, Papasotiriou, Athens, Greece. (in Greek)

Mimikou, M. and Kaemaki, S., 1985, Regionalization of flow duration characteristics, *Journal of Hydrology*, 82, 77–91.

Mimikou, M. and Rao, A.R., 1983, Regional monthly rainfall-runoff model, *Journal of Water Resources Planning and Management*, 109(1), 113–134.

Moutafis, N., 2003, *Floods and Flood Protection Works*, School of Civil Engineering, NTUA, Athens, Greece.

Nicolaou, S., 1971, *Hydrodynamic Works*, School of Civil Engineering, NTUA, Athens, Greece.

Pegram, G.G.S., 1980, An autoregressive model for multilag Markov chains, *Journal of Applied Probability*, 17(2), 350–362.

Quimbo, R., 1967, Stochastic model of daily river flow sequences, Hydrology paper no. 18, Colorado State University, Fort Collins, CO.

Reddy, P.J.R., 1997, *Stochastic Hydrology*, Laxmi Publications Ltd., New Delhi, India.

Rippl, W., 1883, The capacity of storage reservoirs for water supply, *Proceedings of the Institution of Civil Engineers*, 71, 270–278.

Rogers, W.F. and H.A. Zia, 1982, Linear and nonlinear runoff from large drainage basins, *Journal of Hydrology*, 55, 267–278.

Ross, S.M., 1970, *Applied Probability Models with Optimization Application*, Holden-Day Co., San Francisco, CA.

Schultz, G.A., 1976, Determination of deficiencies of the Rippl-diagram method for reservoir sizing by use of synthetically generated runoff data, in *Proceedings of the XIIth Congress of ICOLD*, Mexico City.

Shaw, M., 1994, *Hydrology in Practice*, 3rd edn., Chapman & Hall, London, U.K.

Wallis, J.R. and Matalas, N.C., 1971, Correlogram analysis revisited, *Water Resources Research*, 8(4), 1448–1459.

Yevjevich, V., 1967, An objective approach to definitions and investigations of continental hydrologic droughts, Hydrology paper no. 23, Colorado State University, Fort Collins, CO.

Chapter 7

Urban hydrology and stormwater management

7.1 INTRODUCTION

7.1.1 Definitions of urban hydrology and stormwater management

A significant part of engineering hydrology is dedicated to the so-called urban hydrology or urban stormwater hydrology. These two terms are synonymous and are closely related to various other terms used, such as *urban drainage, urban storm drainage, urban stormwater management* and *urban surface water management*.

Urban hydrology deals with the hydrological cycle within the urban environment. As part of engineering hydrology, it studies the following:

- The precipitation–losses–runoff processes within urban or urbanizing watersheds, emphasizing on the generation of stormwater runoff and its transportation both on the ground and street surface and in the sewer conveyance system, as well as through its appurtenances
- The quality of the stormwater runoff, its fate and its impacts on receiving waters
- The technical aspects of the planning, design, construction, operation, maintenance and effectiveness evaluation of various measures to manage stormwater runoff quantity and quality

In this sense, urban hydrology uses various hydrological methods for runoff quantity computations (the rational method and the SCS curve number method are the most commonly used); some of them are specifically used in urban watersheds.

Urban stormwater management is a more comprehensive term than urban hydrology. It also involves the implementation of various measures to resolve quantity and quality problems related to population growth in urban areas, urban expansion and/or urbanization of natural areas. Although urban stormwater management does not necessarily involve construction of specific structures, it most often does. Typical designs include specific cross sections for the street surface conveyance systems; curbs, gutters and catch basin inlets; storm sewers and their appurtenances; open channels; detention and retention basins with their inlet/outlet structures; and special structures such as energy dissipators. Either specific modifications of existing structures or specific designs are made for controlling water quality of urban stormwater runoff. Finally, urban hydrology is essential in providing technical support in implementing non-structural measures in urban stormwater management, such as flooding area delineation and awareness/education of the public.

According to Walesh (1989), the objectives of urban stormwater management include the following:

- The protection of life and public health, particularly related to lessening of flooding and drainage of standing waters
- The protection of damage to public and private property resulting from flooding
- The minimization of disruption of human activities, as a result of flooding
- The protection of receiving water quality impairment (both of surface and groundwater) from pollutants carried by runoff (non–point source pollution)
- The enhancement of aesthetics and, generally, quality of life in urban areas by incorporating natural features into the surface drainage system (e.g. open grassed swales and channels, grass filters, ponds, wetlands), which add recreational, aesthetic and ecological values.

7.1.2 History

Even though advanced masonry or clay ceramic urban storm and sanitary sewers have been used 4000 years ago in ancient Greece, e.g. during the Minoan civilization (Angelakis et al., 2005), sanitary sewer use in western Europe and the United States was not common until the mid-nineteenth century; wastewater was then freely discharged onto the ground surface, even though storm sewers were in operation. This practice resulted in the occurrence of infectious diseases, such as typhoid fever, dysentery and cholera. The medical doctors could not address the problem, since this was mostly due to infection from consuming contaminated water. The engineers provided the solution to the problem, around the middle of the nineteenth century, by constructing combined sewers. This practice continued until early in the twentieth century, when combined sewer overflow (CSO) and resulting environmental problems led to the construction of separate sewer lines. Nevertheless, old cities or parts of modern cities still operate combined sewers even today; attempts are made, however, to resolve CSO problems. Even though separate sewer lines and also wastewater treatment plants (WWTPs) have been constructed, at least in developed countries, pollution of surface water bodies continues to be a problem. This has identified non–point source pollution, from both agricultural and urban areas, as the main cause of water quality degradation of aquatic systems and the main problem currently addressed in urban stormwater management (Walesh, 1989).

The evolution of urban stormwater management is depicted in Table 7.1. In short, the table presents early human activities to drain wetlands and use floodplains for agricultural and urban development, resulting in the reduction of the natural flood dissipation features of streams and rivers. Later activities include the construction of major dams for flood control and hydropower generation, which provided flood protection of major rivers, even though this, in most cases, caused significant erosion and ecological impacts in downstream reaches and the coastal environment. Further urban growth, with street paving and curb and gutter construction, resulted in the shortening of time of concentration and the increase in hydrograph peaks. As a result, local flooding problems occurred, leading to the necessity for enlargement of sewer lines and/or the use of detention basins. In the 1990s, stormwater-related quality problems of receiving waters were identified, leading to the necessity for stormwater quality management. Today, the tendency is to promote 'natural' flood and pollution control measures, which also include recreational values, and present more aesthetic and ecological methods in urban storm water management.

Table 7.1 Historical development in urban stormwater management

Milestone	Indicative time scale	Impact-result
Natural environment – wetlands: villages and small cities	150 years ago	Limited
Drainage of wetlands – agricultural and urban development on floodplains	100 years ago	Reduction of natural flood dissipation ability – ecological impacts
Dam construction for flood control and hydropower		Flood protection – erosion and ecological impacts
Urban growth – pavement of roads, street and gutter drainage, combined sewers, separate storm sewers	1970s	Shortening of time of concentration, increase of peaks, local flooding – quality problems
Enlargement of sewers – use of detention basins		Stormwater quantity management – quality problems
Quality problems emerge	1990s	Stormwater quality management
Use of more 'natural' systems – proper maintenance	Today	Effective stormwater management

Source: Adapted from Debo and Reese, 2002.

7.1.3 Description of the urban stormwater drainage system

The urban stormwater drainage system aims to

- Provide the means for enhancing water infiltration and minimizing runoff quantity
- Effectively and safely carry the stormwater runoff quantity
- Provide storage for excess runoff to avoid local flooding
- Provide means of retaining or minimizing pollutants carried by runoff to enter receiving waters

To address these four aims, the stormwater drainage system comprises various components categorized into five major categories:

1. Those where runoff is produced, such as rooftops, parking lots and pervious areas.
2. Those providing increased infiltration of rainwater: less impervious areas result in less runoff volume and lower peaks. As a consequence, the drainage system costs less and fewer pollutants reach receiving waters. Incorporating more grassed areas in the urban environment, constructing surface drainage systems (e.g. swales) and building permeable hard surface (e.g. porous pavements, infiltration trenches) all lead to increase in infiltration.
3. Those used to safely convey runoff, such as open swales and ditches, culverts, street cross sections, inlets and catch basins, and storm sewers. Regarding the sewer system, as mentioned, three types of lines may exist in the urban environment: separate storm sewers (they only carry runoff) and sanitary sewers (they carry wastewater), which is the preferred and current practice, and combined sewers (they carry both wastewater and runoff).
4. Those used to store runoff, such as detention and retention basins and storage tanks.
5. Those used to control sediments and other pollutants carried by runoff, such as sedimentation basins, erosion control structures, dry and wet basins and constructed wetlands.

7.1.4 Impacts of urbanization

Urbanization of an area, i.e. the transformation of a natural area into a city, has a major impact on both the quantity and quality of runoff if measures are not taken (Novotny and Olem, 1994). Regarding the quantity of runoff, the common effect is the increase in both the volume and the peaks of runoff hydrographs, with a consequent increase of water stages in drainage channels, increased risk of flooding and expansion of inundation areas. Increases in volume (and peak) are a result of changing land cover by paving roads and constructing buildings and parking lots, i.e. by increasing the impervious area of the urban basin compared to the natural basin, something that decreases infiltration and other losses. This is demonstrated in the hydrographs of Figure 7.1a. Two of the hydrographs correspond to an undeveloped (natural or agricultural) watershed (A) and an urbanized watershed with a surface drainage system using open swales or grass side ditches (B). The figure shows that both the volume of runoff (area under the hydrograph curve) and the peak are increased due to reduced infiltration. However, the time of the peak and the time base of the two hydrographs are comparable (because the time of concentration in the two cases is comparable). A further increase in peak (hydrograph C; Figure 7.1a) is a result of shortening the time of concentration in the urbanized case, and this is mainly an outcome of concentrating runoff in concrete open channels and/or storm sewers, where water flows with increased velocities compared to overland flow and flow in swales or natural channels. All these can be easily understood by also comparing the hydrographs of Figure 7.1b, which can also be used to understand the effect on hydrograph peak and shape in two urbanized watersheds. The two hydrographs now correspond to two hypothetical watersheds A and B where all watershed characteristics are the same except one that is different, as summarized in Table 7.2.

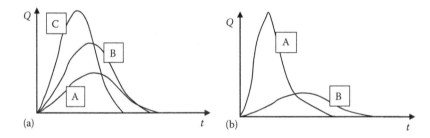

Figure 7.1 (a) Impact of urbanization: (A) undeveloped, (B) urbanized with surface drainage through swales or side ditches, (C) urbanized with drainage in concrete sewers and open channels; (b) comparison of the resulting hydrographs from the same rainfall in two similar watersheds differing in only one characteristic, as summarized in Table 7.2.

Table 7.2 Comparison of two hypothetical watersheds of Figure 7.1b

Feature of comparison	Watershed A	Watershed B
Urban versus natural	Urbanized, drainage on paved roads, in storm sewers and/or lined channels Urban watershed	Natural or agricultural
Total impervious area	High	Low
Hydraulically connected impervious area	High	Low
Watershed slope	Steep	Mild
Storm sewer or channel slope	Steep	Mild
Storm sewer or channel roughness	Low	High

In all cases, the rainfall is considered the same, and the shape and area of the two watersheds are the same.

Similar to runoff quantity, urbanization has significant adverse effects on the quality of the water bodies receiving the runoff (Novotny and Olem, 1994). This is a result of changes in pollutant composition and content in the atmosphere above the urban area, hydrology and drainage, soils, land uses and anthropogenic activities, among others. Pollutants come from various sources such as the industry, municipal wastewater, solid wastes and air pollutant emissions. During the construction, relatively high amounts of solids are produced from erosion and earth movement (estimated at 100 tons/ha/year). These solids end into water bodies causing turbidity and possibly pollution (they may also carry adsorbed compounds) and, consequently, adverse ecological impacts (Novotny and Olem, 1994).

In addition to changing the shape of hydrographs and the resulting drainage/flooding problems, pollution is the other major impact of urbanization. Stormwater runoff pollution is of the non–point source or diffuse type. The characteristics of this type of pollution make it difficult to address, because its sources are widespread over the drainage area, and pollutant loads cannot be easily quantified. More specifically, non–point source pollution has the following characteristics (Novotny and Olem, 1994):

- It enters the aquatic systems at various points, at intermittent time intervals, closely related to the variation to meteorological factors.
- The pollutants are produced in a wide area and are carried through surface runoff for some time before reaching the receiving surface water body or groundwater.
- The pollution is hard to control at the point it enters the water body, because of the large number of entry points and the significantly high volumes/rates of runoff, which would require very large and expensive structures.
- Treatment of runoff analogous to wastewater treatment is a very uncommon option. Non–point source pollution is controlled through measures applied throughout the watershed, aiming to minimize pollutants at the source and/or retain them in special structures. These measures, called best management practices (BMPs), are based on managing land use, anthropogenic activities and surface runoff, and either involve construction of a facility or are of the non-structural type.
- Restoration of the receiving waters quality is a much more expensive and probably a completely unrealistic option compared to the use of BMPs.
- Non–point source pollution depends on meteorology, extreme meteorological or climatic events, geographic and geologic conditions, and varies from place to place and throughout time. Therefore, implementation of control methods is site specific and cannot directly be transferred from one area to another.

Main urban non–point source pollutants include the suspended sediments, nutrients, heavy metals and various toxic substances. These pollutants are results of traffic and other anthropogenic activities and wet or dry deposition from the atmosphere; they are accumulated on the ground surface, washed off by rainfall and transported by runoff. The quantities of pollutants also depend on the drainage system. For example, it is estimated that the pollutant load from residential areas with natural drainage without sewers (e.g. using grassed swales) is significantly less than that from similar watersheds with the same land use and land cover but with drainage by sewers (Novotny and Olem, 1994).

As mentioned, the urban drainage can be separate (i.e. runoff and wastewater flow in separate lines) or combined sewers (the two flow in the same line). The first case is the modern and preferred practice, but there are still combined sewers in operation in various cities of the world. Combined sewers operate normally during dry weather, carrying sewage

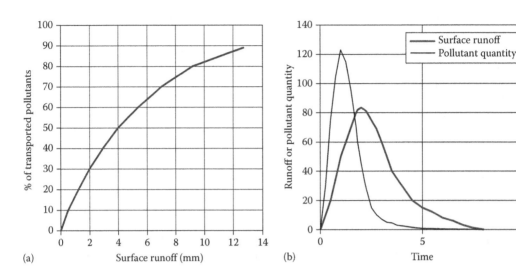

Figure 7.2 (a) Percent of transported pollutants as a function of surface runoff depth; (b) hydrograph and pollutograph qualitatively indicating first-flush phenomenon.

and nuisance water to the WWTP; in this case, no environmental problem occurs because wastewater is treated. However, during wet weather (when the extra sewer collects surface runoff and wastewater), occurrence of two distinct phases is possible: first, at the beginning of surface runoff or when rainfall intensity and duration are small, the combined flow rate is small and mixed domestic sewage and runoff end up in the WWTP and can be treated. Thus, there is no environmental problem in this case. Second, in the case of high-intensity rainfall, the discharge may exceed the design flow rate (capacity) of the WWTP; in this case, a hydraulic control structure (flow divider) prevents entry to the treatment plant, leading flow into the receiving waters. As a result, mixed runoff and wastewater are transferred directly (i.e. without treatment) in the receiving water body, a phenomenon called 'combined sewer overflow' (CSO). CSO can be quite dangerous for ecosystems, fish and humans, since high pollutant and microbial quantities and concentrations enter the water body in relatively short time. In developed countries, during CSO, beaches close to bathing, and fish and mollusc consumption is forbidden. Research has shown that nitrogen and phosphorus loadings from urban areas with combined sewers can be double compared to those from similar areas with separate sewers (Novotny and Olem, 1994).

USEPA (1974) first proposed that pollutants in surface runoff are not transported uniformly during the hydrograph. The largest portion of pollutants is transported during the first 13–25 mm of runoff, as schematically shown in Figure 7.2. This phenomenon is called first flush. Further research (e.g. Ellis, 1986) concluded that this phenomenon is not present in every rainfall event; however, it is a conceptually rational phenomenon and is used in runoff quality management.

7.1.5 Factors influencing urban runoff quality

Several factors influence the quality of surface runoff including (Chui et al., 1982; Tsihrintzis and Hamid, 1997a) rainfall quantity and quality characteristics such as depth, intensity, duration, number of days with no rain and constituent concentrations, traffic volume, land use and cover, geologic characteristics, geographic location, street maintenance practices (e.g. sweeping, flushing) and the type and geometry of the drainage system (i.e. surface drainage, separate or combined sewers).

The number of dry days before the rainfall determines pollutant accumulation on the surface (Huber and Dickinson, 1988). The rainfall intensity affects the volume and discharge of surface runoff, which then determine the rate of washoff, the concentration of the accumulated pollutants and their transport to the receiving aquatic system. In the case of highway runoff, accumulation and washoff of traffic-related pollutants is a function of traffic volume. Land use and land cover are the most important factors determining the type of pollutants in surface runoff. Human activities in the watershed, e.g. land development and construction, contribute to additional pollutant loadings in runoff (Gupta et al., 1981; Boyd and Gardner, 1990). The percent imperviousness of the watershed is an important factor affecting runoff volume and pollutant quantities (Griffin et al., 1980). Geographic, geological and meteorological conditions of the study area should also be taken into consideration.

7.1.6 Pollutant generation processes

The primary processes leading to production of pollutants and quality problems of receiving waters are the following (Boyd and Gardner, 1990; Tsihrintzis and Hamid, 1997a):

Atmospheric scrubbing: Atmospheric particulates from various sources (i.e. dust, soot, aerosols and emitted gases) come down with precipitation and become part of the runoff (Randall et al., 1981).

Scour and erosion: They result from rainfall drop impact over exposed land surfaces, which loosens the soil structure, and subsequent entrainment and transport by runoff. In addition, pollutants adsorbed on soil grains (e.g. fertilizers and pesticides) become part of the runoff, ending in the receiving waters.

Surface washoff: Stormwater washes off all the accumulated pollutants into the receiving water. In urban areas, especially on roads and highways, the washoff volume is relatively high because of larger impervious surface such as parking lots, streets and sidewalks.

Deposition: Sediments and pollutants eroded from pervious surfaces and carried by runoff may deposit by adhering to the surrounding surfaces, settling out and through processes that remove them from the main runoff stream.

Transport and transformation: The transport process moves pollutants between locations and the transformation process alters pollutant characteristics as a result of physical, chemical and/or biological mechanisms. Typical transport processes include advection and diffusion, and typical transformation processes include biodegradation and photolysis (Ferrara, 1986).

7.1.7 Types and sources of pollutants

The following pollutants are typically found in stormwater runoff (Tsihrintzis and Hamid, 1997a):

Suspended solids: Street dust and eroded sediments may make the water turbid or cloudy. Sediment particles may deposit in the bottom, smothering incubating fish eggs, filling spaces between rocks used for cover and disturbing aquatic habitat. In addition, pollutants tend to adhere to sediment particles, particularly to fine-sized ones. Waller and Hart (1986) reported that solid loads in urban stormwater runoff are much higher than loads found in treated sewage.

Heavy metals: Vehicles are a major source of contribution of metals to the environment. Concentration of heavy metals in stormwater runoff is a multiple of that in sanitary sewage. Norman (1991) measured several metals in the Ohio river, such as lead, zinc, copper, chromium, arsenic, cadmium, nickel, antimony and selenium. Certain heavy metals (e.g. copper, cadmium, lead and zinc) are more soluble in water than others and may cause toxic effects at concentrations exceeding threshold values. Industrial and commercial land uses contribute most heavy metals in runoff. These include lead and copper, which are predominantly associated with particulates, and thus, their concentration relates well to that of suspended solids. It is estimated that 40%–75% of metals are derived from highway runoff sources (Hunter et al., 1979). The most toxic compounds originate from traffic. Sources include lead from the use of leaded fuel, lead oxide and zinc from tire wear and copper, and chromium and nickel from wear of the plating, bearings, brake linings and other moving parts of a vehicle.

Chlorides: Salts are applied during the early phase of snowfall to prevent bonding of snow with the pavement. Typical salt quantities in the United States range from 180 to 550 kg per two-lane street per mile (Hoffman et al., 1981). Repeated applications at these rates over the entire winter season end in high amounts of chlorides, which are washed off into the aquatic systems. Streams crossing salted highways have been found to contain higher chloride concentrations downstream of the crossing than upstream, with peak levels occurring between winter and early spring.

Others constituents: Oils, grease and other related hydrocarbons are frequently found in highway runoff. It is estimated that 4.2×10^9 L of automobile and industrial lubricants are lost to the environment annually in the United States, either directly by disposal to sewers and application to land or indirectly by spillage or leaking from vehicles (Hunter et al., 1979; Bomboi et al., 1990). Petroleum-derived hydrocarbons are regularly released into the environment in proportions related to surrounding urbanization and development (Stenstrom et al., 1984). Parking lots are believed to have the highest loading factor, approximately 25 times higher than that of residential areas. Hydrocarbons are emitted from vehicular exhaust systems, are suspended in the atmosphere, get scrubbed by rainfall and become part of the runoff. The primary hydrocarbon loadings (particulates and lubricant oils) in runoff originate from vehicle exhaust systems. Unresolved complex mixtures (UCMs), constituting approximately 80% of the total content of aliphatic hydrocarbons in urban runoff, have also been found in most samples (Boyer and Laitinen, 1975; Bomboi and Hernandez, 1991).

It is estimated that some 50% of solids and 70% of the total polycyclic aromatic hydrocarbons (PAHs) entering aquatic systems originate from highway runoff (Ellis, 1986). Increased PAH and *n*-alkane loadings in urban runoff correlated well with increased urbanization and traffic density (Bomboi and Hernandez, 1991; Hewitt and Rashed, 1992).

The annual BOD_5 loading into a water body from urban runoff is approximately equivalent to that of a secondary WWTP effluent (Gupta et al., 1981). The common sources of BOD_5 are usually vegetation, litter and garbage, and animal waste. Typical mean BOD_5 concentrations of 12 mg/L were observed in stormwater discharges from the Nationwide Urban Runoff Program (NURP) studies (USEPA, 1983).

7.1.8 Impacts to the receiving waters

The impact of urban runoff pollutants on the water quality of aquatic systems depends on its prior water quality and the rates at which these pollutants enter the system. Loadings entering in relatively short periods of time (shock loadings) usually bring dramatic changes in water quality (Tsihrintzis and Hamid, 1997a). These changes may become permanent if shock loadings are frequent during the year. As these pollutants travel along a river, they may settle down at various distances from the original point of introduction, and slowly start affecting the local environment. Toxic substances that dissolve slowly, i.e. heavy metals and PCBs, usually show such characteristics (Gjessing et al., 1984).

The various compounds have different types of impacts on the water quality of the aquatic system (Tsihrintzis and Hamid, 1997a). For example, lead bioaccumulates on the bottom and may retard fish growth and reduce photosynthesis. At certain concentrations, zinc and copper are toxic to fish and aquatic micro-invertebrates. Cadmium and chromium have shown mutagenic and carcinogenic effects. Biochemical oxygen–demanding material can deplete the oxygen level in the aquatic ecosystem, resulting in decrease in the fish population. Excess nutrients, such as nitrogen and phosphorus, cause algal blooms which block sunlight and consume oxygen as they decompose. Oil and grease disposed of from vehicles is toxic to aquatic organisms and affects fish reproduction. Total suspended solids in water increase turbidity level, affecting fish survival (Pitt and Bozeman, 1980; Ferrara, 1986; Schueler, 1987).

7.2 URBAN RUNOFF QUANTITY COMPUTATIONS

Typical methods commonly used in urban runoff hydrologic computations and sewer sizing are the rational method, the SCS curve number method and the unit hydrograph method. The emphasis will be given here to the first two, since they are the most commonly used. The unit hydrograph method has been presented in Section 3.7.

7.2.1 Rational method

The rational method is based on the following formula:

$$Q_p = \frac{1}{360} C_f \, C \, I A \tag{7.1}$$

where
   Q_p is the flow peak of the hydrograph (m³/s)
   C is the surface runoff coefficient
   C_f is the correction to the surface runoff coefficient accounting for the design return
      period
   I is the rainfall intensity (mm/h)
   A is the surface area of the drainage basin (ha)
   (1/360) is a unit conversion factor.

The surface runoff coefficient C is conceptually defined as the ratio of total runoff over total rainfall for a given event. Therefore, it only assumes values less or equal to 1.0. The value of C depends on various factors, such as the percent imperviousness of the watershed, the soil cover, the land use, the slope, the antecedent rainfall, the depression storage, the soil type, the soil moisture, the shape of the watershed and the rainfall intensity, among others. However, the main factors that determine the C-value in practice are the first three, i.e. imperviousness, cover and land use. To guide the hydrologist, several authorities have produced tables containing typical values of runoff coefficient (Table 7.3); these values are valid for storm

Table 7.3 Typical values of surface runoff coefficient C

Land use description	Surface runoff coefficient C	
	Low	High
Commercial		
Urban centre – downtown	0.70	0.95
Neighbourhood areas	0.50	0.70
Residential		
Single family	0.30	0.50
Multiunits, detached	0.40	0.60
Multiunits, attached	0.60	0.75
Suburban	0.25	0.40
Condominium, apartments	0.50	0.70
Industrial		
Light	0.50	0.80
Heavy	0.60	0.90
Parks, cemeteries	0.10	0.25
Construction sites, railroad yards	0.20	0.35
Natural – undeveloped	0.10	0.30
Streets		
Asphalt, concrete	0.70	0.95
Brick, stone	0.70	0.80
Roof	0.70	0.95
Lawns, sandy subsoil		
Flat, <2%	0.05	0.10
Mild slope, 2%–7%	0.10	0.15
Steep slope, >7%	0.15	0.20
Lawns, compacted subsoil		
Flat, <2%	0.13	0.17
Mild slope, 2%–7%	0.18	0.22
Steep slope, >7%	0.25	0.35

Source: ASCE, Design and construction of sanitary sewers, Manual of Practice No. 37, 1970.

Note: The values of C are valid for return periods of less than 10 years. For larger return periods, C_f is greater than 1.0 and is used for correction of C.

events up to a return period of 10 years. The C_f coefficient is used to correct the C values taken from such tables for less frequent storm events. It assumes the values 1.00, 1.10, 1.20 and 1.25 for the 1- to 10-, 25-, 50- and 100-years return periods, respectively. However, to have physical meaning, the product $C_f \cdot C$ should always be less or equal to 1.0.

When using the rational formula, the rainfall intensity is obtained from *idf* curves (see Section 4.8), where the intensity is a function of both the rainfall duration and the return period. Typical such curves are shown in Figure 7.3. The design duration is selected as equal to the time of concentration at the particular point of the drainage basin where the calculation of the peak discharge is made. This assumption results in maximum intensity and it is adequate for calculating the peak discharge. However, it may not be conservative for designing storage structures.

The time of concentration of an urban drainage basin is qualitatively defined as the time needed for a drop of rainwater to move from the 'hydraulically' most remote point of the basin to the point of concentration (i.e. the point of interest where a peak estimate is attempted).

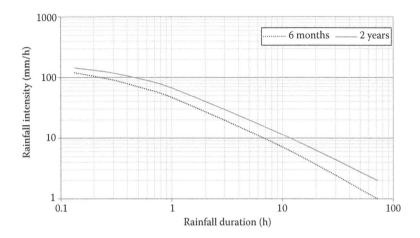

Figure 7.3 Typical *idf* curve.

This time comprises three distinct times: the time of overland flow (T_o), the time of street flow (T_s) and the time of conduit flow, i.e. in sewers or open channels (T_L), as follows:

$$T_c = T_o + T_s + T_L \tag{7.2}$$

Overland flow is defined as flow on a relatively flat plain surface, characterized by small depth and large width, and relatively parallel streamlines. In a natural watershed, it occurs at the upper mountaintop nearly flat areas, where gullies and small channels have not been formed. In an urban area, it occurs on rooftops, parking lots and similar flat paved or unpaved surfaces. The overland time of concentration T_o is calculated from empirical formulas developed for this purpose (Gupta, 1989). The most common ones are the following (in all T_o is computed in hours):

1. The following is the *Kerby (1959) equation*:

$$T_o = 3.03 \left(\frac{\delta L^{1.5}}{H^{0.5}} \right)^{0.467} \tag{7.3}$$

where
   L is the length of flow travel path (km), and the equation is valid for $L < 0.4$ km
   H is the elevation difference between the upper and lower points along the travel path of overland flow (m)
   δ is the coefficient of soil cover, which assumes the following values:

δ value	Description
0.02	Smooth pavement
0.10	Bare compacted soil
0.30	Rough bare soil or sparse grass
0.40	Medium dense grass
0.80	Dense grass, trees

2. The following is the *Izzard (1944) equation*:

$$T_o = \frac{\left(0.024I^{0.33} + \dfrac{878\varepsilon}{I^{0.67}}\right)L^{0.67}}{C^{0.67}H^{0.335}}$$

(7.4)

where

I is the rainfall intensity (mm/h), and Equation 7.4 is valid for $IL < 3.8$
C is the runoff coefficient
ε is the coefficient of soil cover, which assumes the following values:

ε value	Description
0.007	Smooth asphalt
0.012	Concrete pavement
0.017	Tar and gravel pavement
0.046	Sparse grass
0.060	Dense grass

3. The following is the *Bransby Williams (1922) equation*:

$$T_o = \frac{0.96L^{1.2}}{H^{0.2}A^{0.1}}$$

(7.5)

where A is the surface area of the overland flow area (km²).
4. The following is the *Federal Aviation Agency (1970) equation*:

$$T_o = \frac{3.64(1.1 - C)L^{0.83}}{H^{0.33}}$$

(7.6)

5. The following is the Manning kinematic wave equation (Overton and Meadows, 1976; Gupta, 1989):

$$T_o = \frac{0.0913(NL)^{0.8}}{P_{2-year}^{0.5}S^{0.4}}$$

(7.7)

where N is the Manning roughness coefficient for overland flow. Values for this coefficient are presented in Table 7.4. As seen, these values are significantly high compared to respective values for open channel flow (see following Table 7.5); S is the slope along the path of overland flow (m/m); and P_{2-year} is the depth of the rainfall with duration 24 h and return period 2 years (mm). L in Equation 7.7 is expressed in (m) and is limited to less than about 90 m.

In the case when the equation used contains the rainfall intensity I (e.g. Izzard's equation, Equation 7.4), for the computation of T_o, one has to also use the *idf* diagram (Figure 7.3) in a trial-and-error procedure, according to the following steps: (1) One assumes a value for T_o; (2) from the *idf* diagram, for a given return period, and using the assumed T_o value for duration, one can get a value for the intensity I; (3) using this I value in Equation 7.4,

Table 7.4 Values of Manning roughness coefficient N for overland flow

Condition	N value		
	Min	Normal	Max
Concrete	0.010	0.011	0.013
Asphalt	0.010	0.012	0.015
Bare sand	0.010	0.010	0.016
Gravel	0.012	0.012	0.030
Bare clay loam	0.012	0.012	0.033
Turf	0.39	0.45	0.63
Short grass	0.10	0.15	0.20
Dense grass	0.17	0.24	0.30
Trees, forest	0.30	0.45	0.48

Source: Wanielista, M. et al., *Hydrology: Water Quantity and Quality Control*, 2nd edn., John Wiley & Sons, New York, 1997.

Table 7.5 Values of Manning roughness coefficient n for conduits and channels

Conduit material	n value		
	Min	Normal	Max
Brass, smooth	0.009	0.010	0.013
Steel, welded	0.010	0.012	0.014
Steel, riveted	0.013	0.016	0.017
Cast iron, coated	0.010	0.013	0.014
Cast iron, uncoated	0.011	0.014	0.016
Wrought iron, black	0.012	0.014	0.015
Wrought iron, galvanized	0.013	0.016	0.017
Corrugated metal, small corrugations	0.020	0.022	0.025
Corrugated metal, large corrugations	0.030	0.032	0.035
Smooth wall, spiral aluminium	0.010	0.012	0.014
Concrete pipe, straight	0.010	0.012	0.013
Concrete pipe with curves	0.011	0.013	0.014
Storm sewer, straight	0.013	0.015	0.017
Sanitary sewer	0.012	0.013	0.016

Source: Wanielista, M. et al., *Hydrology: Water Quantity and Quality Control*, 2nd edn., John Wiley & Sons, New York, 1997.

one gets a value for T_o; and (4) this new T_o value is compared to the one assumed in the first step, and if not different, one stops; otherwise the procedure is repeated from the first step.

Travel times on street T_s and in conduit T_L are computed using the Manning equation:

$$Q_p = \frac{1}{n} A_L R^{2/3} S^{1/2} \tag{7.8}$$

or

$$V = \frac{1}{n} R^{2/3} S^{1/2} \tag{7.9}$$

These are combined with the equation

$$T_s \quad \text{or} \quad T_L = \frac{L}{V} \tag{7.10}$$

to get

$$T_s \quad \text{or} \quad T_L = \frac{L}{(1/n)R^{2/3}S^{1/2}} \tag{7.11}$$

where
    Q_p is the peak discharge flowing in the conduit (m³/s)
    V is the mean velocity in the street or conduit (m/s)
    L is the length of travel in the street or conduit (m)
    S is the slope of the street or conduit (m/m)
    A_L is the flow section area in the street or conduit perpendicular to the flow direction (m²)
    R is the hydraulic radius defined as the quotient of the flow section area over the wetted
        perimeter P ($=A_L/P$) (m)

The wetted perimeter is defined as the length of the perimeter of the street section or conduit section which is in contact with water; n is the Manning roughness coefficient for open channel flow. Values for n are taken from tables as a function of the conduit material (Table 7.5).

The use of Manning equation is straightforward, based on geometry, for the cases of normal conduit sections, such as rectangular and trapezoidal. Figure 7.4 presents geometric definitions for these two sections. The needed geometry equations to apply Manning equation for these two cases are summarized in Table 7.6, where B is the width (m) of the base of the section, d is the flow depth (m) and z_1 and z_2 are the slopes (i.e. ratio of horizontal to vertical units) of the sides of the trapezoidal section. The triangular section is a special case of the trapezoidal section where $B = 0$.

Particularly for the case of circular pipes, the geometry is more cumbersome and the solution of Manning equation can be aided by the use of Table 7.7, where d (m) is the depth of

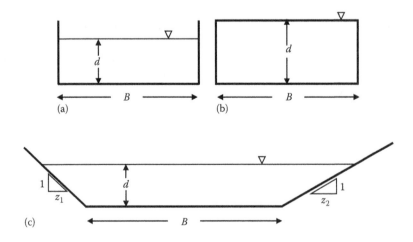

Figure 7.4 Common geometries for channel sections. (a,b) Rectangular section and (c) trapezoidal section.

Table 7.6 Geometry equations useful for the solution of Manning equation for rectangular and trapezoidal channel sections

Section	Flow area A_L (m²)	Wetted perimeter P (m)
Rectangular – open (Figure 7.4a)	$A_L = Bd$	$P = B + 2d$
Rectangular – closed (Figure 7.4b)	$A_L = Bd$	$P = 2B + 2d$
Trapezoidal (Figure 7.4c)	$A_L = \dfrac{2B + (z_1 + z_2)d}{2}d$	$P = B + d\left(\sqrt{1+z_1^2} + \sqrt{1+z_2^2}\right)$

Note: Parameter definitions in Figure 7.4.

flow and D (m) is the pipe diameter. The use of this table is explained in the following. The Manning Equation 7.8 can be written as follows:

$$\frac{A_L R^{2/3}}{D^{8/3}} = \frac{Q_p n}{S^{1/2} D^{8/3}} \tag{7.12}$$

When the flow Q_p, the Manning roughness coefficient n, the slope S and the diameter D are known, then the right-hand side (RHS) of Equation 7.12 is known, and thus, the left-hand side (LHS), i.e. parameter $(A_L R^{2/3})/D^{8/3}$, can be computed. Table 7.7 can then be used to compute both parameters d/D and R/D, from which one can get the flow depth d and the hydraulic radius R. Similarly, if d or R is known, the table can be used to compute any unknown parameter.

Finally, the Manning equation can be used in any channel section, for example, an irregular section (Figure 7.5). In this case, the cross section should be defined by the known coordinates (x,y) of the various points defining the section. To apply Manning equation, one has to create a table where the flow area A_L (m²), wetted perimeter P (m), hydraulic radius R (m) and flow Q (m³/s) are computed as function of the vertical coordinate y (m) assuming various depths d (m) from $y = 0$ up to the upper limit of the cross section (full section). This procedure creates a rating curve for the section, where the flow rate Q (m³/s) is related to various water depths d; then, the solution of Manning equation is straightforward through the use of the table or rating curve. In the case of the cross section of Figure 7.5, and assuming a slope $S = 0.0001$ and Manning roughness $n = 0.050$, one gets the values shown in Table 7.8.

In addition to the computation of the peak discharge, the rational method provides the possibility to derive a hydrograph, which is required when there is need to compute a pollutograph, or to design storage and/or pollution control structures. The rational method hydrograph has a triangular shape with a time base equal to $2T_c$ and a height equal to Q_p.

7.2.2 SCS method

This method was developed by the Soil Conservation Service (SCS; currently: Natural Resources Conservation Service or NRCS) of the U.S. Department of Agriculture (USDA-SCS, 1985; USDA-NRCS, 2015) and is one of the most widely used hydrological methods (Ponce and Hawkins, 1996). According to this method, surface runoff is computed for four hydrologic soil categories (A, B, C and D) based on their drainage properties (soil hydrologic classification), as described in Table 7.9.

Regarding texture, soils are divided in three main categories: sand, silt and clay. SCS provides a more detailed classification in 11 categories, according to the percent content in sand, silt and clay in the soil sample, as shown in Figure 7.6. Table 7.10 presents soil categories and their correspondence to hydrologic soil classification and minimum infiltration rates.

Table 7.7 Solution aid of Manning equation for circular pipes

d/D	$\dfrac{A_L R^{2/3}}{D^{8/3}}$	R/D	d/D	$\dfrac{A_L R^{2/3}}{D^{8/3}}$	R/D
0.00	0.0000	0.0000	0.50	0.1558	0.2500
0.01	0.0000	0.0066	0.51	0.1611	0.2530
0.02	0.0004	0.0131	0.52	0.1665	0.2560
0.03	0.0008	0.0196	0.53	0.1718	0.2589
0.04	0.0011	0.0261	0.54	0.1772	0.2619
0.05	0.0015	0.0326	0.55	0.1825	0.2649
0.06	0.0025	0.0388	0.56	0.1878	0.2674
0.07	0.0035	0.0450	0.57	0.1932	0.2700
0.08	0.0045	0.0511	0.58	0.1985	0.2725
0.09	0.0055	0.0573	0.59	0.2039	0.2751
0.10	0.0065	0.0635	0.60	0.2092	0.2776
0.11	0.0082	0.0694	0.61	0.2145	0.2797
0.12	0.0100	0.0753	0.62	0.2198	0.2818
0.13	0.0117	0.0811	0.63	0.2252	0.2839
0.14	0.0135	0.0870	0.64	0.2305	0.2860
0.15	0.0152	0.0929	0.65	0.2358	0.2881
0.16	0.0176	0.0984	0.66	0.2408	0.2897
0.17	0.0200	0.1040	0.67	0.2458	0.2913
0.18	0.0225	0.1095	0.68	0.2508	0.2930
0.19	0.0249	0.1151	0.69	0.2558	0.2946
0.20	0.0273	0.1206	0.70	0.2608	0.2962
0.21	0.0304	0.1258	0.71	0.2654	0.2973
0.22	0.0335	0.1310	0.72	0.2701	0.2984
0.23	0.0365	0.1362	0.73	0.2747	0.2995
0.24	0.0396	0.1414	0.74	0.2794	0.3006
0.25	0.0427	0.1466	0.75	0.2840	0.3017
0.26	0.0464	0.1515	0.76	0.2881	0.3022
0.27	0.0500	0.1563	0.77	0.2922	0.3027
0.28	0.0537	0.1612	0.78	0.2963	0.3032
0.29	0.0573	0.1660	0.79	0.3004	0.3037
0.30	0.0610	0.1709	0.80	0.3045	0.3042
0.31	0.0652	0.1754	0.81	0.3078	0.3040
0.32	0.0694	0.1799	0.82	0.3112	0.3038
0.33	0.0736	0.1845	0.83	0.3145	0.3037
0.34	0.0778	0.1890	0.84	0.3179	0.3035
0.35	0.0820	0.1935	0.85	0.3212	0.3033
0.36	0.0866	0.1976	0.86	0.3240	0.3020
0.37	0.0912	0.2018	0.87	0.3268	0.3007
0.38	0.0958	0.2059	0.88	0.3296	0.2993
0.39	0.1004	0.2101	0.90	0.3324	0.2980
0.40	0.1050	0.2142	0.91	0.3331	0.2959
0.41	0.1100	0.2180	0.92	0.3339	0.2938
0.42	0.1149	0.2218	0.93	0.3346	0.2917

(Continued)

Table 7.7 (Continued) Solution aid of Manning equation for circular pipes

d/D	$\dfrac{A_L R^{2/3}}{D^{8/3}}$	R/D	d/D	$\dfrac{A_L R^{2/3}}{D^{8/3}}$	R/D
0.43	0.1199	0.2255	0.94	0.3353	0.2896
0.44	0.1248	0.2293	0.95	0.3349	0.2864
0.45	0.1298	0.2331	0.96	0.3303	0.2791
0.46	0.1350	0.2365	0.97	0.3256	0.2718
0.47	0.1402	0.2399	0.98	0.3210	0.2646
0.48	0.1454	0.2432	0.99	0.3163	0.2573
0.49	0.1506	0.2466	1.00	0.3117	0.2500

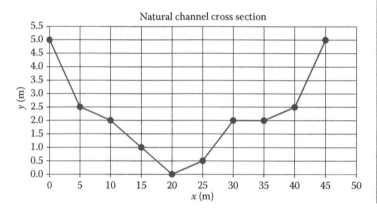

Point no.	x	y
1	0	5.0
2	5	2.5
3	10	2.0
4	15	1.0
5	20	0.0
6	25	0.5
7	30	2.0
8	35	2.0
9	40	2.5
10	45	5.0

Figure 7.5 Typical irregular open channel section; (x,y) coordinates are also shown.

Table 7.8 Computed flows as a function of depth (rating curve) for the cross section of Figure 7.5, for S = 0.0001 and n = 0.050

Depth	Flow area	Wetted perimeter	Hydraulic radius	Discharge
d (m)	A_L (m²)	P (m)	R (m)	Q (m³/s)
0.0	0.000	0.000	0.000	0.000
0.5	1.875	7.575	0.248	0.148
1.0	6.667	11.865	0.562	0.908
1.5	13.542	16.155	0.838	2.408
2.0	22.500	20.445	1.101	4.797
2.5	37.500	35.495	1.056	7.780
3.0	55.500	37.731	1.471	14.357
3.5	74.500	39.967	1.864	22.568
4.0	94.500	42.203	2.239	32.348
4.5	115.500	44.439	2.599	43.667
5.0	137.500	46.675	2.946	56.513

Table 7.9 Hydrologic soil classification according to SCS

Category	Description
A	Soils characterized by low possibility to produce surface runoff and high permeability. Mostly sand with gravel and very low contents of fine sand
B	Soils characterized by adequately low possibility to produce surface runoff and above mean permeability. Mostly sandy soils but finer than those of category A
C	Soils characterized by adequately high possibility to produce surface runoff and below mean permeability. Soils with significant clay content
D	Soils characterized by significantly high possibility to produce surface runoff and below mean permeability. Almost impermeable clay soils

Sources: U.S. Department of Agriculture, Soil Conservation Service (USDA-SCS), *National Engineering Handbook*, Section 4, Hydrology, USDA-SCS, Washington, DC, 1985; U.S. Department of Agriculture, Natural Resources Conservation Service (USDA-NRCS), *National Engineering Handbook*, 2012, http://www.nrcs.usda.gov/wps/portal/nrcs/detailfull/national/water/?cid=stelprdb1043063.

A combination of the SCS method (USDA-SCS, 1985) is now presented combined with the Santa Barbara Unit Hydrograph (SBUH) method (Stubchaer, 1975), as presented by Tsihrintzis and Sidan (1998). The SCS equation computes the cumulative surface runoff depth as a function of time during the rainfall event, according to the following equation:

$$\Sigma R(t) = \frac{[\Sigma P(t) - \lambda S]^2}{\Sigma P(t) + (1 - \lambda)S} \quad \text{for } \Sigma P(t) > \lambda S \tag{7.13}$$

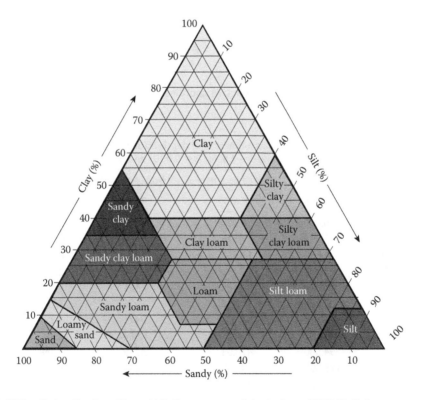

Figure 7.6 SCS soil classification. (From U.S. Department of Agriculture (USDA), Soil survey manual #18, U.S. Department of Agriculture (USDA), Washington, DC, 1951.)

Table 7.10 SCS soil texture classification

Texture category	Minimum infiltration rate (mm/h)	SCS hydrologic soil category
Sand	210.1	A
Loamy sand	61.2	A
Sandy loam	25.9	B
Loam	13.2	B
Silty loam	6.9	C
Sandy clay loam	4.3	C
Clay loam	2.3	D
Silty clay loam	1.5	D
Sandy clay	1.3	D
Silty clay	1.0	D
Clay	0.5	D

Source: Wanielista, M. et al., *Hydrology: Water Quantity and Quality Control*, 2nd edn., John Wiley & Sons, New York, 1997.

and

$$\Sigma R(t) = 0 \quad \text{for } \Sigma P(t) \le \lambda S \tag{7.14}$$

where

$\Sigma R(t)$ is the cumulative surface runoff depth as a function of time t from the beginning of the rainfall (mm)

$\Sigma P(t)$ is the cumulative rainfall depth as a function of time, i.e. the rainfall mass curve (mm)

S is the ultimate storage capacity or maximum retention of the watershed and the soil (mm)

The term $I_a = \lambda S$ represents the initial abstraction, i.e. interception, depression storage and infiltration occurring before the beginning of runoff (mm), and parameter λ is the ratio of the initial abstraction I_a to the maximum retention S, i.e. the portion of the total loss that is due to initial abstraction. Since the numerator of Equation 7.13 is always positive, this equation would predict a runoff depth even when the rainfall depth is less than the initial abstraction, i.e. $\Sigma P(t) \le \lambda S$. Equation 7.14 then states that this has no physical meaning, i.e. there is no runoff when the rainfall depth is less than the initial abstraction.

A typical value for parameter λ is 0.2 (USDA-SCS, 1985). This value resulted from measured data of I_a and S in experimental watersheds (USDA-SCS, 1985). However, when plotted, this empirical data showed significant spreading, and actually, 50% of the data points fell within the range $0.095 \le \lambda \le 0.38$. Several other studies (e.g. McKurk et al., 1980; Cazier and Hawkins, 1984; Bosznay, 1989; Ponce and Hawkins, 1996) have shown that λ can actually vary between 0 and 0.3.

The maximum retention S in Equation 7.13 is expressed by the following empirical equation:

$$S = \frac{25400}{CN} - 254 \tag{7.15}$$

where S in mm and CN is the curve number, an empirical parameter that depends on the soil type, the land use, the surface condition and the antecedent moisture condition (USDA-SCS, 1985; Wanielista, 1990; Ponce and Hawkins, 1996). CN varies between 0 (for a theoretical infinitely abstracting watershed) and 100 (for an impermeable watershed). Values for this parameter are provided in Tables 7.11 through 7.14.

Table 7.11 CN values for urban land uses[a]

Cover description cover type and hydrologic condition	Average percent impervious area[b]	CN for hydrologic soil group			
		A	B	C	D
Fully developed urban areas (vegetation established)					
Open space (lawns, parks, golf courses, cemeteries, etc.)[c]					
Poor condition (grass cover < 50%)		68	79	86	89
Fair condition (grass cover 50%–75%)		49	69	79	84
Good condition (grass cover > 75%)		39	61	74	80
Impervious areas					
Paved parking lots, roofs, driveways, etc. (excluding right-of-way)		98	98	98	98
Streets and roads					
Paved; curbs and storm sewers (excluding right-of-way)		98	98	98	98
Paved; open ditches (including right-of-way)		83	89	92	93
Gravel (including right-of-way)		76	85	89	91
Dirt (including right-of-way)		72	82	87	89
Western desert urban areas					
Natural desert landscaping (pervious areas only)[d]		63	77	85	88
Artificial desert landscaping (impervious weed barrier, desert shrub with 1- to 2 in. sand or gravel mulch and basin borders)		96	96	96	96
Urban districts					
Commercial and business	85	89	92	94	95
Industrial	72	81	88	91	93
Residential districts by average lot size					
1/8 acre or less (town houses)	65	77	85	90	92
1/4 acre	38	61	75	83	87
1/3 acre	30	57	72	81	86
1/2 acre	25	54	70	80	85
1 acre	20	51	68	79	81
2 acres	12	46	65	77	82
Developing urban areas					
Newly graded areas (pervious areas only, no vegetation)		77	86	91	94

Sources: U.S. Department of Agriculture, Soil Conservation Service (USDA-SCS), National Engineering Handbook, Section 4, Hydrology, USDA-SCS, Washington, DC, 1985; U.S. Department of Agriculture, Natural Resources Conservation Service (USDA-NRCS), National Engineering Handbook, 2012, http://www.nrcs.usda.gov/wps/portal/nrcs/detailfull/national/water/?cid=stelprdb1043063; U.S. Department of Agriculture, Natural Resources Conservation Service (USDA-NRCS), Urban hydrology for small watersheds, TR-55, Washington, DC, 1986.

[a] Average runoff condition, and $I_a = 0.2 S$.

[b] The average percent impervious area shown was used to develop the composite CNs. Other assumptions are as follows: impervious areas are directly connected to the drainage system, impervious areas have a CN of 98, and pervious areas are considered equivalent to open space in good hydrologic condition.

[c] CNs shown are equivalent to those of pasture. Composite CNs may be computed for other combinations of open space type.

[d] Composite CNs for natural desert landscaping should be computed based on the impervious area percentage (CN = 98) and the pervious area CN. The pervious area CNs are assumed equivalent to desert shrub in poor hydrologic condition.

Table 7.12 CN values for agricultural watersheds[a]

Cover type	Cover description treatment[b]	Hydrologic condition[c]	CN for hydrologic soil group			
			A	B	C	D
Fallow	Bare soil	–	77	86	91	94
	Crop residue cover (CR)	Poor	76	85	90	93
		Good	74	83	88	90
Row crops	Straight row (SR)	Poor	72	81	88	91
		Good	67	78	85	89
	SR + CR	Poor	71	80	87	90
		Good	64	75	82	85
	Contoured (C)	Poor	70	79	84	88
		Good	65	75	82	86
	C + CR	Poor	69	78	83	87
		Good	64	74	81	85
	Contoured & terraced (C&T)	Poor	66	74	80	82
		Good	62	71	78	81
	C&T + CR	Poor	65	73	79	81
		Good	61	70	77	80
Small grain	SR	Poor	65	76	84	88
		Good	63	75	83	87
	SR + CR	Poor	64	75	83	86
		Good	60	72	80	84
	C	Poor	63	74	82	85
		Good	61	73	81	84
	C + CR	Poor	62	73	81	84
		Good	60	72	80	83
	C&T	Poor	61	72	79	82
		Good	59	70	78	81
	C&T + CR	Poor	60	71	78	81
		Good	58	69	77	80
Close-seeded or broadcast legumes or rotation meadow	SR	Poor	66	77	85	89
		Good	58	72	81	85
	C	Poor	64	75	83	85
		Good	55	69	78	83
	C&T	Poor	63	73	80	83
		Good	51	67	76	80
Pasture, grassland or range-continuous forage for grazing[d]		Poor	68	79	86	89
		Fair	49	69	79	81
		Good	39	61	74	80
Meadow-continuous grass, protected from grazing and generally mowed for hay		Good	30	58	71	78
Brush-brush-forbs-grass mixture with brush the major element[e]		Poor	48	67	77	83
		Fair	35	56	70	77
		Good	30[f]	48	65	73

(Continued)

Table 7.12 (Continued) CN values for agricultural watersheds[a]

Cover type	Cover description treatment[b]	Hydrologic condition[c]	CN for hydrologic soil group			
			A	B	C	D
Woods–grass combination (orchard		Poor	57	73	82	86
or tree farm)[g]		Fair	43	65	76	82
		Good	32	58	72	79
Woods[h]		Poor	45	66	77	83
		Fair	36	60	73	79
		Good	30	55	70	77
Farmstead–buildings, lanes, driveways and surrounding lots		–	59	74	82	86
Roads (including right-of-way)						
Dirt		–	72	82	87	89
Gravel		–	76	85	89	91

Sources: U.S. Department of Agriculture, Soil Conservation Service (USDA-SCS), *National Engineering Handbook*, Section 4, Hydrology, USDA-SCS, Washington, DC, 1985; U.S. Department of Agriculture, Natural Resources Conservation Service (USDA-NRCS), *National Engineering Handbook*, 2012, http://www.nrcs.usda.gov/wps/portal/nrcs/detailfull/national/water/?cid=stelprdb1043063; U.S. Department of Agriculture, Natural Resources Conservation Service (USDA-NRCS), Urban hydrology for small watersheds, TR-55, Washington, DC, 1986.

[a] Average runoff condition, and $I_a = 0.2S$.

[b] Crop residue cover applies only if residue is on at least 5% of the surface throughout the year.

[c] Hydrologic condition is based on combinations of factors that affect infiltration and runoff, including (a) density and canopy of vegetative areas, (b) amount of year-round cover, (c) amount of grass or close-seeded legumes, (d) percent of residue cover on the land surface (good $\geq$ 20%), and (e) degree of surface toughness.

Poor: Factors impair infiltration and tend to increase runoff.

Good: Factors encourage average and better than average infiltration and tend to decrease runoff.

For conservation tillage poor hydrologic condition, 5%–20% of the surface is covered with residue (less than 750 pounds per acre for row crops or 300 pounds per acre for small grain).

For conservation tillage good hydrologic condition, more than 20% of the surface is covered with residue (>750 pounds per acre for row crops or 300 pounds per acre for small grain).

[d] *Poor*: <50% ground cover or heavily grazed with no mulch.

Fair: 50%–75% ground cover and not heavily grazed.

Good: >75% ground cover and lightly or only occasionally grazed.

[e] *Poor*: <50% ground cover.

Fair: 50%–75% ground cover.

Good: >75% ground cover.

[f] If actual curve number is <30, use CN = 30 for runoff computation.

[g] CNs shown were computed for areas with 50% woods and 50% grass (pasture) cover. Other combinations of conditions may be computed from the CNs for woods and pasture.

[h] *Poor*: Forest litter, small trees and brush are destroyed by heavy grazing or regular burning.

Fair: Woods are grazed, but not burned, and some forest litter covers the soil.

Good: Woods are protected from grazing, and litter and brush adequately cover the soil.

Tables 7.11 through 7.13 are used to select values that closely describe the conditions of the watershed. Table 7.14 is used to adjust the selected values according to the antecedent moisture conditions (AMCs) which are determined according to the guideline on the footnote of the table. Normal condition is AMC II, with AMC I being the dry condition and AMC III the wet case.

Equations 7.13 and 7.15 can be combined into the following equation to compute runoff:

$$\Sigma R(t) = \frac{[CN(\Sigma P(t) + 254\lambda) - 25,400\lambda]^2}{CN\{CN\ [\Sigma P - 254(1-\lambda)] + 25,400(1-\lambda)\}} \quad \text{for } \Sigma P(t) > \frac{25,400\lambda}{CN} - 254\lambda \quad (7.16)$$

Table 7.13 CN values for permeable surfaces in urban areas

Description	Hydrologic soil group			
	A	B	C	D
Bare ground	77	86	91	94
Gardens or row crop	72	81	88	91
Grass in good condition (cover exceeding 75% of the pervious area)	39	61	74	80
Grass in fair condition (cover 50%–75% of the pervious area)	49	69	79	84
Grass in poor condition (cover less than 50% of the pervious area)	68	79	86	89
Woods	36	60	73	79

Source: U.S. Department of Agriculture, Soil Conservation Service (USDA-SCS), *National Engineering Handbook*, Section 4, Hydrology, USDA-SCS, Washington, DC, 1972.

Table 7.14 Adjustment of CN value according to antecedent soil moisture conditions

Antecedent moisture condition (AMC)		
AMC II (normal)	AMC I (dry)	AMC III (wet)
100	100	100
95	87	98
90	78	96
85	70	94
80	63	91
75	57	88
70	51	85
65	45	82
60	40	78
55	35	74
50	31	70
45	26	65
40	22	60
35	18	55
30	15	50

Source: U.S. Department of Agriculture, Soil Conservation Service (USDA-SCS), *National Engineering Handbook*, Section 4, Hydrology, Washington, DC, 1972.

Note: AMC I: Relatively dry soils, rainfall in the preceding 5 days < 12.5 mm. AMC II: Normal case, rainfall in the preceding 5 days between 12.5 and 38 mm. AMC III: Relatively wet soils, rainfall in the preceding 5 days >38 mm.

and

$$\Sigma R(t) = 0 \quad \text{for } \Sigma P(t) \leq \frac{254,00\lambda}{CN} - 254\lambda \tag{7.17}$$

This equation shows that runoff is a function of both CN and λ. The instantaneous (incremental) runoff depth $R(t)$ (mm) at any time t (i.e. for a time period Δt) can be computed from the cumulative one (Equations 7.13, 7.14 or 7.16, 7.17) using

$$R(t) = \Sigma R(t) - \Sigma R(t - \Delta t) \tag{7.18}$$

and the instantaneous discharge (hydrograph) $Q_i(t)$ (m³/s) is given by the following equation:

$$Q_i(t) = \frac{1}{360} \frac{R(t)}{\Delta t} A \tag{7.19}$$

where

    A is the watershed surface area (ha)
    Δt is the time step of the hyetograph (h)

The final hydrograph is produced after routing the instantaneous hydrograph through the drainage basin. One appropriate method (Tsihrintzis and Sidan, 1998) is the SBUH method which was first presented by Stubchaer (1975). It is presented here in a modified form where the SCS method is used to compute the runoff depth, as presented in Equations 7.13 through 7.19. The instantaneous hydrograph routing is done using the following equation (Stubchaer, 1975):

$$Q_f(t) = Q_f(t - \Delta t) + K[Q_i(t - \Delta t) + Q_i(t) - 2Q_f(t - \Delta t)] \tag{7.20}$$

where

    $Q_f(t)$ is the routed hydrograph through the urban watershed (m³/s)
    K is the routing constant, which, according to Stubchaer (1975), is defined as

$$K = \frac{\Delta t}{2T_c + \Delta t} \tag{7.21}$$

where T_c is the time of concentration defined using the methods presented in previous sections.

7.3 URBAN RUNOFF QUALITY COMPUTATIONS

7.3.1 USEPA method for the prediction of annual unit pollutant loadings

Several authors have produced annual unit loadings of pollutants per unit area of the urban watershed. Marsalek (1978) provides values for BOD, nitrogen, phosphorus suspended solids and seven metals for urban areas with both separate and combined sewers, and for four land uses, leading to very low, low, medium or high pollutant production rates. Sonzogni et al. (1980) provide annual unit loadings for suspended solids, nitrogen phosphorus and four metals for four specific land uses and one general urban land use.

Similar work was presented by Heaney et al. (1977) for separate and combined sewers and is used by USEPA (1979) in a method to compute mean annual pollutant loadings in surface runoff. The method uses the following equation:

$$m_x = 0.04422\, \beta_x\, P_{year}\, \varphi \sigma \tag{7.22}$$

where

m_x is the annual unit loading of pollutant X (kg/ha/year)
β_x is the pollutant X loading coefficient, taken from Table 7.15
P_{year} is the mean annual rainfall depth (mm)
φ is the population density coefficient (Equations 7.23 through 7.25)
σ is the mechanical street sweeping coefficient (Equations 7.26 and 7.27)

The population density coefficient is given by the following equations. For residential areas,

$$\varphi = 0.142 + 0.218\,(0.4047\,P_d)^{0.54} \tag{7.23}$$

where P_d is the population density (capita/ha). For commercial/industrial land uses,

$$\varphi = 1.0 \tag{7.24}$$

and for other urban areas (e.g. parks, cemeteries, educational land uses),

$$\varphi = 0.142 \tag{7.25}$$

The mechanical street sweeping coefficient σ depends on the frequency of sweeping, i.e. the time interval T_s (days) between street sweepings. For $T_s > 20$ days,

$$\sigma = 1.0 \tag{7.26}$$

and for $T_s \leq 20$ days,

$$\sigma = \frac{T_s}{20} \tag{7.27}$$

Table 7.15 Loading coefficient β_x (Equation 7.22)

Land use	BOD_5	Suspended solids	Volatile solids	PO_4	N
Separate sewer systems					
Residential	0.799	16.3	9.4	0.0336	0.131
Commercial area	3.200	22.2	14.0	0.0757	0.296
Industrial area	1.210	29.1	14.3	0.0705	0.277
Other uses	0.113	2.7	2.6	0.0099	0.060
Combined sewer systems					
Residential	3.290	67.2	38.9	0.1390	0.540
Commercial area	13.200	91.8	57.9	0.3120	1.220
Industrial area	5.000	120.0	59.2	0.2910	1.140
Other uses	0.467	11.1	10.8	0.0411	0.250

Source: U.S. Environmental Protection Agency (USEPA), *1978 Needs Survey – Continuous Stormwater Pollution Simulation System, User's Manual*, EPA 430/9-79-004, U.S. Environmental Protection Agency, Washington, DC, 1979.

In addition to the mean annual unit loading, the method can also predict the mean annual concentration C_x of pollutant X. To do so, one needs the mean annual runoff depth, for which Heaney et al. (1977) propose the following equation:

$$R_{year} = 25.4 \left\{ \left[0.15 + 0.75 \left(\frac{I}{100} \right) \right] \left(\frac{P_{year}}{25.4} \right) - 5.234 \left(\frac{D_s}{24.4} \right)^{0.5957} \right\}$$ (7.28)

where

R_{year} is the mean annual surface runoff depth (mm)

I is the percent imperviousness of the area (%)

D_s is the depression storage (mm), which can be estimated using known methods or, alternatively, the following equation:

$$D_s = 25.4 \left[0.25 - 0.1875 \left(\frac{I}{100} \right) \right]$$ (7.29)

7.3.2 USGS method for the prediction of mean annual pollutant quantity

The USGS method is based on the development of regression equations from the statistical analysis of 2813 rainfall events in 173 stations in 30 urban areas in the United States (Driver and Tasker, 1990), resulting in the following equation:

$$M_x = 0.4545 \, N \, c_e 10^{\left[a + b \sqrt{\frac{A}{2.59}} + cI + d \left(\frac{P_{year}}{25.4} \right) + e^{(1.8T+32)} + fy \right]}$$ (7.30)

where

M_x is the annual quantity of pollutant X (kg/year)

N is the mean annual number of rainfalls events (a rainfall event is defined as the one having a minimum depth of 1.3 mm; events are separated by a 6 h period of no rain);

c_e, a, b, c, d, e, f coefficients depending on the pollutant type (taken from Table 7.16)

A is the surface area of the urban watershed (km^2)

I is the percent imperviousness (%)

P_{year} is the mean annual rainfall depth (mm)

T is the mean minimum temperature of January (°C)

y is the land use coefficient ($y = 1$ when the commercial and industrial land uses cover more than 75% of the urban watershed; $y = 0$ for any other case)

Table 7.17 provides upper and lower limits for parameters A, I, P_{year} and T; the method is applicable when these parameters are within these limits.

7.3.3 Simple method of Washington, DC

This method was presented by Schueler (1987) in the manual of methods for the management of surface runoff and pollution control of Washington, DC. The method calculates the quantity of a pollutant for a year or a certain time period, according to the following equation:

$$M_x = 10^{-4} P \, p \, C_x \, C \, A$$ (7.31)

Table 7.16 Values of coefficients *a*, *b*, *c*, *d*, *e*, *f* and c_e of Equation 7.30

Pollutant	a	b	c	d	e	f	c_e
COD	1.1174	2.0069	0.0051	–	–	–	1.298
Suspended solids	1.5430	1.5906	–	0.0264	–0.0297	–	1.521
Dissolved solids	1.8449	2.5468	–	–	–0.0232	–	1.251
Total nitrogen	–0.2433	1.6383	0.0061	–	–	–0.4442	1.345
Total ammonia	–0.7282	1.6123	0.0064	0.0226	–0.0210	–0.4345	1.277
Total phosphorus	–1.3884	2.0825	–	0.0234	0.0213	–	1.314
Dissolved phosphorus	–1.3661	1.3955	–	–	–	–	1.469
Cu	–1.4824	1.8281	–	–	–0.0141	–	1.403
Pd	–1.9679	1.9037	0.0070	0.0128	–	–	1.365
Zn	–1.6302	2.0392	0.0072	–	–	–	1.322

Source: Driver, N.E. and Tasker, G.D., Techniques for estimation of storm-runoff loads, volumes, and selected constituent concentrations in urban watersheds in the United States, U.S. Geological Survey Water Supply Paper 2363, U.S. Department of Interior, Washington, DC, 1990.

Table 7.17 Upper and lower limits of parameters *A*, *I*, P_{year} and *T* for which Equation 7.30 is valid

Pollutant	A (km²)	I (%)	P_{year} (mm)	T (°C)
COD	0.049–1.831	4–100	212.9–1574.8	–16 to 14.8
Suspended solids	0.049–1.831	4–100	212.9–1254.3	–16 to 10.1
Dissolved solids	0.052–1.166	19–99	260.1–955.3	–11.4 to 2.1
Total nitrogen	0.049–2.150	4–100	300.5–1574.8	–16 to 14.8
Ammonia	0.049–1.831	4–100	212.9–1574.8	–16 to 14.8
Total phosphorus	0.049–2.150	4–100	212.9–1574.8	–16 to 14.8
Dissolved phosphorus	0.052–1.831	5–99	212.9–1173.0	–11.8 to 2.1
Cu	0.036–2.150	6–99	212.9–1574.8	–9.3 to 14.8
Pd	0.049–2.150	4–100	212.9–1574.8	–16 to 14.8
Zn	0.049–2.150	13–100	212.9–1574.8	–11.4 to 14.8

Source: Driver, N.E. and Tasker, G.D., Techniques for estimation of storm-runoff loads, volumes, and selected constituent concentrations in urban watersheds in the United States, U.S. Geological Survey Water Supply Paper 2363, U.S. Department of Interior, Washington, DC, 1990.

where

M_x is the quantity of pollutant *X* for a certain time period (kg)

P is the rainfall depth for the given time period (mm) (when the calculation is for the mean annual pollutant quantity, then the mean annual rainfall depth P_{year} is used)

p is the annual percentage of rainfall events that result in surface runoff (%) (in the case when the calculation is for a single rainfall event, this parameter assumes the value 1.0)

C_x is the mean pollutant concentration in surface runoff (mg/L), which result either from measurements in the area or by using Table 7.18, which contains typical values

A is the surface area of the urban watershed (ha) (the method is valid for areas less than 260 ha)

C is the surface runoff coefficient for the study area, which, as presented, is taken from tables (e.g. Table 7.3) or computed using the following equation as a function of the imperviousness *I* (%) of the urban watershed:

$$C = 0.05 + 0.009I \tag{7.32}$$

Table 7.18 Typical pollutant concentration in urban areas

Pollutant	Concentrations (mg/L) in following land uses					
	New suburban areas (Washington, DC)	Old urban areas (Baltimore, MD)	Central commercial zone (Washington, DC)	Mean from studies in various urban areas in the United States (NURP study by USEPA)	Hardwood Forest (North Virginia)	Highways in the United States (NURP study by USEPA)
Phosphorus						
Total	0.26	1.08	–	0.46	0.15	–
Orthophosphate	0.12	0.26	1.01	–	0.02	–
Dissolved	0.16	–	–	0.16	0.04	0.59
Organic	0.10	0.82	–	0.13	0.11	–
Nitrogen						
Total	2.00	13.6	2.17	3.31	0.78	–
Nitrite	0.48	8.9	0.84	0.96	0.17	–
Ammonia	0.26	1.1	–	–	0.07	–
Organic	1.25	–	–	–	0.54	–
TKN	1.51	7.2	1.49	2.35	0.61	2.72
COD	35.6	163.0	–	90.8	>40.0	124.0
BOD$_5$	5.1	–	36.0	11.9	–	–
Metals						
Zn	0.037	0.397	0.250	0.176	–	0.380
Pb	0.018	0.389	0.370	0.180	–	0.550
Cu	–	0.105	–	0.047	–	–

Source: Schueler, T.R., Controlling Urban Runoff: A Practical Manual for Planning and Designing Urban BMPs, Washington Metropolitan Water Resources Planning Board, Washington, DC, July 1987.

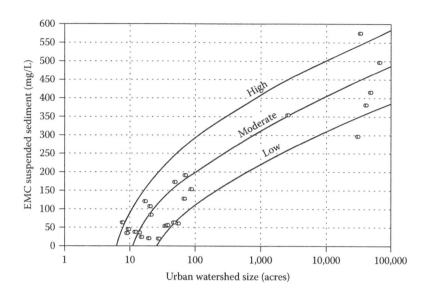

Figure 7.7 Sediment event mean concentration in mg/L as a function of the urban area size in acres (I acre = 0.4047 ha). (From Schueler, T.R., *Controlling Urban Runoff: A Practical Manual for Planning and Designing Urban BMPs*, Washington Metropolitan Water Resources Planning Board, Washington, DC, July 1987.)

Table 7.19 Criteria for using each curve in Figure 7.7 (Schueler, 1987)

Criterion	Sediment event mean concentration (EMC)		
	Low	Moderate	High
Channel stability condition in erosion	Vegetated swales or sewers	Intermediate situation	Open channels, cut banks alternating with channel sandbars, fallen trees
Capacity of sediment storage in the channel	Small deposits in storm drains, stabilized land use	Intermediate situation	Large silt or clay deposits, evidence of recent or ongoing construction, water becomes murky after disturbing bottom
Flow velocity in the stream	Low slope, low imperviousness	Intermediate situation	High slope, high watershed imperviousness

The concentration of suspended solids in surface runoff cannot be computed using Equation 7.31 (Schueler, 1987). The following method is used instead, which has resulted from statistical analyses of data from 25 urban watersheds in the area of Washington, DC, which varied in area from 2 ha to 405 km². The mean concentrations of suspended solids for each rainfall event were found in these watersheds and the graph in Figure 7.7 was created, which contains three curves (for low, medium and high suspended solids concentration) and shows that the mean concentration depends on watershed size (Schueler, 1987). The selection of curve in Figure 7.7 is in accordance with the description of the drainage system condition in Table 7.19.

7.3.4 Pollutant accumulation on street surface

In the previous sections, methods were presented to compute mean pollutant concentrations in runoff. Since the time interval of computation was annual or event based, an assumption was made that pollutant production follows a linear trend. However, studies have shown that pollutant production and accumulation on the street surface is not a linear function. Actually, it has been found that pollutant accumulation reaches an equilibrium quantity with time, which neither increases nor decreases, unless it rains or the street is mechanically swept. The explanation of the equilibrium is that the rate of accumulation reaches after a time the rate of removal because of wind and traffic transport. Various pollutant accumulation equations have been proposed. For example, the following are used in the USEPA Storm Water Management Model (SWMM) (Huber and Dickinson, 1988):

Linear ($b = 1$) or hyperbolic ($b \neq 1$)

$$M(t) = a t^b \qquad (7.33)$$

Exponential

$$M(t) = a(1 - e^{-bt}) \qquad (7.34)$$

Michaelis–Menton

$$M(t) = \frac{at}{b+t} \qquad (7.35)$$

where
 $M(t)$ is the mass of pollutant on the street surface at time t
 a, b are empirical coefficients, which are site specific

To apply Equations 7.33 through 7.35, one has to know the values of coefficients a and b. These can be derived at a specific site by curve fitting on collected data with time or through calibration of a stormwater quality model that uses them (Tsihrintzis and Hamid, 1997a,b, 1998).

Alternatively to Equations 7.33 through 7.35, the following method can also be used to predict pollutant accumulation. It is based on mass conservation (USEPA, 1979; Novotny and Olem, 1994). During any time interval Δt, the change of mass of a pollutant on the street surface is equal to the mass of pollutant that enters the area minus the mass that escapes from the area. The pollutants that enter the area come from (1) solids, tree leaves, etc.; (2) dry (without rain) deposition (settlement) of pollutants and solids suspended in the atmosphere; and (3) deposition from traffic as a result of pollutants from automobile emissions and products of friction between the various mechanical parts of automobiles. Pollutants escape from the area as a result of turbulence created by traffic and wind. Based on these, the following equation was proposed by USEPA (1977):

$$M(t) = M(t - \Delta t)\left(1 - \frac{\xi}{100}\right) + m\Delta t \qquad (7.36)$$

where
 $M(t)$ is the pollutant quantity at time t (kg/ha)
 $M(t - \Delta t)$ is the pollutant quantity at time $t - \Delta t$ (kg/ha)
 t and Δt are the time and time interval (Δt is usually taken as 1 day), respectively
 m is the mean rate of pollutant accumulation during Δt (kg/ha/time interval Δt)
 ξ is the percent of pollutant escape during the time interval Δt (%)

When $\xi = 0$, there is no pollutant escape, the equation is linear and accumulation is unlimited. When $\xi \neq 0$, the equation is non-linear. USEPA (1979) proposes to separate parameter ξ into two components: escape due to natural and chemical reasons and escape due to human management practices (e.g. street sweeping). Therefore, we get the following equation:

$$\xi = \xi_d + \xi_r \tag{7.37}$$

where

ξ is the total percent of pollutant escape (%)
ξ_d is the percent of pollutant escape due to natural and chemical phenomena (%), with values for typical pollutants 6.67% for BOD and TKN, and 0.0 for suspended solids and lead
ξ_r is the percent of pollutant escape due to human management practices (%)

This parameter can be used to quantify the effect of these practices on pollutant accumulation. In the case of street sweeping, ξ_r represents the effectiveness of sweeping in pollutant removal (which is discussed in the following paragraph).

Novotny and Olem (1994) present the following equation for ξ_d:

$$\xi_{d,day} = 1.16e^{-0.08H}(V + W) \tag{7.38}$$

where $\xi_{d,day}$ is daily percent escape of dust and solids (%) which usually ranges between 20% and 40%. For $\Delta t < 1$ day, a proportion of $\xi_{d,day}$ is used; H is the curb height (cm), usually ranging between 15 and 18 cm; V is the traffic speed (km/h); and W is the wind speed (km/h).

The parameter m in Equation 7.36 is either calibrated with measurements in the urban watershed or it is calculated based on the methods presented in previous sections (7.3.1 through 7.3.3). The preferred method is the USEPA method (Heaney et al., 1977), presented in Section 7.3.1, but any method providing mean annual pollutant quantities can be used, as follows:

$$m = \frac{\Delta t}{365} m_x \tag{7.39}$$

where

Δt is the time interval (days)
m_x is the annual unit loading of pollutant X (kg/ha/year)

7.3.5 Washoff of accumulated pollutants

Washoff of accumulated pollutants by surface runoff can be expressed by equations similar to Equations 7.33 through 7.35. Mostly an exponential equation is used, something based on the hypothesis that washoff is a process that follows first-order kinetics with time, described by the following differential equation (Novotny and Olem, 1994):

$$\frac{dM(t)}{dt} = -k\,r\,M(t) \tag{7.40}$$

where

$M(t)$ is the quantity of pollutant remaining on the street surface as a function of time t (kg)
k is the washoff coefficient (mm^{-1})
r is the surface runoff rate or intensity (mm/h)

The solution of Equation 7.40 is of the form

$$M(t) = M(t - \Delta t)e^{-kr\Delta t} \tag{7.41}$$

where $M(t)$ and $M(t - \Delta t)$ are the pollutant masses remaining on the street surface at times t and $(t - \Delta t)$ (kg). From Equation 7.41, the pollutant quantity $\Delta M(t)$ that is washed off during the time interval Δt can be calculated as follows:

$$\Delta M(t) = M(t - \Delta t) - M(t) = M(t - \Delta t)(1 - e^{-kr\Delta t}) = M(t - \Delta t)(1 - e^{-k\Delta R}) \tag{7.42}$$

where ΔR is the incremental surface runoff depth for the time interval Δt (mm). The Hydrologic Engineering Center of the U.S. Army (USCOE, 1977) found that in the case of solids, not all of them are available for washoff for all runoff depths, based on Equation 7.42. This is due to differences in size, texture and specific weight. To correct Equation 7.42, they proposed to use a coefficient of availability γ on the RHS part of this equation. Then, Equation 7.42 becomes

$$\Delta M(t) = M(t - \Delta t) - M(t) = \gamma M(t - \Delta t)(1 - e^{-kr\Delta t}) = \gamma M(t - \Delta t)(1 - e^{-k\Delta R}) \tag{7.43}$$

where

$$\gamma = 0.057 + 0.04(\Delta R)^{1.1} \quad \text{for } \Delta R < 17.7 \text{ mm} \tag{7.44}$$

and

$$\gamma = 1 \quad \text{for } \Delta R \geq 17.7 \text{ mm} \tag{7.45}$$

When using the coefficient of availability γ, Equation 7.41 becomes

$$M(t) = M(t - \Delta t)(1 - \gamma + \gamma e^{-k\Delta R}) \tag{7.46}$$

Novotny and Olem (1994) has calculated the value of the washoff coefficient k as 0.19 mm^{-1} where r and t are expressed in mm/h and h, respectively. According to USEPA (1979), the washoff coefficient k is a function of both watershed imperviousness and slope, and varies from 0.06 mm^{-1} for flat pervious surfaces to 0.18 mm^{-1} for inclined impervious surfaces. USEPA (1979) provides a graph to compute the appropriate value of k for watersheds served with separate storm sewers. Since most pollutants are produced and washed off from impervious inclined surfaces, the proposed value by Novotny and Olem (1994) (0.19 mm^{-1}) is nearly the same as the maximum value given by USEPA (1979) (0.18 mm^{-1}). It has to be noticed that higher values should be used for areas with combined sewers.

The volume ΔV of surface runoff during the time interval Δt can be computed from the following equation:

$$\Delta V(t) = 10\Delta R(t)A \tag{7.47}$$

where
ΔV is the surface runoff volume (m^3)
A is the surface area of the watershed (ha)
10 is a unit conversion factor

Therefore, the concentration of pollutant X originating from washoff in surface runoff can be computed from the following equation:

$$C_x(t) = 1000 \frac{\Delta M_x(t)}{\Delta V(t)} \tag{7.48}$$

where

$C_x(t)$ is the concentration of pollutant X (mg/L)

$\Delta M_x(t)$ results from Equations 7.42 and 7.43

Equations 7.42 through 7.48 can be used to produce a loadograph and a pollutograph for the washed-off pollutants.

7.4 SURFACE RUNOFF QUANTITY AND QUALITY MANAGEMENT

7.4.1 General

Because of its nature, the management of non–point source pollution from urban areas, in general, does not follow the traditional methods used in the treatment of point source pollution. Several methods have been developed, with some requiring construction (structural methods) and others not (non-structural methods). These methods are called best management practices' (BMPs), because the goal is not to achieve treatment so that the effluent is below a certain concentration for the pollutants (something impossible since the concentrations vary with time, magnitude of rainfall event, antecedent dry days and several other factors), but to achieve retention of pollutants to the maximum extent possible. This implies minimization or elimination of pollutants entering the aquatic system.

BMPs can be described as measures that slow, retain and/or absorb pollutants produced from non–point sources and associated with surface runoff (Mandelker, 1989). BMP selection criteria depend on climatic, geographic and economic factors. Beale (1992) pointed out three main options to be considered in the control of urban runoff: to reduce flows entering the drainage system, to increase the capacity of the drainage system and to attenuate the flow entering the drainage system. According to Schueler et al. (1992a), an effective BMP system design comprises six basic components, namely runoff attenuation, runoff conveyance, runoff pre-treatment, runoff treatment, system maintenance and secondary impact mitigation.

Runoff attenuation: Most BMPs indirectly control pollution by minimizing the volume of runoff. The runoff volume can be primarily contained by minimizing either total basin imperviousness or the 'hydraulically connected impervious area' (HCIA) of the watershed, and delineating and protecting the environmental reserve areas. Runoff attenuation serves to enhance the performance of the remaining components of the BMP system. The HCIA of the watershed is the one that directly drains into the drainage system and is not isolated from it. This is depicted in Figure 7.8. The street surface and areas directly draining in the street form the HCIA. Roofs surrounded by impervious areas are part of the total impervious area of the watershed but not of the HCIA.

Runoff conveyance: Runoff should be safely transported to the BMP. Swales, exfiltration drains and parallel pipe systems are some of these methods.

Runoff pre-treatment: To assure longevity and proper operation, some BMPs require pre-treatment. For example, in some BMPs, coarse sediments should be captured or trapped before entering the BMP to preserve storage volume and/or prevent clogging.

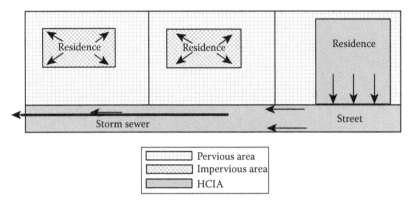

Figure 7.8 Definition of hydraulically connected impervious area.

Examples of some pre-treatment methods include stilling basins, grass filter strips and filter cloth barriers for infiltration systems. For industrial or transportation applications, water quality inlets or settling basins should be used to remove oily wastes prior to entering into the BMP system.

Runoff treatment: This is the heart of the BMP system. In general, there are four basic treatment options available which include filtration, detention, retention and infiltration. An effective BMP system may use two, three or even all four of these options.

System maintenance: It is necessary to maintain the long-term performance of the BMP system. Since the system accumulates significant quantities of sediment and possibly toxic pollutants, the disposal or containment of these residuals should be carefully planned. The design of the BMP must include easy and permanent access, as well as environmentally and economically sound removal procedures.

Secondary impact mitigation: This determines whether the BMP system creates any secondary impacts on the downstream community. For example, secondary impacts from ponds include the discharge of hypoxic and/or thermally enriched water downstream, the removal of downstream riparian cover and possibly filling or alteration of wetland areas adjacent to the pond. For infiltration systems, they include possible risk of groundwater contamination and system failure.

BMPs can be either structural or non-structural. The first kind requires the construction of a structure where surface runoff is collected, treated or retained. The non-structural BMPs include practices and policies which reduce pollution usually at the source, e.g. environmental education, reduction of fertilized use, mechanical street sweeping, regular maintenance of the sewer system and reduction of salt use for deicing. To effectively reduce non–point source pollution, the structural BMPs are combined with non-structural ones, and in most cases, a combination of various BMPs is used in an urban watershed.

Several government agencies in the United States and various authors provide manuals and guidelines for effective BMP description, design, installation and maintenance. These include, among others, the Maryland Department of the Environment (1984, 1987); Maryland Department of Natural Resources (1985, 1987); Schueler (1987); City of Seattle (1989); King County (1990); Minnesota Pollution Control Agency (1992); Urbonas and Roesner (1986); Roesner, Urbonas and Sonnen (1988); Torno (1989); Stahre and Urbonas (1990); Novotny and Olem (1994); Metropolitan Washington Council of Governments (1992); Tsihrintzis and Hamid (1997a). Table 7.20 presents main BMPs by dividing them into four main categories and various sub-categories.

Table 7.20 Various categories and types of BMPs

No.	Category	Description		
1	Pollution control at the source	Environmental education and awareness		
2		Reduction of pollutants from atmospheric scrubbing (dry deposition)	Reduction of pollutants ending to the atmosphere	
3		Streets and impermeable surfaces	Street and impermeable surface flushing	
4			Street sweeping	
5		Permeable surfaces	Erosion control measures	
6			Covering with grass, mulch and special meshes	
7			Control of use of chemicals (e.g. fertilizers, pesticides)	
8		Hydrological modification of the surface drainage system (increase of permeability, increase of storage, reduction of impermeability)	Infiltration	Porous pavements
9				Infiltration trenches
10				Infiltration inlets
11			Increase of surface retention	Retention on the roof
12				Rainwater harvesting
13			Reduction of hydraulically connected impervious areas	Grass filters
14			Environmental corridors, protection zones and vegetation development	Grass swales
15	Pollution control in the sewer and the drainage ditches	Before the entrance to the sewer	Oil/grease separators	
16			Water quality inlets	
17			Grids	
18		In the sewer or channel	Channel stabilization structures	
19			Vegetated canals	
20			Riprap, gabions and other methods of channel protection	
21		Storage, retention/detention in the storm sewer		
22		Storm sewer washing		
23		Control of illicit sanitary sewer connections		
24	Pollution control with surface detention/ retention and storage	Flood control – dry basins	Detention basins	In series
25				In parallel
26		Water quality control	Basin for infiltration of first flush (in parallel with the storm sewer)	
27			Extended dry basin	
28			Wet basin	
29			Wetlands	
30		Detention basins for combined sewer		In series
31				In parallel
32				Inside the aquatic system
33	Runoff treatment			

Source: Adapted from Novotny, V. and Olem, H., *Water Quality – Prevention, Identification, and Management of Diffuse Pollution*, Van Nostrand Reinhold, New York, 1994.

7.4.2 Pollution control at the source

The control of the source of pollution, i.e. the reduction or minimization or elimination of pollutant production, is the most important action towards the reduction of urban non–point source pollution. The BMPs in this category (No. 1–14; Table 7.20) are mostly applicable in the case of separate sewer systems; their applicability in combined sewers is limited (Novotny and Olem, 1994). In this main BMP category, as Table 7.20 shows, there are five main sub-categories of BMPs and several other sub-categories.

BMP 1 – Environmental education and awareness: It is a very important non-structural BMP. It may comprise, among others, solid waste control programmes on the streets, public environmental awareness and education programmes, presentations in schools by experts, organization of environmental excursions and cleaning campaigns of aquatic systems, advertisements in mass media, production and distribution of pamphlets, installation of warning signs at storm sewer inlets, promotion of programmes for used oil recycling, promotion of regular maintenance of cars, promotion of use of public transportation media, invitation of the population to participate in environmental programmes through ecological organizations and application of a policy imposing fines.

BMP 2 – Reduction of pollutants from atmospheric scrubbing (dry deposition): This implies the control of atmospheric pollution resulting mostly from emissions from the industry and the traffic. Examples include the use of natural gas instead of other fossil fuels in the industry and the use of renewable energy sources.

BMP 3 – Street and impermeable surface flushing: This BMP and the following one are mostly used for aesthetic reasons, i.e. to remove solids and dust from impervious surfaces. Street flushing is more common in Europe while street sweeping is more common in the United States and is now expanding in Europe. From the aesthetic point of view, there is no preferred method. However, if one considers the fact that some pollutants are also washed off during flushing, this practice should be applied only in combined sewer systems, when pollutants are transported to the WWTP and not directly to the receiving waters, impacting water quality (Novotny and Olem, 1994). For street flushing, special vehicles are used, equipped with a water tank and special orifices.

BMP 4 – Street sweeping: Street sweeping is more appropriate for separate sewer systems because accumulated pollutants are removed before being washed off by rain (Novotny and Olem, 1994). Therefore, they are not discharged into the receiving waters. For street sweeping, special vehicles are used. There are two types, one equipped with two rotating brushes (mechanical type) and the other with brushes and vacuum (vacuum type). The second type is generally more effective; however, the effectiveness is usually less than 50% in all cases, particularly for fine particles of diameter less than 3.2 mm. Other factors affecting removal is the speed of the vehicle, the number of times it passes from the area, the size of the particle and the content of particles per street length (Clark and Cobbins, 1963; Sartor and Boyd, 1972; Pitt, 1979; Novotny and Olem, 1994).

BMP5 – Erosion control measures: BMPs in the fourth sub-category of control of pollution at the source comprise pollutants from permeable surfaces. Main pollutants are transported sediments, a result of erosion, but also chemical substances adsorbed on sediments. It has to be emphasized that erosion is a process that may produce significant amount of pollutants from bare soils in urban areas, particularly when construction activities take place, and needs attention. The following method is often used in urban areas to control erosion from small surfaces.

BMP 6 – Covering with grass, mulch and special meshes: Erosion is significantly reduced when the bare soil is seeded with grass or covered with mulch. Mulch results from tree trimming and branch cutting, straw or from vegetation remnants from various agricultural activities. When applied on bare soil, mulch may reduce raindrop impact, increase roughness and reduce runoff velocity, increase depression storage and reduce runoff volume and increase soil moisture, enhancing vegetation growth. It is effective in reducing erosion on relatively mild slopes up to 15%, while its ability is enhanced when used together with various metal, plastic or paper meshes or nets or grass seeding.

BMP 7 – *Control of use of chemicals*: Under this measure, fertilizer and pesticide use in lawns and parks in the urban landscape should be reduced to the absolutely minimum. The advice of an expert on the minimum chemical requirements is necessary. In addition, alternatives to chemicals should be used, such as the use of compost, reuse of wastewater for irrigation and use of wastewater sludge. All these of course should be used with caution. Finally, in areas with intensive snowfall in the winter, salts and sand should be used with caution.

BMP 8 – *Porous pavements*: This BMP belongs to the general category of hydrological changes, and more specifically, it is one of the four methods of increasing infiltration. The main aim is to reduce the volume of runoff and, subsequently, to reduce pollutant loadings in separate sewer systems or overflow frequency in combined sewer systems. Porous pavements are the first BMP in this category. They are constructed so as to allow runoff to infiltrate through them to the subsoil. This is achieved either by not using fine material in asphalt construction which makes it permeable, by special construction of concrete or by using special paving stones which allow water to percolate. In general, the infiltration capability of porous pavements is significantly high, even exceeding the intensity of very infrequent storms (Novotny and Olem, 1994).

Porous pavements are mainly used in open parking areas or in secondary streets. They are effective in areas with mild slope and quite permeable subsoil, where the water table and the underlying rock are deep. In terms of construction effort and cost, they are comparable to traditional pavements. Their advantages over traditional pavements include the following: They reduce or even can eliminate runoff, they result in smaller sewer size downstream and they contribute to groundwater recharge. Their main disadvantage is the high risk of clogging, with a high cost for repair and danger of flooding. For this reason, care must be exercised so that no solids and sediments end into the porous pavement. In addition, maintenance is a must, for example, regular sweeping. Finally, to minimize the risk of flooding, an escape route should be designed in the area of the porous pavement for water to overflow in case of clogging.

Regarding drainage, there are two types of porous pavements: total infiltration and partial infiltration. The first type is depicted schematically in Figure 7.9. The function of this structure is the following: due to the significantly high permeability of the porous pavement, rainfall or runoff water enters fast from the ground surface through the pavement to the underground reservoir. This contains rock of diameter 3.8–7.6 cm and its thickness is a design parameter. The rock is separated from the upper porous pavement and the natural subsoil beneath by a gravel filter and/or a geotextile material (Figure 7.9). Water is stored in the voids of the rock and leaves the structure through seepage to the subsoil. To reduce the risk of sediments entering the top of the porous pavement, a grass strip can be designed to surround the porous pavement, as shown in the plan view of Figure 7.9. Finally, an overflow path is provided at the lower point of the pavement to allow for floodwater to escape in case of pavement clogging. The underground reservoir can be designed to hold the hydrograph of the design flood.

The partial infiltration system is depicted in Figure 7.10. It is similar to the total infiltration system in function; the only difference is that it is equipped with an overflow pipe which allows for the underground reservoir to operate up to a certain level. This pipe connects to the downstream storm sewer and conveys any excess water into it. Usually, storage in the underground reservoir is for the first flush up to the 2-year rainfall event.

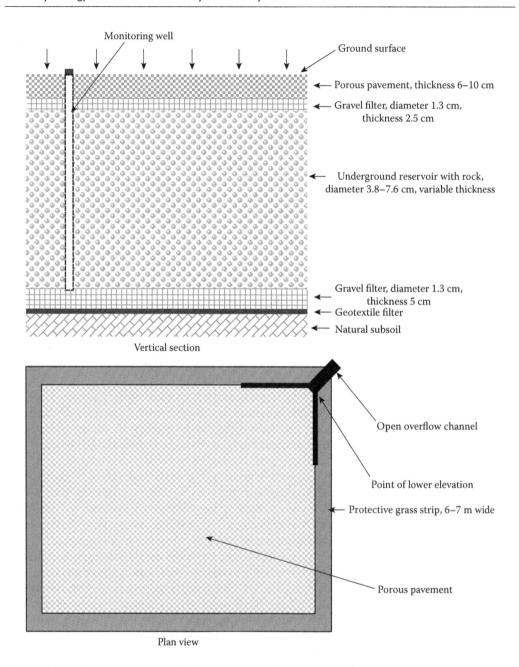

Figure 7.9 Typical vertical section and plan view of a total infiltration porous pavement.

In cases where there are sediments and/or oil in the runoff entering the porous pavement, pre-treatment may be needed, as mentioned. This may comprise a grass strip and, possibly, a sand filter (see also following BMP) surrounding the porous pavement. A way to do this is depicted in the schematic diagram of Figure 7.11, where one sees a section of the grass strip, a small weir and the sand/gravel filter. The operation of this structure is self-explanatory.

To size the depth of the underground reservoir, the main parameter is the volume of the voids of the rock. This volume can be determined from the porosity, which is defined

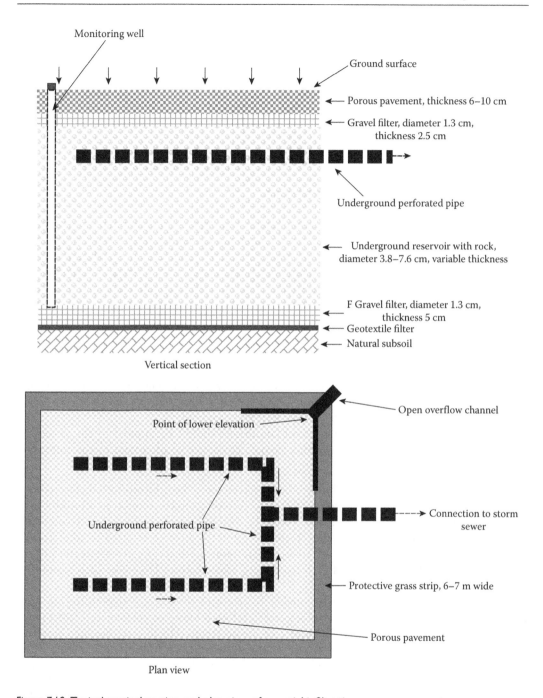

Figure 7.10 Typical vertical section and plan view of a partial infiltration porous pavement.

as the ratio of the voids volume to the total volume of the rock. Typical porosities for various material textures can be found in any geotechnical engineering or groundwater hydraulics textbook (i.e. 0.25–0.40 for gravel). It is recommended that the porosity of the material is always determined *in situ* before construction by standard experimental procedures.

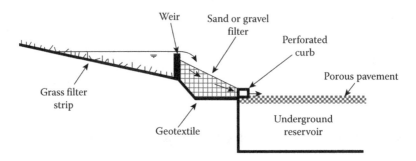

Figure 7.11 Pre-treatment before the porous pavement.

The porous pavements are very effective in removing dissolved pollutants or fine solids. The underground reservoir is mainly used for runoff storage, while pollutant retention takes place only after infiltration in the subsoil. Main treatment processes taking place in porous pavements include the following (Schueler, 1987): (1) sorption of pollutants, which takes place in the upper 30 cm of the soil under the underground reservoir and concerns mostly dissolved constituents (e.g. orthophosphate, zinc); (2) trapping of solid particles, in which case the soil acts as a filter; and (3) biological reactions by various aerobic bacteria under the reservoir, which are enhanced as the dry days between rainfall events increase, i.e. flooding is less frequent.

Pollutant removal of porous pavements is satisfying for metals, solids and nitrogen, and to a lesser degree for phosphorus. According to Schueler (1987), long-term removal percentages range 82%–95% for sediments and solids, 65% for total phosphorus, 80%–85% for total nitrogen, 82% for COD, 99% for zinc and 98% for lead.

Several criteria have to be followed and conditions should be applied in the design of porous pavements and other infiltration systems. These are summarized in the following Table 7.21.

BMP 9 – Infiltration trenches: Similar to porous pavements, the infiltration trenches remove dissolved pollutants and solid particles effectively. A main problem, similar to all infiltration systems, is clogging; therefore, their use should be avoided in areas where coarse solids are a problem. Alternatively, in such cases, protective grass strips can be used to filter runoff before reaching the trench. To use a trench, a minimum infiltration capacity of the subsoil should exist; therefore, subsoils of SCS hydrologic classification D (Table 7.10) should be avoided.

There are two types of trenches, namely, surface and subsurface, depending on how runoff enters the trench. Typical cross sections and plan views of surface and subsurface trenches are presented in Figures 7.12 and 7.13, respectively. The main difference between the two types is that one is open to the air at the upper part and runoff enters from there (Figure 7.12), while other kind is underground and runoff enters either directly through a sewer that ends into the trench or through a perforated pipe that empties a collection box or a water quality inlet or an oil/grease separator, where the sewer ends (Figure 7.13; see also the following BMPs). Similar to porous pavements, infiltration trenches can be of total or partial infiltration, depending on collecting the entire design storm or a portion of it. In the case of surface trench, overflow can occur from the upper part, while the subsurface trench should be equipped with a perforated pipe that collects overflow and leads it to the downstream sewer, as discussed in porous pavements. In several cases, the trench is designed only for the volume of the first flush (i.e. 13–25 mm of runoff).

Table 7.21 Rules, criteria and conditions for optimum design and best performance of porous pavements and other infiltration systems

No.	For best performance
1	Most of the annual runoff volume should be infiltrated to the subsoil.
2	Larger in surface and less deep systems are preferred.
3	The underground reservoir should empty through infiltration at most in 3 days (preferably in 2 days), something that allows for the development of aerobic bacteria. For faster drainage, only soils of SCS hydrologic classification A–C should be used (Tables 7.9 and 7.10). Soils of category D (Table 7.10) are not appropriate. A minimum runoff retention time of 6–12 h in the underground reservoir should also be allowed.
4	Regular sweeping (with vacuum vehicles) should take place. High-pressure flushing should take place at least four times a year.
5	Area slope for porous pavements and surface trenches should be less than 5%. For subsurface trenches, slopes can be up to 20%.
6	A minimum distance of 0.6–1.2 m should exist between the bottom of the underground reservoir and the underlying bedrock and/or the highest stage of the water table.
7	To minimize contamination risk, a minimum distance should be maintained between potable water wells and infiltration structures. Infiltration structures should be located downstream of groundwater flow direction.
8	To avoid building foundation settlement problems, a minimum distance of 3 m should be maintained between infiltration structures and building foundations.
9	An observation well (Figures 7.9 and 7.10) should be placed in the underground reservoir. This allows for checking drainage time and clogging, and also sampling for chemical analyses.
10	To avoid damage and malfunction, the existence of the infiltration structure should be identified by placing signs on the surface. The size should identify the type of the structure and should prohibit surface paving and discharge of sand, oil and other pollutants.
11	Porous pavements are recommended for parking areas and secondary roads. Recommended drainage areas sizes are 0.1–4 ha. For infiltration trenches, recommended areas are up to 2 ha.

Source: Schueler, T.R., *Controlling Urban Runoff: A Practical Manual for Planning and Designing Urban BMPs*, Washington Metropolitan Water Resources Planning Board, Washington, DC, July 1987.

A particular type of trench can be used to drain specifically roofs or impermeable surfaces, to offer direct infiltration of runoff from such areas. A schematic presentation of such a system is shown in Figure 7.14.

The main advantages of infiltration trenches include reduction of both peak and volume of runoff (usually design storm with a return period of 2–10 years is used) and recharge of groundwater (it is expected that 60%–90% of annual runoff may end into groundwater when these systems are used). Similar to porous pavements, pollutant removal only includes dissolved constituents and fine solid particles with processes similar to those mentioned for porous pavements. Removals have been presented by Schueler (1987). They vary from 75% to 99% for sediments and solids, 50% to 75% for total phosphorus, 45% to 70% for total nitrogen, 75% to 99% for trace metals and 70% to 90% for BOD (Schueler, 1987). Lower removals correspond to infiltration of first flush (13 mm) and higher ones to total infiltration. Design criteria and rules presented in Table 7.21 for porous pavements also apply for infiltration trenches.

BMP 10 – Infiltration inlets: There are no standard designs for these systems, but a typical design is presented in Figure 7.15. They are made of reinforced concrete. Their base is open to allow for runoff infiltration. Because of the small storage capacity, they are usually designed for first flush (i.e. 13–25 cm of runoff), while the rest of runoff overflows through a weir to the sewer (Figure 7.15). The design/sizing of the system comprises of four parts: (1) the inlet, (2) the underground reservoir, (3) the weir and (4) the sewer.

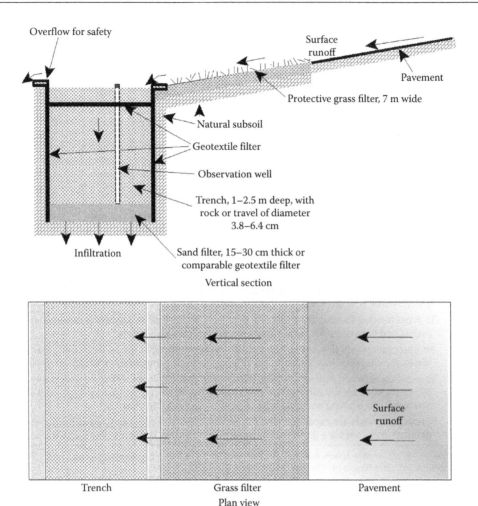

Figure 7.12 Typical section and plan view of a surface infiltration trench. (Adapted from Schueler, T.R., *Controlling Urban Runoff: A Practical Manual for Planning and Designing Urban BMPs,* Washington Metropolitan Water Resources Planning Board, Washington, DC, July 1987.)

BMP 11 – Retention on the roof: This BMP aims to reduce the runoff volume. It is applied on flat roofs, where the gutter inlets are raised above the roof floor to have water retained on the roof and eventually evaporated. A usual retention is for the first flush, i.e. 13 cm of runoff, which in this case coincides with rainfall. For rainfall events exceeding this depth, excess water overflows to the gutters. This BMP is effective in reducing runoff volume; in terms of runoff quality, the improvement is minor, since only pollutants from atmospheric scrubbing and/or wet depositions are retained. Furthermore, to apply this BMP, the roof should be properly designed/checked to sustain additional loads from ponding water and avoid leakage into the building.

BMP 12 – Rainwater harvesting: Rainwater harvesting is the collection and storage of rainwater for potable or non-potable in-house uses (Gikas and Tsihrintzis, 2012; Tsihrintzis and Baltas, 2014); it is a relatively inexpensive method to reduce the consumption of potable water supplied by public sources, and may contribute to the reduction of runoff volumes and peaks. Untreated harvested rainwater can be used for non-potable uses, such as toilet

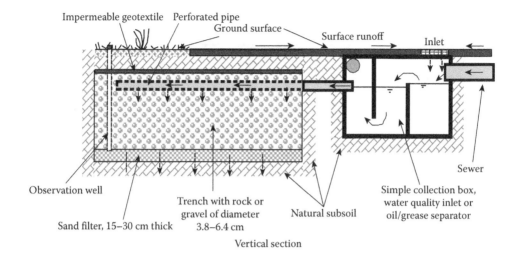

Vertical section

Figure 7.13 Typical section of a subsurface infiltration trench. (Adapted from Schueler, T.R., *Controlling Urban Runoff: A Practical Manual for Planning and Designing Urban BMPs*, Washington Metropolitan Water Resources Planning Board, Washington, DC, July 1987.)

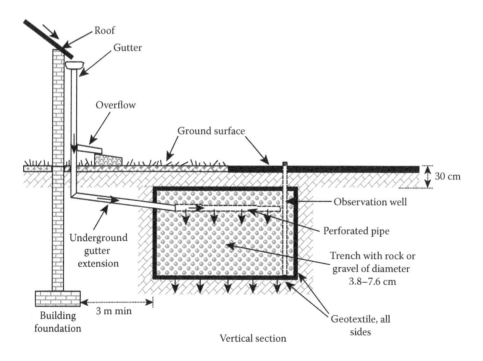

Vertical section

Figure 7.14 Typical infiltration trench for a roof. (Adapted from Schueler, T.R., *Controlling Urban Runoff: A Practical Manual for Planning and Designing Urban BMPs*, Washington Metropolitan Water Resources Planning Board, Washington, DC, July 1987.)

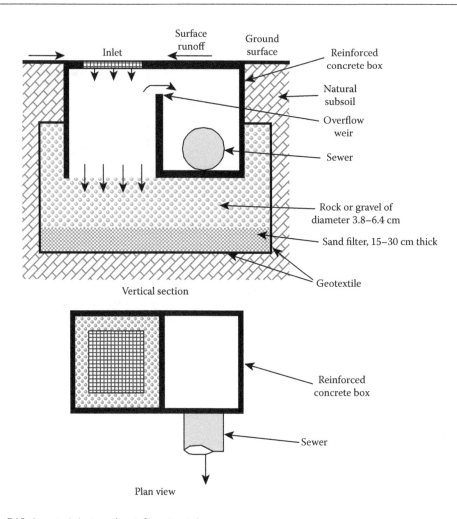

Figure 7.15 A typical design of an infiltration inlet.

flushing, washing clothes, other household uses and garden irrigation, while potable uses are also common in several countries (e.g. Australia), but they may require appropriate treatment of harvested rainwater, depending on its quality (Tsihrintzis and Baltas, 2014). A layout similar to the one presented in Figure 7.14 is used for a rainwater-harvesting system; however, the trench should be replaced by a collection tank which receives and stores the rainwater. Various descriptions and sizing techniques of a rainwater-harvesting tank have been presented (Tsihrintzis and Baltas, 2014). The quality of harvested rainwater and health risk assessment have been studied by various authors (Gikas and Tsihrintzis, 2012).

BMP 13 – Grass filters: Grass filters were presented with porous pavements (Figures 7.9 to 7.11) and infiltration trenches (Figure 7.12), where they are used for runoff pre-treatment, i.e. retention of solids to avoid clogging. They can also be used to reduce the hydraulically connected impervious area or along streams to retain mostly solids and, possibly, other pollutants from reaching the water (Figure 7.16).

They are flat slightly sloping surfaces, where flow is shallow and evenly distributed perpendicular to the flow path, i.e. overland flow. Their good performance is based on this property. Performance gets worst when rills and gutters are formed due to erosion, and flow

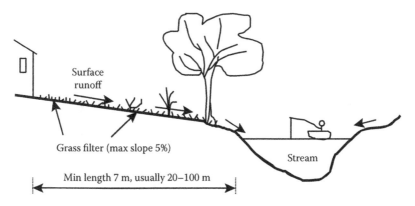

Figure 7.16 Grass filter along a stream.

concentrates in certain paths. To avoid erosion, a certain type of erosion-resistant grass is used, e.g. Bermuda grass. In addition to grass, filters can contain trees and bushes and, generally, be better designed to fit the local landscape.

Their function can be summarized as follows (Novotny and Olem, 1994): (1) They reduce the runoff velocity, resulting in an increase in the time of concentration and a reduction of peak; (2) they reduce the imperviousness of the urban watershed, resulting in a reduction of runoff volume and an increase of infiltration; and (3) they remove solids, organic matter and metals. Their treatment performance depends on several parameters, including filter length along the flow path, slope, watershed area, runoff velocity, solid particle diameter and soil infiltration capacity. The two main parameters are the length and the slope. An absolutely minimum length is 7 m, while lengths of 20–100 m are often used (Figure 7.16). The slope should not exceed 5% for pollutant removal (Novotny and Olem, 1994), but even higher slopes should be covered by grass for erosion protection; (4) dissolved constituents are removed through infiltration, which is more effective when slope is mild and there are trees and bushes; (5) maintenance is recommended which includes grass cutting (two to three times a year) and repair of eroded areas; (6) grass filters along streams can be part of environmental corridors which increase biodiversity and aesthetics of the landscape, and can support park and recreational activities (walking, jogging, biking, nature observation, etc.) in the urban area. To function as wildlife corridors, the length of the grass filter (Figure 7.16) should increase significantly (200–1000 m).

BMP 14 – Grass swales: Grass swales can be placed: (1) along street sides in suburban areas, when there is area available to replace curbs; (2) along highways at the right side to collect runoff; and (3) along highways at the middle to separate the two directions of traffic (Figure 7.17). Swales usually are small, open, shallow canals of nearly triangular, trapezoidal (Figure 7.17) or parabolic section of very mild bank slope.

They offer (1) reduction of runoff peak and volume through two mechanisms, reduction of velocity and, thus, increase of time of concentration, and infiltration of part of runoff to the subsoil. The infiltrating volume can be increased if small check dams or weirs are placed at several locations across the swale to allow for water ponding and increase of depth upstream of the check dam (Figure 7.17); (2) in terms of water quality improvement, similarly to grass filters, their function is restricted to reducing solid particles and the portion of dissolved pollutants contained in the infiltrating water; and (3) they are a more economic solution, in terms of construction, compared to standard curb and gutter and sewer option.

Design rules/steps for grass swales include the following (Novotny and Olem, 1994): (1) The slope along the swale should be very mild, preferably nearly horizontal. In any case,

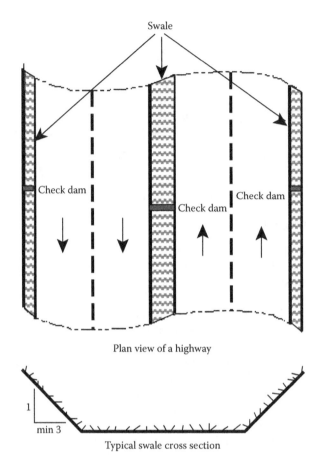

Figure 7.17 A schematic plan view and typical cross section of highway swales.

the maximum allowed slope should be 5%. (2) The bank slope should be milder than 3:1 horizontal to vertical (Figure 7.17). (3) The grass species used should be resistant to erosion. (4) Swales should be used only with subsoils of SCS hydrologic category A and B (Table 7.9). (5) To increase infiltration, check dams or weirs can be used. These can be either concrete, wooden or made of earth. For safety reasons and to avoid mosquito development, the ponding duration should not exceed 24 h and the depth should be less than 50 cm. (6) The design velocity should be less than 0.9 m/s to avoid erosion. (7) Maintenance is necessary and should include grass cutting once a year, repair of bare areas by replanting, minimization or elimination of use of fertilizers and collection of garbage. (8) To increase ecological and aesthetic value, trees and bushes can be planted.

7.4.3 Pollution control in the sewer and the drainage ditches

BMP 15 – Oil/grease separators: They are placed at the beginning of storm sewers or at inlets. They are concrete boxes used to separate oil and grease floating on runoff and solids carried by runoff (Schueler, 1987). They are generally small in volume (usually the effective volume is about 11 m^3); therefore, they do not provide storage and consequent reduction of hydrograph peak. They retain relatively large solid particles (i.e. sand), oil and grease, while fine particles and dissolved constituents pass downstream with runoff.

For these reasons, they are mostly used for pre-treatment or in combination with other BMPs (e.g. Figure 7.13). They are placed in areas where hydrocarbons and oil derivatives from vehicles are expected, e.g. parking lots, gas stations and streets. The drainage area to one oil/grease separator should not exceed 0.4 ha. Maintenance is mandatory (recommended every 6 months) and requires removal of collected sediments and pollutants which may be toxic. If these are not removed, they get resuspended and carried by the runoff of an intense rainfall. To remove solids, special vehicles with vacuum suction are used. Therefore, manholes are needed at the top of the structure. The oil/grease is removed by special materials which can absorb it without absorbing water. This material needs regular replacement.

A typical design is shown in Figure 7.18. The structure comprises of three compartments (Schueler, 1987). The first one retains solids which deposit, the second one retains floating oil/grease and the third one connects to the downstream storm sewer. The first and second compartments are separated by a vertical wall and communicate through two circular openings, 15 cm in diameter, and placed at about 1.2 m from the bottom of the structure. A grid (13 mm in size) is placed in front of the two openings to hold floating solids and protect the orifices from clogging. The oil and grease float and are retained in both compartments. To trap it, the second compartment communicates with the third through a pipe, which has a Γ-shape to convey water only from the bottom (30 cm above the bottom). The two vertical walls between compartments do not meet the ceiling of the structure but allow for a 10 cm space for overflow. In some cases, holes (15 cm in diameter) can also be drilled at the bottom to drain the structure to the subsoil. However, these are not usually effective, since they quickly clog with sediments.

BMP 16 – Water quality inlets: These sewer inlets are specifically designed to retain coarse solid particles and solid waste, and oil/grease (Schueler, 1987). In addition, they control odours from sewers to reach the atmosphere. There is no standard design, but it depends on local practice; therefore, Figure 7.19 only shows the concept. The storage water volume varies from 0.08 to 2.21 m³, but about half of it is usually taken by sediment. Due to the

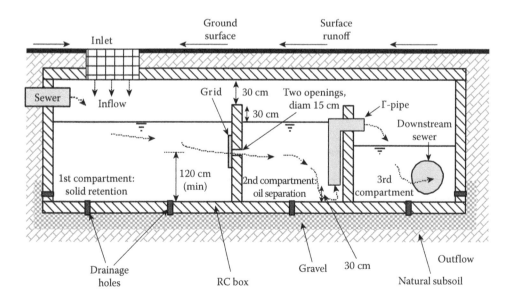

Figure 7.18 A schematic diagram of oil/grease separator. (Adapted from Schueler, T.R., *Controlling Urban Runoff: A Practical Manual for Planning and Designing Urban BMPs*, Washington Metropolitan Water Resources Planning Board, Washington, DC, July 1987.)

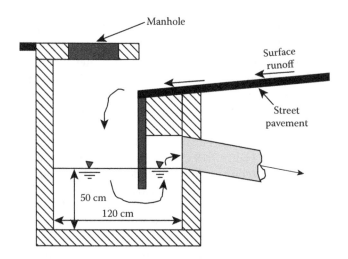

Figure 7.19 A schematic diagram of a water quality inlet. (Adapted from Schueler, T.R., *Controlling Urban Runoff: A Practical Manual for Planning and Designing Urban BMPs*, Washington Metropolitan Water Resources Planning Board, Washington, DC, July 1987.)

small volume, their effectiveness in water quality is limited. There is a need for regular maintenance and cleaning at least twice a year, in which case the removal of solids and heavy metals can reach 25%.

BMP 17 – Grids: Grids are usually placed at the outlet of main storm sewers to the receiving waters. A typical design is depicted in Figure 7.20. Their function is to collect mostly floating solid waste (e.g. foam cups) and debris (branches, wood) from spreading at the surface of the aquatic system. They need regular maintenance which consists of cleaning and removing the collected material to avoid clogging and, possibly, upstream flooding.

BMP 18 – Channel stabilization structures: Such structures are used to protect open channels from erosion. They are presented in detail in various textbooks (e.g. Goldman et al., 1986; Simons and Senturk, 1992; Vanoni, 2006) and are also briefly discussed in Chapter 8.

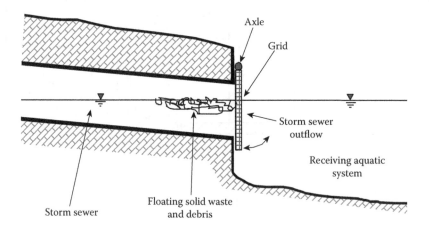

Figure 7.20 Typical grid at the end of a storm sewer.

BMP 19 – Vegetated canals: A method of protection of channels and canals from erosion is to use grass and vegetation. The specific theory and design is presented in various textbooks (e.g. Chow, 1959; French, 1985).

BMP 20 – Riprap, gabions and other methods of channel protection: Such methods to protect channels from erosion are presented in various textbooks (e.g. Goldman et al., 1986).

BMP 21 – Storage, retention/detention in the storm sewer: BMPs under this category offer indirect improvement to runoff quality by reducing volume and/or flow peak. Four methods are used:

1. Temporary or permanent storage of runoff in underground storage tanks (Figure 7.21). Most often, reinforced concrete or metal pipes of large diameter are used, which are placed in parallel to the storm sewer and communicate with it through orifices or weirs. The entire or portion of the runoff is diverted and stored. The drainage of the underground storage tanks is (a) through infiltration to the subsoil, in which case the tanks should be perforated, and (b) by drainage back to the storm sewer after the flood passes.
2. Temporary storage in a detention tank within the storm sewer, which is created by enlarging the sewer (Figure 7.22). Sizing of the detention tank is based on flood routing theory (Puls method), presented later in this chapter (see also BMPs 26 or 27).

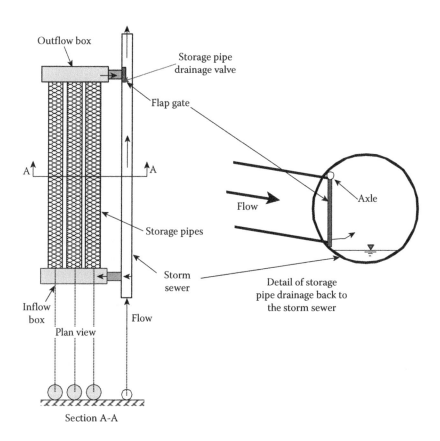

Figure 7.21 Storage in underground pipes.

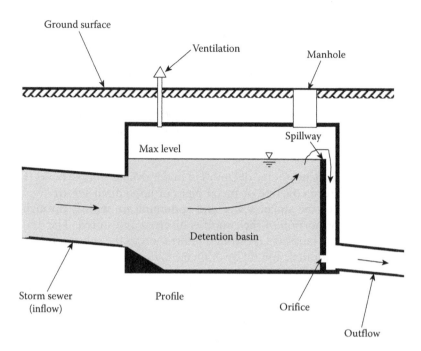

Figure 7.22 Underground detention basin.

3. Storage in a tunnel which can be constructed under the sewer at a greater depth (Figure 7.23). This solution is useful when there is space limitation for a regular surface or subsurface detention pond. Since the tunnel is usually excavated in rock, drainage is possible only through pumping of the stored water back to the storm sewer after the flood has passed. This is also a solution that can be used in solving CSO problems (see following BMPs 30 to 32).
4. Storage within the storm sewer or canal. This is important when the sewer is large. The computations follow the Muskingum method theory, presented in the following.

If the hydrographs at the upstream and downstream sides of a storm sewer reach are measured simultaneously, one will observe that these differ in flow peak, time base and generally shape, even if there is no inflow or outflow of water from the storm sewer reach. The hydrograph peak at the downstream end of the reach is lower than that at the upstream end and the time base is longer. This is a result of water storage in the storm sewer. The hydrograph at the downstream section can be computed if the one at the upstream section and the geometry of the conduit are known, using flow routing. A common method used is the Muskingum method.

In a canal or storm sewer reach flowing open, the water surface has a slope and the water stored in the reach is a function of water stage at both the upstream and downstream sections of the reach (Figure 7.24). The figure shows that the total stored volume comprises two parts: prism storage S_p (defined as the volume of water between conduit invert and the line parallel to the invert from the downstream water surface) and wedge storage S_w (defined as the water stored above the parallel line to the bed and the free water surface). Prism storage is a function of the downstream water stage and, therefore, outflow O from

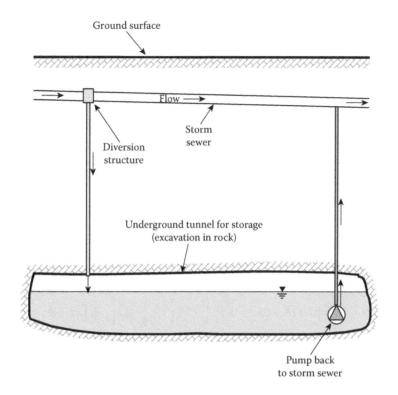

Figure 7.23 Use of a tunnel for storage.

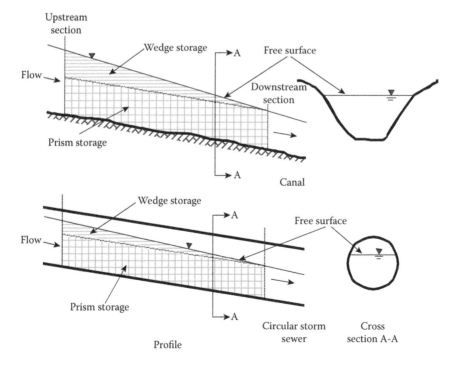

Figure 7.24 Definition of wedge and prism storage in a canal and a circular storm sewer.

the downstream section of the reach. Wedge storage depends on the difference between inflow I and outflow O from the reach. Therefore, the total volume can be written as follows:

$$S = S_p + S_w = f_1(O) + f_2(I - O) \tag{7.49}$$

where
  I and O are inflow and outflow from the reach (m³/s)
  f_1, f_2 imply functions of

The Muskingum method assumes that functions f_1 and f_2 are linear:

$$f_1(O) = KO \quad \text{and} \quad f_2(I - O) = \beta(I - O) \tag{7.50}$$

In this case, Equation 7.49 gives

$$S = KO + \beta(I - O) = \beta I + (K - \beta)O = K\left[\left(\frac{\beta}{K}\right)I + \left(1 - \frac{\beta}{K}\right)O\right] = K[xI + (1 - x)O] \tag{7.51}$$

where $x = \beta/K$ is a dimensionless coefficient, which shows the relative weight of the inflow and outflow in determining the storage in the conduit reach. This coefficient takes values between 0.0 and 0.5. The 0.0 value implies that inflow is not important and the 0.5 value implies that inflow and outflow are equally important. In the Muskingum method, to solve the flow routing problem, Equation 7.51 is combined with the water budget equation for the reach. Let us consider a time interval Δt. If $I_m(t)$ and $O_m(t)$ are the mean inflow and outflow rates in the conduit reach during Δt, the water budget can be written as follows:

$$I_m(t)\Delta t - O_m(t)\Delta t = \Delta S(t) \tag{7.52}$$

where

$$I_m(t) = \frac{I(t) + I(t - \Delta t)}{2} \tag{7.53}$$

$$O_m(t) = \frac{O(t) + O(t - \Delta t)}{2} \tag{7.54}$$

$$\Delta S(t) = S(t) - S(t - \Delta t) \tag{7.55}$$

where
  $I(t)$, $O(t)$, and $S(t)$ are inflow rate, outflow rate and storage within the reach at time t
  $I(t - \Delta t)$, $O(t - \Delta t)$ and $S(t - \Delta t)$ are the respective quantities at time $t - \Delta t$

The combination of Equations 7.51 through 7.55 leads to the following:

$$O(t) = C_0 I(t) + C_1 I(t - \Delta t) + C_2 O(t - \Delta t) \tag{7.56}$$

where coefficients C_0, C_1 and C_2 are defined as follows:

$$C_0 = \frac{0.5\Delta t - Kx}{K(1-x) + 0.5\Delta t} \tag{7.57}$$

$$C_1 = \frac{0.5\Delta t + Kx}{K(1-x) + 0.5\Delta t} \tag{7.58}$$

$$C_2 = \frac{K(1-x) - 0.5\Delta t}{K(1-x) + 0.5\Delta t} \tag{7.59}$$

It is obvious that

$$C_0 + C_1 + C_2 = 1 \tag{7.60}$$

Equation 7.56 can be used to compute step by step, in an iterative procedure, the outflow hydrograph $O(t)$ at the downstream end of the reach for a given inflow hydrograph $I(t)$, if K and x (and thus, C_0, C_1, C_2) are known, and the value in the previous time step $O(t - \Delta t)$ has been computed (or is known for $t = 0$). First, the values of coefficients C_0, C_1 and C_2 have to be computed using Equations 7.57 through 7.59.

To determine the values of parameters K and x, the following procedure can be used:

1. Hydrographs are measured, simultaneously, at the upstream and downstream sections of the conduit reach.
2. For each time interval Δt, the change in storage ΔS is computed using Equation 7.52.
3. A value is assumed for parameter x (in the interval $[0,0.5]$), and the value of the quantity $[xI + (1 - x)O]$ is computed (in RHS of Equation 7.51).
4. The storage $S(t)$ is computed by accumulating ΔS, $S = \Sigma \Delta S$.
5. A plot of S (from step 4) versus $[xI + (1 - x)O]$ (from step 3) is prepared.
6. According to Equation 7.51, the plot should represent a straight line of slope K, if the selected value for parameter x (step 3) is correct.
7. If the data deviate from the straight line, the assumption of the x value is incorrect, and another guess is needed. Then, steps 3–7 are repeated to get the plot where the points fall most closely to the straight line.
8. The straight regression line is fitted, and its slope is computed which gives the K value.

BMP 22 – Storm sewer washing: Several times solids deposit in storm sewers and also a biological slime develops (particularly in combined systems). Sewer washing may be important, particularly of combined systems, to remove solids and/or slime. This should be taking place during dry weather, forcing the wash water to end into the WWTP (Novotny and Olem, 1994). High-pressure water jets can be used for effective washing.

BMP 23 – Control of illicit sanitary sewer connections: Illicit connections of house plumping into separate storm sewers are a fact in many cities. This has adverse impacts on receiving waters since untreated sewage ends there. It is then important for public agencies to find and minimize/eliminate illicit connections (either through discharge and/or water quality monitoring at various locations along the storm sewer, or by using specialized cameras). Furthermore, public awareness on illicit connection adverse impacts and imposing of strict fines may also be necessary in some instances.

7.4.4 Pollution control with surface detention/retention and storage

BMP 24 – Flood detention basins in series (or online) with the storm sewer: They offer water quality advantages only indirectly by reducing the hydrograph peak, which is the main reason of their construction, while the total volume of the hydrograph is unaffected. As a result, they reduce the downstream flood risk, the velocity in the conduit and the erosion risk. Several times, their construction offers opportunities for aesthetics improvements through the existence of ponding water, development of parks and recreation areas and support of wetland vegetation.

Figure 7.25 presents a schematic plan view, indicating how such a basin is located in relation to the storm sewer, 'in series', i.e. the storm sewer ends and empties into the detention basin and the outlet of the detention basin connects to the downstream sewer. The outlet of the detention basin should be of restricted capacity compared to the inflowing sewer, in order for incoming water to pond and be stored in the basin, i.e. as the inflowing hydrograph comes into the basin, the water-level in the basin rises and reaches a maximum level, and as the inflowing hydrograph recedes, the water-level in the basin falls. As a result of water storage, the outflow hydrograph from the basin is modified compared to the inflowing hydrograph, and more specifically, the outflow hydrograph has a lower peak and a larger time base. Since there is no loss or gain of water, the inflowing and outflowing hydrographs have the same volume.

The computation of the outflowing hydrograph is based on the theory or flood routing through a reservoir (Puls method). To solve this problem, the inflowing hydrograph and the geometry of the detention basin and the outflow structure, i.e. the storage volume and the outflow rate as a function of water-level in the basin, have to be known. Similar to the Muskingum method, let us define $I(t)$ and $O(t)$ as the inflow and outflow rates to and from the basin and $S(t)$ as the volume of water stored at time t. Then it is generally valid that both outflow rate and storage in the basin depend on water-level in the basin:

$$O = f_3(H) \quad \text{and} \quad S = f_4(H) \tag{7.61}$$

where
f_3 and f_4 imply functions of
H is the water-level in the detention basin

Similar to the Muskingum method, the water budget Equation 7.52 holds. If Equations 7.53 through 7.55 are used again to express mean inflow and outflow, and change of stored water in the basin, then Equation 7.52 reduces to

$$\left[\frac{S(t)}{\Delta t} + \frac{O(t)}{2} \right] = \left[\frac{S(t - \Delta t)}{\Delta t} + \frac{O(t - \Delta t)}{2} \right] + \frac{I(t) + I(t - \Delta t)}{2} - O(t - \Delta t) \tag{7.62}$$

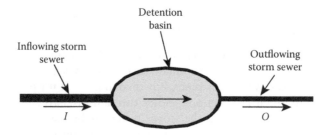

Figure 7.25 Detention basin in series (plan view).

If we define the following parameter $\Gamma(t)$ as

$$\Gamma(t) = \frac{S(t)}{\Delta t} + \frac{O(t)}{2} \qquad (7.63)$$

Then, Equation 7.62 reads

$$\Gamma(t) = \Gamma(t - \Delta t) + I_m(t) - O(t - \Delta t) \qquad (7.64)$$

According to Equations 7.61 and 7.63

$$O(t) = f_s(\Gamma(t)) \qquad (7.65)$$

and

$$\Gamma(t) = f_s^{-1}(O(t)) \qquad (7.66)$$

where f_s is some function which depends on detention basin geometry and the properties of the outflow structure. Equations 7.64 and 7.65 or 7.66 form a system of two equations with two unknowns. The unknowns are $\Gamma(t)$ and $O(t)$ if the values of $\Gamma(t - \Delta t)$, $O(t - \Delta t)$ and $I_m(t)$ are known. This implies that the system can be solved in a step-by-step iterative procedure where the values of the previous time step are used to compute those of the next time step Δt. In addition, the geometry expressed by function f_s and the first value of $O(0)$ at $t = 0$ should be known. The procedure is applied very quickly in a table and can be very fast if a spreadsheet is used.

BMP 27 – Flood detention basins in parallel (or offline) with the storm sewer: They have the main advantages of the previously described basins, consisting mainly of reducing the hydrograph peak. The location of the basin (in plan view) in relation to the storm sewer is shown schematically in Figure 7.26. In this case, the storm sewer does not enter the basin as in the previous case (Figure 7.25).

If the aim is to reduce the peak of the hydrograph (i.e. flood control), there is a side weir placed along the storm sewer (Figure 7.27); in other words, the side of the storm sewer has an opening at a certain distance above the invert, which allows passage of water, when it exceeds a certain depth (i.e. discharge) to spill out into the basin. Figure 7.28 shows that low flows ($Q < Q^*$) pass the basin and continue downstream in the main storm sewer, while part

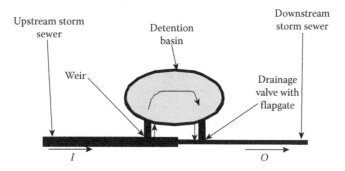

Figure 7.26 Detention basin in parallel (plan view).

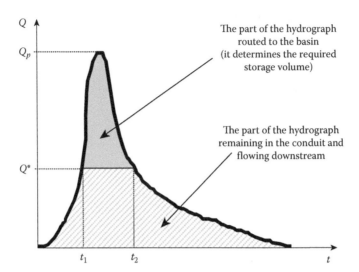

Figure 7.27 Reduction of the peak of the hydrograph in a detention basin placed in parallel.

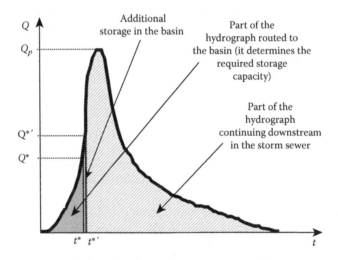

Figure 7.28 Determination of hydrograph routed to an infiltration basin in parallel.

of high flows ($Q > Q^*$, $t_1 < t < t_2$) is transferred to the detention basin. The detention basin is either designed to infiltrate the stored water or this can return to the main storm sewer, after the flood passes, through a pipe that connects to the storm sewer at a lower point (Figure 7.27). A detail of this structure is similar to that presented in Figure 7.21.

BMP 26 – Basin for infiltration of first flush in parallel with the storm sewer: The basin is again placed in parallel with the storm sewer (Figure 7.26). However, the hydraulic structure that transfers water to the basin is different from the one described in the previous BMP. The aim here is not to reduce the hydrograph peak (Figure 7.27) but to transfer to the basin the first flush, i.e. the initial part of the hydrograph, or about 13–25 mm of runoff (Figure 7.28). The difference is seen easily by comparing Figures 7.27 and 7.28. A typical hydraulic structure design to route the part of the hydrograph to the basin is depicted in Figure 7.29. Instead of the side weir of the previous BMP, which is used to transfer flows higher of a certain flow,

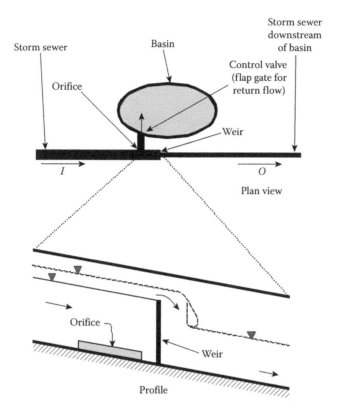

Figure 7.29 Basin plan view and hydraulic structure for an infiltration basin for first flush.

a weir is placed across the storm sewer width to block flows below a certain value ($Q < Q^*$). In addition, a side or bottom orifice has to be placed in the storm sewer upstream of the weir to receive this flow and transfer it into the basin. The concept is presented in Figure 7.29. The bed of the basin is constructed at the same level as the storm sewer invert at the point of connection; in this case, the maximum water depth in the basin will be the difference of the level of the crest of the weir and the storm sewer invert. There should also be a return flow control valve (flap gate towards the basin) which permits water to enter the basin from the storm sewer but does not allow flow in the opposite direction; thus, water is trapped in the basin.

BMP 27 – Extended dry basin: As mentioned, water quality benefits from using detention basins are indirect and only come from reducing the peak of the hydrograph. The main reason is that such structures are mostly designed having in mind flood control; therefore, the design is based on infrequent storm events. As a result, runoff from frequent events carrying pollutants passes through the basin relatively fast and pollutants are not retained.

Extended dry basins are detention basins properly configured and designed to offer water quality improvement of runoff. Water of frequent storms is retained for about 24 h, something that leads to about 90% removal of solids. However, nitrogen and phosphorus are not removed. A configuration of an extended dry basin is shown in Figure 7.30. The hydraulic structure of the outflow is designed to provide the required hydraulic retention time (HRT) for the runoff of a frequent flood to remain in the basin, but also it provides the storage for a larger event. This is done as follows: (1) The basin has two operational parts, the upper and the lower ones. The lower one is intermittently flooded and retains water for at least 24 h to improve water quality. The upper part is used for flood detention of less frequent events.

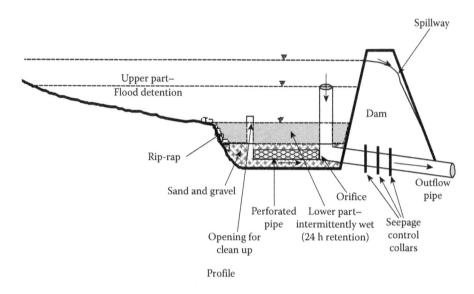

Profile

Figure 7.30 Extended dry pond.

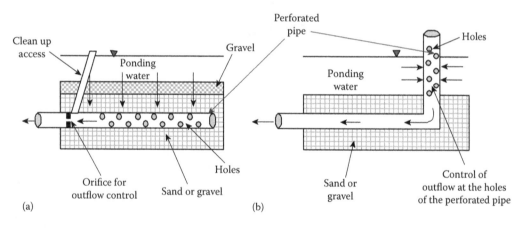

Figure 7.31 Alternative outflow hydraulic structure details from the lower part of the extended dry pond: (a) Control by orifice and (b) control by perforated standpipe.

(2) The outflow hydraulic structure drains slowly the lower part (in 24 h or more). For this, a common design is to use a perforated vertical or horizontal pipe (Figure 7.31) placed in gravel, or an orifice, or a combination of both.

The storage capacity of the lower part can be determined based on either of the following criteria (Schueler, 1987): (1) runoff from a storm event of a 1-year return period and duration of 24 h with an HRT of 24 h, (2) runoff from a storm event of a 2-year return period and duration of 24 h with an HRT of 24 h and (3) first flush (i.e. 13–25 mm or runoff) with an HRT of 24 h. The upper part of the basin is dimensioned using the methods presented in BMPs 24 or 25.

The main mechanism of the extended dry pond in pollutant removal is deposition. Therefore, the effectiveness is on solid removal, while the removal of dissolved pollutants is limited. According to Schueler (1987), the removal of solids depends on HRT and varies from 60%–70% for HRT of 6 h to 80%–90% for HRT of 48 h. Greater HRTs can result

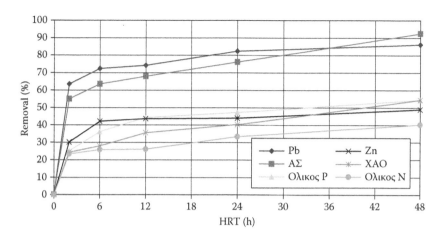

Figure 7.32 Percent removal of various pollutants as a function of HRT. (From Schueler, T.R., *Controlling Urban Runoff: A Practical Manual for Planning and Designing Urban BMPs*, Washington Metropolitan Water Resources Planning Board, Washington, DC, July 1987.)

in almost complete removal. Observed long-term removal was 65% for HRT between 6 and 12 h. Other pollutants are removed in lower percentages. For example, a maximum of 40%–50% was observed for phosphorus, nitrogen, BOD and COD for HRT of 32–48 h. Metals are removed at high percentages when easily adsorbed to solids (like Pb) and to a lesser degree when they are in dissolved form (like Zn). Finally, extended dry ponds remove bacteria by an order of magnitude and hydrocarbons by 60%–70% at HRT of 32 h. Figure 7.32 presents removals for various constituents in relation to the HRT.

The design of the extended dry pond is based on the following criteria (Schueler, 1987):

1. The lower level should be designed for at least the first flush ($R = 13$–25 mm). Better performance can be achieved when designing to detain the runoff from the 1-year or 2-year storm of a 24 h duration. In general, the lower level should detain 50%–90% of the annual rainfall events.
2. Based on Figure 7.32, the residence time, i.e. the time it takes for the lower part to empty, should be at least 24 h up to 48 h. As mentioned, a special design of the outlet structure is needed, so that drainage is slow and takes the required time. One option is the one schematically shown in Figure 7.31a; in this case, one orifice controls the outflow. Therefore, the design parameter is the diameter of this orifice. The other option would be to have a structure similar to the one shown in Figure 7.31b, where design parameters are the diameter of the holes of the perforated pipe, the level of each hole and the number of holes. The first case is easier to design but may suffer from clogging problems, since it is embedded underneath the basin bed, and sediments and debris may be a problem. The other option is more difficult to design, but it may work better with lower clogging risk.
3. Planting of the lower part of the basin and creation of a wetland may lead to increased nitrogen and phosphorus removals. Common depths, in this case, should not exceed 15–30 cm.
4. Riprap should be placed on the banks and wherever there is a chance for erosion.
5. The basin banks should be between 3:1 and 20:1.
6. A clear zone 8 m wide should surround the basin. Trees around the basin area increase aesthetics and ecological values.

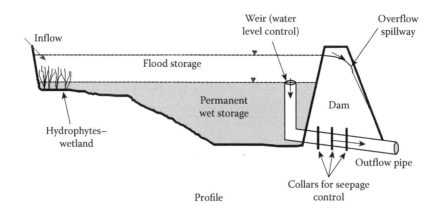

Figure 7.33 A Schematic diagram of a longitudinal section of a wet basin.

7. An access road should be provided (3 m wide) for maintenance, which should take place at least twice a year, and should include repairs, cleaning of outlet structure and sediment removal, among others.
8. Basin catchment area should be of the order of 4–8 ha.

BMP 28 – Wet basin: Wet basins (Figure 7.33) are good alternatives to extended dry basins, because they are more effective in removing sediments, BOD and trace metals. They also remove, to a large degree, dissolved pollutants, such as nitrate and orthophosphate. Finally, due to the permanent water, they offer more ecological and aesthetics opportunities, e.g. they can be used in parks. Depending on the design, they can reduce the hydrograph peak and probably, to a lesser degree, reduce the volume of runoff. The minimum size of a drainage basin to include a wet pond is 8 ha.

The mechanisms acting in wet ponds include (1) deposition: the pond acts as a settling basin for solids–solid deposition is the main process enhanced by the existence of permanent water and, possibly, by growing hydrophytes; (2) biological processes: these occur mainly through growing plants and/or algae and contribute to dissolved constituent removal (i.e. nutrients); and (3) decomposition by bacteria: these occur in the bottom sediments and concern decomposition of phosphorus, nitrogen and organic compounds. Typical percentages of pollutant removals in wet ponds are as follows (Schueler, 1987): 54% for solids, 20% for COD, 51% for Zn, 65% for Pb, 20% for organic nitrogen, 20% for organic phosphorus, 60% for dissolved nitrogen and 80% for dissolved phosphorus.

The performance of the wet basin in removing pollutants depends on its size compared to the catchment area. Figures 7.34 and 7.35 show this dependence. In Figure 7.34, pollutant removal is related to the ratio of basin wet volume (V_b) to runoff volume (V_R) from the mean precipitation over the entire watershed (Schueler, 1987). Alternatively, Driscoll (1988) presented the relation of pollutant removal as a function of the percentage of the basin wet surface area (A_b) to the watershed area A (Figure 7.35). In both graphs, one can see that curves reach equilibrium, i.e. pollutant removals do not increase significantly after a certain basin size, and specifically, when the basin volume exceeds about four times the runoff of the mean precipitation, or the area of the basin exceeds about 1–2% of the watershed surface area.

The effectiveness of the wet basin in removing solids can be evaluated by treating it as a settling basin, i.e. one has to calculate the settling velocity of the expected solid particles based on Stokes law; a particle of a certain size settles when the vertical component of the flow velocity is less than the particle settling velocity.

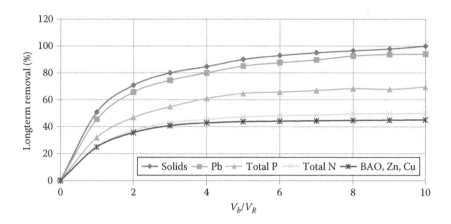

Figure 7.34 Wet basin performance as a function of the ratio of wet basin volume to the volume of runoff from the mean precipitation over the entire watershed. (Schueler, T.R., *Controlling Urban Runoff: A Practical Manual for Planning and Designing Urban BMPs*, Washington Metropolitan Water Resources Planning Board, Washington, DC, July 1987.)

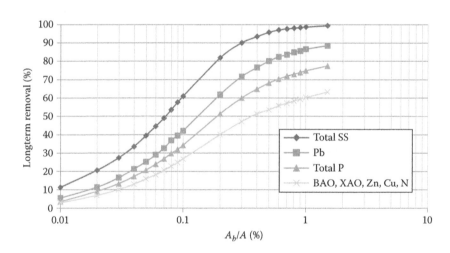

Figure 7.35 Removal of various pollutants as a function of the ratio of the surface area of the wet basin to the drainage area. (From Driscoll, E.D., Long-term performance of water quality ponds, in *Design of Urban Runoff Quality Controls, Proceedings of the Engineering Foundation Conference*, ASCE, New York, 1988, pp. 145–162.)

The design of the wet basins should follow certain criteria (Schueler, 1987):

1. Long and narrow basins are preferred over short and wide to avoid 'short-circuiting', i.e. flow passing quickly through the basin without replacing the existing water. Minimum ratio of length to width is 3 to 1. When this is not possible to keep, then baffles can be placed in the basin to guide the flow to follow a sinuous path instead of a straight one.
2. Shallow basins are preferred over deep ones to enhance solids deposition. A minimum depth of 0.6 m should be considered to minimize bottom sediment resuspension, and a maximum depth of 2.5 m to avoid thermal stratification. The depth should be shallower (around 0.3 m) at the edges of the basin towards the banks to enhance

hydrophyte growth; this increases efficiency in treatment, contributes to aesthetics improvement and recreation opportunities, offers ecological values and increases safety. A surrounding zone with vegetation, 8–10 m wide, is also recommended.

3. The banks should have a maximum slope 3:1 to avoid erosion, and should be preferably planted with grass. Rip rap should be placed at erosion-prone areas. Areas downstream of pipe outlets should also be protected.
4. Wet basins are recommended at watersheds with a minimum size of 8 ha.
5. Soil types below the wet basin should be of categories C or D. Basins in soils of categories A or B are not recommended unless they are isolated using a geomembrane or a clay layer.
6. Frequent maintenance is mandatory, consisting of grass cutting, removing of solid and other waste, and checking for erosion, settlements and structure stability, clogging of hydraulic structures, good operation, etc. An access road is required.
7. The design of the wet basin, particularly when it is of relatively large size, should also consider the effects of thermal impacts and anoxic conditions on the downstream stream where the basin discharges.

BMP 29 – Wetlands: Wetlands and mostly constructed wetlands have been successfully used in treating stormwater runoff from urban and agricultural areas. They are effective systems in removing organic matter, nitrogen and phosphorus, and heavy metals. The depth of water in these systems varies from 15 to 30 cm. Common vegetation used includes *Typha* sp., *Scirpus* sp. and *Phragmites* sp. There are now published several specialized books on wetlands and constructed wetlands (e.g. Reed et al., 1995; Kadlec and Wallace, 2009; Stefanakis et al., 2014).

BMP 30 – Detention basins for combined sewers (in series): Detention basins for combined sewers are designed either in series or in parallel to the sewer line, similar to those for storm sewers. They are usually sized for the first flush. They are used to control CSO, but also to minimize human health and odour problems; thus, since combined sewers are old systems, in most cases, basins are constructed underground, mostly because usually above-ground space is not available for their construction; nevertheless, open basins can be constructed at locations distant from settlements.

Figure 7.36 presents the operation of a combined sewer and includes the sewer, the WWTP and the aquatic system. A hydraulic structure (flow divider, hydraulic switch) is also shown in Figure 7.36, which is used to direct the flow either to the WWTP or to the aquatic system. In the second case, CSO occurs and untreated wastewater ends in the aquatic system. Combined sewers are usually designed for a 5- to 10-year return period. The diameter of the influent pipe to the WWTP usually has much lesser capacity than the sewer and is usually designed to bring the peak of the sewage flow and probably some nuisance water draining on the streets, i.e. five to six times, the daily mean sewage flow. This flow is called critical flow (Q_c). During dry weather or light precipitation, the flow Q in the sewer is less than Q_c, and the flow divider directs the flow towards the WWTP (Figure 7.36a). After treatment, the flow returns to the combined sewer and, finally, outflows to the aquatic system. During wet weather with Q exceeding Q_c, the flow divider allows only Q_c to go to the WWTP and the remaining flow ends directly and untreated to the aquatic system (Figure 7.36b), which can have adverse impacts on both humans and aquatic life. This is done to avoid flooding of the WWTP and washing out of the bacteria with adverse impact on its operation.

Detention basins for combined sewers in series operate similarly to those for separate sewers (BMP 26). However, in case they are placed underground, they may need pumping of the stored CSO instead of drainage by gravity. A typical layout of a basin placed in series is shown in Figure 7.37. As seen, the basin is placed downstream of the flow divider; therefore,

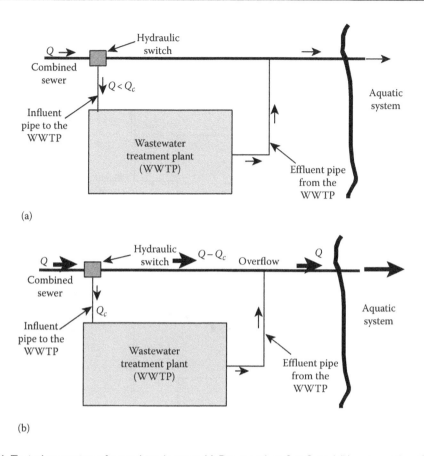

(a)

(b)

Figure 7.36 Typical operation of a combined sewer. (a) Dry weather $Q < Q_c$ and (b) wet weather $Q > Q_c$.

it operates only when $Q > Q_c$. When $Q \leq Q_c$, the system operates as presented in Figure 7.36a, and the basin stays dry since the entire flow is diverted by the flow divider to the WWTP. When $Q > Q_c$, the flow continues passing through the flow divider and reaches the basin. Initially, the basin fills up as shown in Figure 7.37a, and there is no flow in the downstream sewer ending into the aquatic system. Therefore, the basin controls CSO outflow to the aquatic system. Flow Q_c is diverted from the basin to the WWTP, which, after treatment, is discharged into the aquatic system. CSO to the aquatic system starts after the capacity of the basin is exceeded (Figure 7.37b). Nevertheless, there is still water stored above the spillway crest, and, even in this case, there is improvement since the flow peaks are lowered and the time base of the hydrograph is lengthened. Computation can be done using the Puls method for flood routing.

BMP 31 – Detention basins for combined sewers (in parallel): These are placed in parallel to the combined sewer; therefore, a hydraulic structure should be used to divert the flow to the basin. A typical layout is shown in Figure 7.38 which is self-explanatory.

BMP 32 – Detention basins for combined sewer (inside the aquatic system): This method was developed and is used in Sweden (Field, 1990; Field et al., 1990). The advantage is that the structure is inside the aquatic system, solving problems of space availability; furthermore, these are light structures (in contrast to underground basins) and the construction cost is lower.

A typical layout is shown in Figure 7.39. The basin comprises nine compartments communicating with each other. Depending on the area of the watershed and the runoff volume,

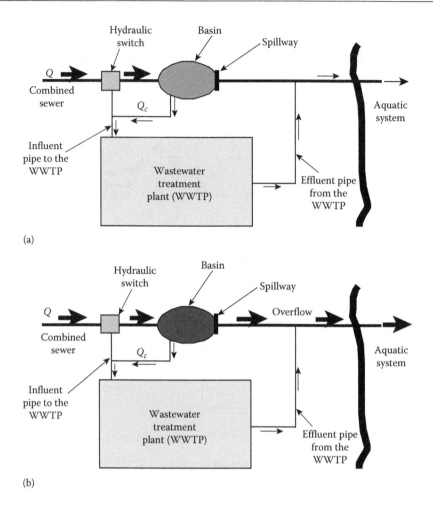

(a)

(b)

Figure 7.37 Detention basin for CSO control placed in series. (a) The basin is filling ($Q > Q_c$) and (b) the basin is full and spills ($Q > Q_c$).

the number and size of compartments can be designed accordingly. The last compartment communicates directly with the aquatic system. When $Q \leq Q_c$, the combined sewer flow ends into the WWTP and the basin is full of clean water from the aquatic system. When $Q > Q_c$, CSO reaches the first compartment of the basin, pushing clean water out into the second compartment. This continues as CSO continues until some or all of the compartments are filled. If the CSO volume is less than the capacity of the basin, CSO remains entrapped in the basin; otherwise, a portion of it remains in the basin and some escapes into the aquatic system. After the end of runoff, entrapped CSO is pumped back to the WWTP for treatment (Figure 7.39) and clean water fills the basin again (Novotny and Olem, 1994).

7.4.5 Runoff treatment

Conventional runoff treatment methods, similar to those used in wastewater treatment, are usually not an option. This is due to the relatively high volume of runoff, which would make application of such methods expensive. To reduce the cost, such methods could be employed only for treatment of the first flush. Alternatively, treatment can follow collection of runoff

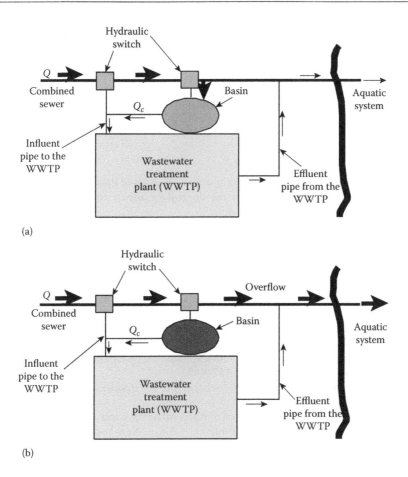

(a)

(b)

Figure 7.38 Detention basin for CSO control placed in parallel. (a) The basin is filling ($Q > Q_c$) and (b) the basin is full ($Q > Q_c$).

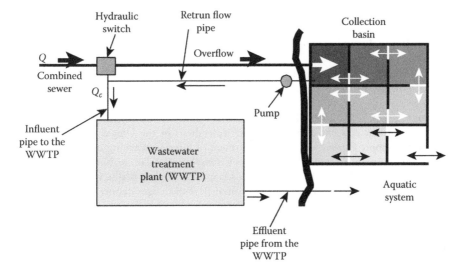

Figure 7.39 Detention basin for combined sewer inside the aquatic system.

in basins, and then pumping at slow rates in the treatment facility. Most of the time, such methods are used to treat collected combined sewer flow, particularly those based on biochemical processes. Most methods can be found in textbooks on wastewater treatment and will not be described here. They include those based on physicochemical processes (e.g. settling, separation) and biochemical methods (e.g. biodiscs, anaerobic ponds, aerated ponds, facultative ponds, constructed wetlands). Particularly, for CSO, disinfection is also used.

REFERENCES

ASCE, 1970, Design and construction of sanitary sewers, Manual of Practice No. 37, Reston, Virginia.

Beale, D.C., 1992, Recent developments in the control of urban runoff, *Journal of Water and Environmental Management*, 6(2), 141–150.

Bomboi, M.T. and Hernandez, A., 1991, Hydrocarbons in urban runoff: Their contribution to the wastewaters, *Water Research*, IAWPRC, 25, 557–565.

Bomboi, M.T., Hernandez, A., Marino, F. and Hontoria, E., 1990, Applications of multivariate analysis for characterization of organic compounds from urban runoff, *The Science of the Total Environment*, 93, 523–536.

Bosznay, M., 1989, Generalization of SCS curve number method, *J. Irrigation Drainage Eng.*, ASCE, 115(1), 139–144.

Boyd, G. and Gardner, N., December 1990, Urban stormwater: An overview for municipalities, *Public Works*, pp. 39–42.

Boyer, K.W. and Laitinen, H.A., 1975, Automobile exhaust particulates: Properties of environmental significance, *Environ. Sci. Technol*, 9(5), 457–469.

Bransby Williams, G., September 1922, Flood discharge and the dimensions of spillways in India, *The Engineer (London)*, 121, 321–322.

Cazier, D.J. and Hawkins, R.H., 1984, Regional application of the curve number method, *Proc. ASCE Irrigation Drainage Division Specialty Conference*, ASCE, New York.

Chow, V.T., 1959, *Open-Channel Hydraulics*, McGraw-Hill, Inc., New York.

Chui, T.W., Mar, B.W. and Hornet, R.R., 1982, A pollutant loading model for highway runoff, *Journal of Environmental Engineering*, ASCE, 108(6), 1193–1210.

City of Seattle, Washington, 30 June 1989, *Water Quality, Best Management Practices Manual*, Seattle, WA.

Clark, D.E. and Cobbins, W.C., 1963, Removal effectiveness of simulated dry fallout from paved areas by motorized vacuumized street sweepers, U.S. Naval Radiological Defense Laboratory, San Francisco, CA.

Debo, T.N. and Reese, A., 2002, *Municipal Stormwater Management*, 2nd Edition, CRC Press, Boca Raton, FL, 1176p.

Driscoll, E.D., 1988, Long-term performance of water quality ponds, in *Design of Urban Runoff Quality Controls*, Proceedings of the Engineering Foundation Conference, ASCE, New York, pp. 145–162.

Driver, N.E. and Tasker, G.D., 1990, Techniques for estimation of storm-runoff loads, volumes, and selected constituent concentrations in urban watersheds in the United States, U.S. Geological Survey Water Supply Paper 2363, U.S. Department of Interior, Washington, DC.

Ellis, J.B., 1986, Pollutional aspects of urban runoff, in *Urban Runoff Pollution*, NATO ASI Series, Vol. G10, H.C. Torno, J. Marsalek and M. Desbordes (eds.), Springer-Verlag, Berlin, Germany, pp. 1–31.

Federal Aviation Agency, 1970, Advisory circular on airport drainage, Report A/C 150-5320-58, U.S. Department of Transportation, Washington, DC.

Ferrara, R.A., 1986, Toxic pollutants: Impact and fate in receiving waters, in *Urban Runoff Pollution*, NATO ASI Series, Vol. G10, H.C. Torno, J. Marsalek and M. Desbordes (eds.), Springer-Verlag, Berlin, Germany, pp. 423–435.

Field, R., 1990, Combined sewer overflows: Control and treatment, in *Control and Treatment of Combined-Sewer Overflows*, P.E. Moffa (ed.), Van Nostrand Reinhold, New York, pp. 119–191.

Field, R., Dunkers, K. and Forndran, A., 1990, Demonstration of in-receiving water storage of combined sewer overflows in marine/estuarine environment by the flow balance method, in *Proceedings of the Fifth International Conference of Urban Storm Drainage*, University of Osaka, Osaka, Japan, 23–27 July 1990, pp. 759–764.

French, R.H., 1985, *Open Channel Hydraulics*, McGraw Hill, New York.

Gikas, G.D. and Tsihrintzis, V.A., 2012, Assessment of water quality of first-flush roof runoff and harvested rainwater, *Journal of Hydrology*, 466–467, 115–126.

Gjessing, E., Lygren, E., Berglind, L., Gulbrandsenm, T. and Skaane, R., 1984, Effects of highway runoff on lake water quality, *Science of the Total Environment*, 33, 245–257.

Goldman, S.J., Jackson, K. and Bursztynsky, T.A., 1986, *Erosion & Sediment Control Handbook*, McGraw-Hill Book Co., New York.

Griffin, D.M., Grizzard, T.J., Randall, C.W., Helsel, D.R. and Hartigan, J.P., 1980, Analysis of non-point pollution export for small catchments, *Journal of Water Pollution Control Federation*, WEF, 52(4), 780–790.

Gupta, M.K., Agnew, R.K. and Kobringer, N.P., 1981, *Constituents of Highway Runoff*, Vol. I, FHWA/RD-81/045, Federal Highway Administration, Washington, DC.

Gupta, R.S., 1989, *Hydrology and Hydraulic Systems*, Prentice Hall, Englewood Cliffs, NJ.

Hartigan, J.P., 1989, Basis for design of wet detention basin BMPs, in *Design of Urban Runoff Quality Controls*, L.A. Roesner, B. Urbonas and M.B. Sonnen (eds.), ASCE, New York, pp. 290–304.

Heaney, J., Huber, W.C., Sheikh, H., Medina, M.A., Doyle, J.R., Peltz, W.A. and Darling, J.E., 1977, *Nationwide Evaluation of Combined Sewer Overflows and Urban Stormwater Discharges*, Vol. 2: *Cost Assessment and Impacts*, EPA-600/2-77-064, Environmental Protection Agency, Washington, DC.

Hewitt, C.N. and Rashed, M.B., 1992, Removal rates of selected pollutants in the runoff waters from a major rural highway, *Water Research*, IAWRPC, 26, 311–319.

Hoffman, R.W., Goldman, C.R., Paulson, S. and Winters, G.R., 1981, Aquatic impacts of deicing salts in the central Sierra Nevada mountains, California, *Water Resources Bulletin*, AWRA, 17(1), 280–285.

Huber, C.W. and Dickinson, R.E., 1988, *Storm Water Management Model, Version 4: Users Manual*, EPA 600/3-88/001a, Environmental Research Laboratory, USEPA, Athens, GA.

Hunter, J.V., Sabatino, T., Gomperts, R., and Mackenzie, M.J., 1979, Contribution of urban runoff to hydrocarbon pollution, *Journal of Water Pollution Control Federation*, WEF, 51, 2129–2138.

Izzard, C.F., 1944, The surface-profile of overland flow, *Transactions of the American Geophysical Union*, 25, 959–968.

Kadlec, R.H. and Wallace, S.D., 2009. *Treatment Wetlands*, 2nd edn., CRC Press, Boca Raton, FL.

Kerby, W.S., June 1959, Time of concentration for overland flow, *Civil Engineering*, 10(6), 362.

King County, Washington, 1990, *Surface Water Design Manual*, Department of Public Works.

Mandelker, D.R., 1989, Controlling nonpoint source water pollution, *Chicago-Kent Law Review*, 65(2), 479–502.

Marsalek, J., 1978, Pollution due to urban runoff: Unit loads and abatement measures, pollution form land use activities reference group, International Joint Commission, Windsor, Ontario, Canada.

Maryland Department of the Environment, Sediment and Stormwater Administration, 1984, Maryland standards and specifications for stormwater management infiltration practices, Baltimore, MD.

Maryland Department of the Environment, Sediment and Stormwater Division, 1987, Design procedures for stormwater management detention structures, Baltimore, MD.

Maryland Department of Natural Resources, Sediment and Stormwater Administration, 1985, Inspector's guidelines manual for stormwater management infiltration practices, Baltimore, MD.

Maryland Department of Natural Resources, Sediment and Stormwater Division, 1987, Guidelines for constructing wetland stormwater basins, Baltimore, MD.

Metropolitan Washington Council of Governments, Department of Environment Programs, 1992, Analysis of urban BMP performance and longevity, Metropolitan Washington Council of Governments, Washington, DC.

Minnesota Pollution Control Agency, 1992, Protecting water quality in urban areas, St Paul, MN.

Norman, C.G., 1991, Urban runoff effects on Ohio river water quality, *Water Environment and Technology*, 3, 44–46.

Novotny, V. and Olem, H., 1994, *Water Quality – Prevention, Identification, and Management of Diffuse Pollution*, Van Nostrand Reinhold, New York.

Overton, D.E. and Meadows, M.E., 1976, *Storm Water Modeling*, Academic Press, New York.

Pitt, R., 1979, *Demonstration of Nonpoint Pollution Abatement through Improved Street Cleaning Practices*, EPA 600/2-79-161, U.S. Environmental Protection Agency, Cincinnati, OH.

Pitt, R. and Bozeman, M., 1980, *Water Quality and Biological Effects of Urban Runoff on Coyote Creek*, Phase I – Preliminary Survey Report, EPA-600/2-80-104, U.S. Environmental Protection Agency, Cincinnati, OH.

Ponce, V.M. and Hawkins, R.H., 1996, Runoff curve number: Has it reached maturity?, *J. Hydrol. Eng.* ASCE, 1(1), 11–19.

Randall, C.W., Grizzard, T.J., Helsel, D.R., and Griffin, D.M., Jr., 1981, Comparison of pollutants mass loads in precipitation and runoff in urban areas, *Urban Stormwater Quality, Management and Planning*, Proceedings of the Second International Conference on Urban Storm Drainage, Urbana, IL, Water Resources Publications, Littleton, CO, pp. 29–38.

Reed, S., Middlebrooks, E. and Crites, R., 1995. *Natural Systems for Waste Management and Treatment*, McGraw Hill, New York.

Roesner, L.A., Urbonas, B. and Sonnen, M.B. (eds.), 1988, *Design of Urban Runoff Quality Controls*, Proceedings of an Engineering Foundation Conference on Current Practice and Design Criteria for Urban Quality Control, ASCE, Potosi, MO, 10–15 July 1988.

Sartor, J.D. and Boyd, G.B., 1972, *Water Pollution Aspects of Street Surface Contamination*, EPA R2-72-081, U.S. Environmental Protection Agency, Washington, DC.

Schueler, T.R., July 1987, *Controlling Urban Runoff: A Practical Manual for Planning and Designing Urban BMPs*, Washington Metropolitan Water Resources Planning Board, Washington, DC.

Schueler, T.R., Galli, J., Herson, L., Kumble, P. and Shepp, D. 1992a, Developing effective BMP system for urban watersheds: Analysis of urban BMP performance and longevity, Metropolitan Washington Council of Governments, Department of Environment Programs, Washington, DC.

Schueler, T.R., Kumble, P.A. and Heraty, M.A., 1992b, *A Current Assessment of Urban Best Management Practices. Techniques for Reducing Non-point Source Pollution in the Coastal Zone*, Technical Guidance Manual, Metropolitan Washington Council of Governments, Office of Wetlands, Oceans and Watersheds, U.S. Environmental Protection Agency, Washington, DC.

Simons, D.B. and Senturk, F., 1992, *Sediment Transport Technology – Water and Sediment Dynamics*, Water Resources Publications, Littleton, CO.

Sonzogni, W.C. et al., 1980, Pollution from land runoff, *Environmental Science and Technology*, 14(2), 148–153.

Stahre, P. and Urbonas, B., 1990, *Stormwater Detention for Drainage, Water Quality and CSO Management*, Prentice Hall, Englewood Cliffs, NJ.

Stefanakis, A., Akratos, C. and Tsihrintzis, V.A., 2014, *Vertical Flow Constructed Wetlands: Eco-Engineering Systems for Wastewater and Sludge Treatment*, Elsevier, 378pp. ISBN: 978-0-12-404612-2.

Stenstrom, M.K., Silverman, G. and Burszlynsky, T.A., 1984, Oil and grease in urban stormwater, *Journal of Environmental Engineering*, ASCE, 110(1), 58–72.

Stubchaer, J.M., 1975, The Santa Barbara urban hydrograph method, *Proc. National Sympos. on Urban Hydrology and Sediment Control*, University of Kentucky, Lexington, July 28–31, pp. 131–141.

Torno, H. (ed.), 1989, Urban stormwater quality enhancement-source control, retrofitting, and combined sewer technology, in *Proceedings of an Engineering Foundation Conference*, Davos Planz, Switzerland, 22–27 October 1989.

Tsihrintzis, V.A. and Baltas, E., 2014, Determination of rainwater harvesting tank size, *Global NEST Journal*, 16(5), 822–831.

Tsihrintzis, V.A. and Hamid, R., April 1997a, Modeling and management of urban stormwater runoff quality: A review, *Water Resources Management*, 11(2), 137–164.

Tsihrintzis, V.A. and Hamid, R., February 1997b, Urban stormwater quantity/quality modeling using the SCS method and empirical equations, *Journal of the American Water Resources Association*, 33(1), 163–176.

Tsihrintzis, V.A. and Hamid, R., 1998, Runoff quality prediction from small urban catchments using SWMM, *Hydrological Processes*, 12(2), 311–329.

Tsihrintzis, V.A. and Sidan, C.B., 1998, Modeling urban runoff processes using the Santa Barbara method, *Water Resources Management*, EWRA, 12(2), 139–166.

Urbonas, B. and Roesner, L.A. (eds.), 1986, Urban runoff quality-impact and quality enhancement technology, in *Proceedings of an Engineering Foundation Conference*, New England College, Henniker, NH, 23–27 June 1986.

U.S. Army Corps of Engineers (USCOE), 1977, *STORM: Storage, Treatment, Overflow Runoff Model – User's Manual*, 723-S8-L7520, Hydrologic Engineering Center, Davis, CA.

U.S. Department of Agriculture (USDA), 1951, Soil survey manual #18, U.S. Department of Agriculture (USDA), Washington, DC.

U.S. Department of Agriculture, Natural Resources Conservation Service (USDA-NRCS), 1986, *Urban Hydrology for Small Watersheds*, TR-55.

U.S. Department of Agriculture, Natural Resources Conservation Service (USDA-NRCS), 2012, *National Engineering Handbook*, http://www.nrcs.usda.gov/wps/portal/nrcs/detailfull/national/water/?cid=stelprdb1043063, accessed January 28, 2016.

U.S. Department of Agriculture, Soil Conservation Service (USDA-SCS), 1972, *National Engineering Handbook*, Section 4, Hydrology, Washington, DC.

U.S. Department of Agriculture, Soil Conservation Service (USDA-SCS), 1985, *National Engineering Handbook*, Section 4, Hydrology, Washington, DC.

U.S. Environmental Protection Agency (USEPA), 1974, *Water Quality Management Planning for Urban Runoff*, EPA 440/9-75-004, U.S. Environmental Protection Agency, Washington, DC.

U.S. Environmental Protection Agency (USEPA), 1979, *1978 Needs Survey – Continuous Stormwater Pollution Simulation System, User's Manual*, EPA 430/9-79-004, U.S. Environmental Protection Agency, Washington, DC.

U.S. Environmental Protection Agency (USEPA), 1983, *Results of the Nationwide Urban Runoff Program – Volumes I, II, III, IV*, Water Planning Division, U.S. Environmental Protection Agency, Washington, DC.

Vanoni, V.A. (ed.), 2006, *Sedimentation Engineering*, ASCE Manuals and Reports on Engineering Practice No. 54, ASCE, Reston, VA.

Walesh, S.G., 1989, *Urban Surface Water Management*, John Wiley & Sons, New York, USA, 518p., ISBN: 0-471-83719-9.

Waller, D. and Hart, W.C., 1986, Solids, nutrients and chlorides in urban runoff, in *NATO ASI Series, Vol. G10, Urban Runoff Pollution*, H.C. Torno, J. Marsalek and M. Desbordes, (eds.), Springer-Verlag, Berlin, Germany, pp. 59–85.

Wanielista, M., 1990, *Hydrology and Water Quantity Control*, John Wiley & Sons, New York, 565p., ISBN: 0-471-62404-7.

Wanielista, M., Kersten, R. and Eaglin, R., 1997, *Hydrology: Water Quantity and Quality Control*, 2nd edn., John Wiley & Sons, New York.

Chapter 8

Sediment transport and erosion

8.1 INTRODUCTION

A part of hydraulic engineering is devoted to the so-called sediment transport science or sediment transport engineering or sediment transport technology. All these are nearly synonymous terms, describing the engineering science that (1) deals with the phenomena and processes taking place not only in fluvial systems (on which we concentrate in this chapter) but also on bare soil subjected to precipitation where sediments are generated, lakes or reservoirs and estuaries receiving river sediment-laden water and the coastal environment where sediments are finally deposited and transported alongshore, and (2) provides engineering solutions to alleviate adverse problems of erosion, sediment transport and deposition. Therefore, *sediment transport* phenomena or processes deal with the interrelationship between the sediment particles and the flowing or nearly stagnant water. Typical processes include (1) the impact of rainfall that drops on soil, soil grain detachment and erosion, (2) sediment entrainment by runoff and transportation, (3) channel aggradation/degradation and (4) sediment deposition. These processes are often described in brief or in a simplified way, as erosion, sediment movement or transport, deposition or sedimentation.

We will deal with some of these in this chapter. Our emphasis will be on presenting the basic theory, addressing sediment-related problems and providing engineering solutions. Of course, the chapter is introductory, does not address all these exhaustively, and presents only representative and commonly used methods; the reason for this is that there are several books written specifically to the subject of erosion and/or sediment transport. Such books are listed in the reference list for the interested reader.

So, what may be these sediment-related problems that need engineering solutions? The following list is not exhaustive but presents the most typical problems requiring the knowledge of erosion and sediment transport theory and practice:

- Uphill soil erosion may cause slope instability.
- Soil erosion of agricultural land may result in loss of fertile soils.
- Urbanization and road or highway construction increase erosion and sediment yields.
- River general degradation near bridges and/or local scour may endanger bridge stability.
- River bank erosion may result in bank failure and loss of land.
- River aggradation may cause local or general flooding.
- Sediment deposition may affect river navigable depth.
- Sediment-laden rivers cause reservoir silting and loss of storage capacity, affecting reservoir useful or operational life.
- Navigable river silting may require dredging.

- Sediment trapping in a reservoir may result in downstream river degradation.
- Sediment trapping in a reservoir may cause coastal sediment starvation and loss of valuable land from coastline erosion and retreat.
- Sediment and turbidity may have adverse ecological impacts on river or lake wildlife (e.g. affect fish).
- Sediments may be the carriers and storage agents of other pollutants (e.g. phosphorus, metals, pesticides), which can be released in time in aquatic systems where sediments are deposited, causing water quality problems.
- Sediments may influence water quality for water uses, such as irrigation, potable supplies and recreation.
- The presence of sediments may affect natural water aesthetics or the perception about its good quality.

Typical examples of engineering designs requiring the knowledge of sediment transport include soil erosion protection measures, design of debris basins, river channelization, hydraulic design of bridges, design of river degradation control and stability measures, channel bank protection measures, flood control design, reservoir design to minimize silting, and in-stream sand and gravel mining planning, among others.

8.2 PROPERTIES OF SEDIMENT

8.2.1 Size and shape

Size is the basic property of a single particle. However, when dealing with sediments, because of the great variation of particles in a sediment sample, average size values over groups of sizes are commonly used. The most well-accepted and commonly used sediment size scale is the one developed by Lane et al. (1947), which is presented in Table 8.1.

To determine size, either sieve analysis (for sand and larger sizes; Table 8.1) or visual-accumulation tube analysis (for smaller than sand sizes) is used. Such analyses result in the determination of the sieve diameter for a single particle in the sample or the group, and the development of the so-called particle size distribution curve (or sediment gradation curve), which presents the variation of the percentages of the weight of the various sizes in a sediment sample that is smaller than a given sieve size as function of the size. Three typical curves are presented in Figure 8.1. The vertical axis of this figure presents the percent of the total sample weight passing, i.e. being finer than a certain sieve size. The horizontal axis of the graph shows this sieve size. Curve A in Figure 8.1 shows a sediment sample with finer material compared to curves B and C. Curves B and C have, on the average, comparable sediment sizes; however, the sizes of curve B are spread over a wider range compared to those of curve C, i.e. the sample of curve C is more uniform in terms of particle sizes.

The sediment gradation curve allows for the definition of diameter D_x, where x implies a certain percentage on the vertical axis of the graph (Figure 8.1). Diameter D_x implies the sieve diameter that allows x percent (%) of the sample by weight to pass, i.e. the sieve diameter for which x percent of the sample is finer. A commonly used diameter is D_{50}, which implies the sieve diameter that allows 50% of the sample to pass. Other diameters used in various sediment transport formulas and other equations are $D_{10}, D_{35}, D_{40}, D_{65}, D_{85}, D_{90}, D_{15.9}, D_{84.1}$ and others. As an example, in Figure 8.1, for curve A, we have approximately $D_{50} = 0.8$ mm, $D_{80} = 1.1$ mm and $D_{10} = 0.5$ mm. Similarly, for curve C, $D_{50} = 2.3$ mm, $D_{80} = 2.65$ mm and $D_{10} = 1.95$ mm.

Table 8.1 Lane's sediment grade sizes

Class name		Range (mm)
Boulders	Very large	4000–2000
	Large	2000–1000
	Medium	1000–500
	Small	500–250
Cobbles	Large	250–125
	Small	125–64
Gravel	Very coarse	64–32
	Coarse	32–16
	Medium	16–8
	Fine	8–4
	Very fine	4–2
Sand	Very coarse	2–1
	Coarse	1–1/2
	Medium	1/2–1/4
	Fine	1/4–1/8
	Very fine	1/8–1/16
Silt	Coarse	1/16–1/32
	Medium	1/32–1/64
	Fine	1/64–1/128
	Very fine	1/128–1/256
Clay	Coarse	1/256–1/512
	Medium	1/512–1/1024
	Fine	1/1024–1/2048
	Very fine	1/2048–1/4096

Source: Lane, E.W., *Trans. Am. Geophys. Union*, 28(6), 936, 1947.

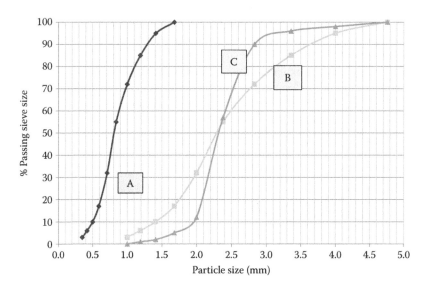

Figure 8.1 Typical sediment gradation curves.

Other common definitions of the size of a single sediment particle include the mean diameter (i.e. the arithmetic average of the longest, intermediate and shortest dimensions of the particle in three perpendicular axes), the nominal diameter (i.e. the diameter of a sphere with the same volume as the particle) and the fall diameter (i.e. the diameter of a sphere of specific gravity 2.65 which falls at the same terminal velocity with the particle in distilled water at 24°C).

In addition to size, an important parameter describing particle characteristics is the shape, which describes how the particle looks, i.e. its form regardless of its size. To quantify shape, there have been several parameters proposed. The most common parameter is the shape factor, most frequently given by the following expression:

$$SF = \frac{c}{\sqrt{ab}} \tag{8.1}$$

where a, b and c are the longest, intermediate and shortest dimensions of the particles, respectively, in three perpendicular axes.

8.2.2 Density, specific weight and specific gravity

The density ρ_s (i.e. particle mass per unit volume [kg/m^3]), the specific weight γ_s (i.e. particle weight per unit volume [N/m^3]) and the specific gravity S_g (i.e. the ratio of particle specific weight γ_s to that of water γ at 4°C) of sediment particles are also important parameters entering sediment transport computations. A usual value for the specific gravity, commonly used in computations, is 2.65; this is the specific gravity of quartz which is the most abundant mineral in river beds and sediments.

8.2.3 Fall velocity

The fall velocity is the terminal velocity of a single particle settling in still distilled water extending infinitely. It can be computed by balancing the particle buoyant weight that tends to move the particle downwards and the drag force resisting the motion. The main parameters then affecting the fall velocity are the particle size, shape, density and surface roughness, and the fluid density and viscosity. Expressions for fall velocity have been originally developed for spherical particles. Then, solutions have been presented considering the various factors affecting fall velocity of non-spherical particles, including effect of particle shape, boundaries, turbulence, sediment concentration and particle size distribution, among others.

The earlier discussions are outside the scope of this chapter and can be found in various specific textbooks (i.e. Simons and Sentürk, 1992; Yang, 1996; Vanoni, 2006). Here, we will present two formulas to compute the fall velocity, which, however, are valid when certain conditions hold.

The first formula, used in various applications in various fields, is the classic Stokes law (Stokes, 1851). It predicts the fall velocity of a spherical particle according to the following formula:

$$V_f = \frac{g}{18v} D^2 \frac{\gamma_s - \gamma}{\gamma} = \frac{g}{18v} D^2 (S_g - 1) \tag{8.2}$$

where

V_f is the fall velocity (m/s)
g is the gravitational acceleration (9.81 m/s²)
D is the particle diameter (m)
υ is the kinematic viscosity (m²/s)
γ_s and γ are the specific weights of particle and water (N/m³), respectively

The formula is valid for values of the particle Reynolds number $Re_p < 0.1$ (where Re_p is defined by $Re_p = V_f D/\upsilon$). This range includes quartz particles ($S_g = 2.65$) in the range of silt and clay ($D < 0.062$ mm; Table 8.1).

The second formula is Rubey's formula (Rubey, 1933), which can be used to compute fall velocities of quartz silt, sand and gravel particles (therefore, it has a wider application, particularly for fluvial engineering applications), and reads

$$V_f = \lambda \sqrt{Dg\frac{\gamma_s - \gamma}{\gamma}} = \lambda \sqrt{Dg\,(S_g - 1)} \tag{8.3}$$

where all the parameters have been defined already and λ is given by

$$\lambda = \begin{cases} 0.79 & \text{for } D > 1 \text{ mm} \\ \left[\dfrac{2}{3} + \dfrac{36\upsilon^2}{gD^3\,(S_g - 1)}\right]^{1/2} - \left[\dfrac{36\upsilon^2}{gD^3\,(S_g - 1)}\right]^{1/2} & \text{for } D \leq 1 \text{ mm} \end{cases} \tag{8.4}$$

8.3 FLOW RESISTANCE

Compared to lined (i.e. rigid wall, non-erodible) open channels, flow resistance in alluvial rivers and streams is difficult to describe and determine. One reason is that the cross-sectional shape and the bed configuration continuously change as the flow transports sediments. Various types of bed forms develop; which type prevails depends on the interrelationship of velocity, Froude number and resistance to flow, among other parameters. While in lined channels the flow resistance is only due to skin roughness, in alluvial channels the so-called form drag provides additional resistance to flow, and the resulting total resistance, also depending on the flow depth, is in most cases greater.

8.3.1 Channel flow resistance

Resistance to flow in open channels is presented in various publications (e.g. Chow, 1959; French, 1985; Simons and Sentürk, 1992; Yen, 1992). Three traditional equations are used for flow resistance computation in open channels: the Darcy–Weisbach, the Chézy and the Manning equations. These, respectively, read

$$V = \sqrt{\frac{8g}{f}}\sqrt{RS} \tag{8.5}$$

$$V = C\sqrt{RS} \tag{8.6}$$

$$V = \frac{1}{n} R^{2/3} S^{1/2} \tag{8.7}$$

where

  V is the mean flow velocity (m/s)
  R is the hydraulic radius of the cross section (m), defined as the ratio of flow area to the wetted perimeter
  S is the slope of the energy grade line (m/m), which in the case of uniform flow coincides with the bed and water surface slope
  f is the Darcy–Weisbach frictional factor
  C is the Chézy roughness coefficient ($m^{1/2}/s$)
  n is the Manning roughness coefficient ($s/m^{1/3}$)

The three resistance coefficients f, C and n depend on the flow Reynolds number Re, the boundary relative roughness and the shape of the channel (Chow, 1959; Tsihrintzis and Madiedo, 2000). f is expressed in terms of the Reynolds number in a Moody-type diagram (e.g. Chow, 1959) for the three flow states (i.e. laminar, transitional and turbulent). Similar diagrams have also been developed for C and n (e.g. Henderson, 1966). However, usually values for C and n are taken from tables as constants (see the following discussion), depending mostly on the channel wall material, condition and geometry (e.g. Cowan, 1956; Chow, 1959; Barnes, 1967; Arcement and Schneider, 1989; Yen, 1992; Tsihrintzis and Madiedo, 2000). The reason for this is that fully developed turbulent flow (i.e. turbulent flow that does not depend on the Reynolds number but only on relative roughness) is assumed, which is the case in almost all practical applications. Finally, a channel does not have a constant resistance coefficient under all occasions and conditions, but satisfactory characterization in terms of frictional factors has been achieved for the case of steady, uniform, sediment-free flows in channels of impervious rigid boundaries, with densely distributed, nearly homogeneous roughness on the wetted perimeter (Chow, 1959; Coyle, 1975; Simons and Sentürk, 1992; Guillen, 1996; Tsihrintzis and Madiedo, 2000).

If we define the shear velocity V_* as

$$V_* = \sqrt{gRS} \tag{8.8}$$

then Equations 8.5 through 8.7 become

$$V = \sqrt{\frac{8}{f}} V_* \tag{8.9}$$

$$V = \frac{C}{\sqrt{g}} V_* \tag{8.10}$$

$$V = \frac{1}{n\sqrt{g}} R^{1/6} V_* \tag{8.11}$$

From these, one can easily determine the relation between the three flow resistance coefficients:

$$\sqrt{\frac{8}{f}} = \frac{C}{\sqrt{g}} = \frac{1}{n\sqrt{g}} R^{1/6} \tag{8.12}$$

The shear velocity is related to the shear stress τ_o (or unit tractive force) exerted by the flow on the channel boundary by the following equation:

$$V_* = \sqrt{\frac{\tau_o}{\rho}} \tag{8.13}$$

Combining Equations 8.8 and 8.13,

$$\tau_o = \rho V_*^2 = \gamma R S \tag{8.14}$$

From the three presented resistance equations, the most commonly used in practical applications is the Manning equation. However, its application is limited to channels, smooth or rough, under fully turbulent conditions. In applying this equation, the proper determination of the Manning roughness coefficient n is very important. Chow (1959) presents the following factors affecting the n value: (1) the surface roughness of the material of the channel section, which is due to the size and shape of grains; (2) the vegetation present, which increases channel roughness, with determining factors the height, density, distribution and type of the vegetation (Tsihrintzis and Madiedo, 2000); (3) the channel irregularity, accounting for factors such as changes in cross-sectional shape and size, and irregularities of the bed along the channel; (4) the channel alignment, where sharp changes of curvature, turns and meandering increase n values; (5) silting and scouring, with both changing the n values of lined and unlined channels; (6) obstructions, such as bridge piers, which increase n values; (7) water depth, with n generally decreasing as depth increases; and (8) size and shape, which are factors with only a minor or no influence on n values.

Cowan (1956) developed a procedure to estimate the n value, based on the following formula, which considers most of the aforementioned factors:

$$n = (n_0 + n_1 + n_2 + n_3 + n_4)m \tag{8.15}$$

where n_0 is the basic n value for a straight uniform channel of a certain cross-sectional material and n_1, n_2, n_3 and n_4 are increases of n_0 accounting for surface irregularities, variation in shape and size of the channel section, obstructions and vegetation, respectively. Factor m accounts for channel meandering. Table 8.2 (Chow, 1959) provides values of factors n_1–n_4 and m based on channel conditions. Cowan's (1956) method applies according to the following limitations (Chow, 1959): (1) Water is sediment free. (2) The method is not applicable for channels with hydraulic radius larger than 5 m. (3) Channels are unlined with a minimum n_0 of 0.020.

A comprehensive set of n values for various channel materials and conditions are presented in Table 8.3 (after Chow, 1959). Photographs of channel sections associated with n values are provided by Chow (1959), Barnes (1967), French (1985), Arcement and Schneider (1989)

Table 8.2 Cowan (1956) procedure for the computation of the Manning roughness coefficient

Parameter in Equation 8.15	Condition	Description	Values
n_o – material of cross section	Earth	Basic n value for straight, uniform, smooth channels in natural materials	0.020
	Rock cut		0.025
	Fine gravel		0.024
	Coarse gravel		0.028
n_1 – degree of irregularity	Smooth	Best attainable for the materials involved	0.000
	Minor	Good dredged channels, slightly eroded or scoured side slopes of canals	0.005
	Moderate	Fair-to-poor dredged channels, moderately sloughed or eroded side slopes of canals	0.010
	Severe	Badly sloughed banks of natural streams, badly sloughed or eroded side slopes of canals, unshaped, jagged and irregular surface of channels in rock	0.020
n_2 – section shape and size variation	Gradual	Gradual change in size or shape of cross section	0.000
	Alternating occasionally	Occasionally alternating large and small sections or occasional shape changes and shifting or main flow from side to side	0.005
	Alternating frequently	Frequently alternating large and small sections or frequent shape changes and shifting or main flow from side to side	0.010–0.015
n_3 – obstructions	Negligible	Presence and characteristics of obstructions, i.e. debris, tree roots, boulders and logs, with factors considered such as the degree of reduction of water area, character of obstructions in inducing turbulence (sharp edge, smooth surfaced) and position/spacing of obstructions in longitudinal and transverse directions	0.000
	Minor		0.010–0.015
	Appreciable		0.020–0.030
	Severe		0.040–0.060
n_4 – vegetation	Low	Dense growths of flexible turf grasses, flow depth 2–3 times the vegetation height; or flexible tree seedlings, flow depth 3–4 times the vegetation height	0.005–0.010
	Medium	Turf grasses, flow depth 1–2 times the vegetation height; or stemmy grasses, weeds or tree seedlings, moderate cover, flow depth 2–3 times the vegetation height; or brushy growths, moderately dense on channel sides slopes, no vegetation on the channel bottom, hydraulic radius greater than 0.6 m	0.010–0.025
	High	Turf grasses, flow depth about vegetation height; or dormant season (similar to willows or cottonwood 8–10 years old, some weeds and brushes, no foliage, hydraulic radius greater than 0.6 m); or growing season (similar to bushy willows 1-year old, some weeds in full foliage, dense on channel sides slopes, no significant vegetation on the channel bottom, hydraulic radius greater than 0.6 m)	0.025–0.050

(Continued)

Table 8.2 (Continued) Cowan (1956) procedure for the computation of the Manning roughness coefficient

Parameter in Equation 8.15	Condition	Description	Values
	Very high	Turf grasses, flow depth less than half of vegetation height; or growing season (similar to bushy willows 1-year old, weeds in full foliage on channel side slopes or dense growth of cattails on the channel bottom, hydraulic radius up to 3–5 m); or growing season (similar to trees with weeds and brush, all in full foliage, hydraulic radius up to 3–5 m)	0.050–0.100
m – meandering	Minor	Ratio (meander length to the straight length of the channel reach) 1.0–1.2	1.000
	Appreciable	Ratio 1.2–1.5	1.150
	Severe	Ratio >1.5	1.300

Source: Adapted from Chow, V.T., *Open-Channel Hydraulics*, McGraw-Hill Inc., New York, 1959.

and Chaudhry (1993), among others, which can help to determine, by comparison, the proper *n* for a given stream, river or floodplain.

The Manning *n* for alluvial channels has been related by various researchers to the sediment particle size of the bed. Obviously, this *n* value refers to the skin or surface roughness of the bed. Typical suggested formulas have the following general form:

$$n = \frac{D_x^{1/6}}{b} \tag{8.16}$$

where

D_x, as mentioned earlier, is a certain representative diameter (m), with *x* representing the percentage of finer material according to the gradation curve of the bed material
b is an empirical coefficient

Table 8.4 summarizes the coefficients and units of various formulas of Equation 8.16 (Simons and Sentürk, 1992). Table 8.5 presents a comparison of *n* value predictions by all these formulas, which is also depicted in Figure 8.2. One can see that although a different representative diameter D_x is used in various formulas, predictions are relatively close to each other, with the exception probably of Keulegan's Equations 8.16d and e. The Meyer–Peter and Müller (1948) formula (Equation 8.16c) compares well with Strickler's (1923) formulas (Equations 8.16a and b) and Keulegan's formula (1949) (Equation 8.16f). The Meyer–Peter and Müller (1948) formula (Equation 8.16c) is a well-tested equation and is recommended for sand and gravel beds of alluvial streams. The Lane and Carlson (1953) formula (Equation 8.17e) predicts slightly increased *n* values and is applicable for cobbles.

8.3.2 Flow resistance in alluvial streams and rivers

As mentioned earlier, in addition to skin resistance discussed in the previous section, excess flow resistance occurs in alluvial channels with sediment transport due to the creation of bed forms in the movable bed. The resulting overall resistance may be considerably higher compared to skin friction without bed forms.

Table 8.3 Chow's (1959) values for the Manning roughness coefficient *n* for excavated earthen and natural streams

Description	Manning n values		
	Min	Normal	Max
A. Excavated or dredged channels			
a. Earth, straight and uniform			
1. Clean, recently completed	0.016	0.018	0.020
2. Clean, after weathering	0.018	0.022	0.025
3. Gravel, uniform section, clean	0.022	0.025	0.030
4. With short grass, few weeds	0.022	0.027	0.033
b. Earth, winding and sluggish			
1. No vegetation	0.023	0.025	0.030
2. Grass, some weeds	0.025	0.030	0.033
3. Dense weeds or aquatic plants in deep channels	0.030	0.035	0.040
4. Earth bottom and rubble sides	0.028	0.030	0.035
5. Stony bottom and weedy banks	0.025	0.035	0.040
6. Cobble bottom and clean sides	0.030	0.040	0.050
c. Dragline – excavated or dredged			
1. No vegetation	0.025	0.028	0.033
2. Light brush or banks	0.035	0.050	0.060
d. Rock cuts			
1. Smooth and uniform	0.025	0.035	0.040
2. Jagged and irregular	0.035	0.040	0.050
e. Channels not maintained, weeds and brush uncut			
1. Dense weeds, high as flow depth	0.050	0.080	0.120
2. Clean bottom, brush on sides	0.040	0.050	0.080
3. Same, highest stage of flow	0.045	0.070	0.110
4. Dense brush, high stage	0.080	0.100	0.140
B. Natural streams			
B.1 Minor streams (top width at flood stage <30 m)			
a. Streams on plain			
1. Clean, straight, full stage, no rifts or deep pools	0.025	0.030	0.033
2. Same as 1 above, but more stones and weeds	0.030	0.035	0.040
3. Clean, winding, some pools and shoals	0.033	0.040	0.045
4. Same as 3 above, but some weeds and stones	0.035	0.045	0.050
5. Same as 4 above, lower stages, more ineffective slopes and sections	0.040	0.048	0.055
6. Same as 4 above, but more stones	0.045	0.050	0.060
7. Sluggish reaches, deep pools	0.050	0.070	0.080
8. Very weedy reaches, deep pools or floodways with heavy stand of timber and underbrush	0.075	0.100	0.150
b. Mountain streams, no vegetation in channel, banks usually steep, trees and brush along banks submerged at high stages			
1. Bottom: gravels, cobbles and few boulders	0.030	0.040	0.050
2. Bottom: cobbles with large boulders	0.040	0.050	0.070

(Continued)

Table 8.3 (Continued) Chow's (1959) values for the Manning roughness coefficient *n* for excavated earthen and natural streams

Description	Manning n values		
	Min	Normal	Max
B.2 Floodplains			
a. Pasture, no brush			
1. Short grass	0.025	0.030	0.035
2. High grass	0.030	0.035	0.050
b. Cultivated areas			
1. No crop	0.020	0.030	0.040
2. Mature row crops	0.025	0.035	0.045
3. Mature field crops	0.030	0.040	0.050
c. Brush			
1. Scattered brush, heavy weeds	0.035	0.050	0.070
2. Light brush and trees, in winter	0.035	0.050	0.060
3. Light brush and trees, in summer	0.040	0.060	0.080
4. Medium-to-dense brush, in winter	0.045	0.070	0.110
5. Medium-to-dense brush, in summer	0.070	0.100	0.160
d. Trees			
1. Dense willows, summer straight	0.110	0.150	0.200
2. Cleared land with tree stumps, no spouts	0.030	0.040	0.050
3. Same as 2 above, but with heavy growth of spouts	0.050	0.060	0.080
4. Heavy stand of timber, a few down trees, little undergrowth, flood stage below branches	0.080	0.100	0.120
5. Same as 4 above, but with flood stage reaching branches	0.100	0.120	0.160
B.3 Major streams (top width at flood stage >30 m) with the *n* value less than that for minor streams of similar description, because banks offer less effective resistance			
a. Regular section with no boulders or brush	0.025	—	0.060
b. Irregular and rough section	0.035	—	0.100

Table 8.4 Coefficients of Equation 8.16 for surface roughness *n* determination based on the sediment particle size according to various studies

Name	D_x	D_x unit	b	Equations	Comments
Strickler (1923)	D_{50}	ft	25.6	8.16a	
	D_{65}	mm	75.75	8.16b	
Meyer–Peter and Müller (1948)	D_{90}	m	26.0	8.16c	Bed with sand and gravel, well tested
Keulegan (1938)	D_{50}	ft	46.9	8.16d	
Keulegan (1949)	D_{90}	ft	49.0	8.16e	
	D_{65}	ft	29.3	8.16f	
Lane and Carlson (1953)	D_{75}	in.	39.0	8.16g	Bed with cobbles

Note: 1 ft = 0.3048 m; 1 in. = 0.0254 m.

Table 8.5 Comparison of predictions of Manning *n* by the various formulas of Equation 8.16

Type	D (mm)	8.16a	8.16b	8.16c	8.16d	8.16e	8.16f	8.16g
Cobbles	250	0.038	0.033	0.031	0.021	0.020	0.033	0.038
	125	0.034	0.030	0.027	0.018	0.018	0.029	0.033
Gravel	64	0.030	0.026	0.024	0.016	0.016	0.026	0.030
	32	0.027	0.024	0.022	0.015	0.014	0.023	0.027
	16	0.024	0.021	0.019	0.013	0.012	0.021	0.024
	8	0.021	0.019	0.017	0.012	0.011	0.019	0.021
	4	0.019	0.017	0.015	0.010	0.010	0.017	0.019
Sand	2	0.017	0.015	0.014	0.009	0.009	0.015	0.017
	1	0.015	0.013	0.012	0.008	0.008	0.013	0.015
	0.5	0.013	0.012	0.011	0.007	0.007	0.012	0.013
	0.25	0.012	0.010	0.010	0.007	0.006	0.010	0.012
	0.125	0.011	0.009	0.009	0.006	0.006	0.009	0.011
	0.0625	0.009	0.008	0.008	0.005	0.005	0.008	0.009

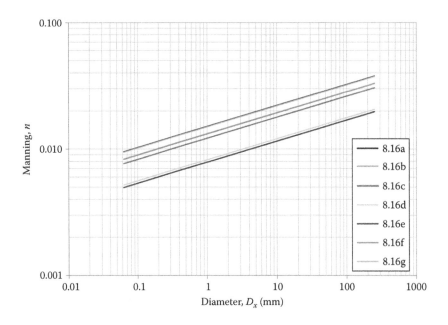

Figure 8.2 Comparison of predictions of Manning *n* by the various formulas of Equation 8.16.

The various forms that can be developed in an alluvial channel with sediment transport are the following (Simons and Sentürk, 1992; Yang, 1996): (1) ripples (i.e. small bed forms occurring as sand waves on the bed with wavelengths of about 30 cm and height of 5 cm); (2) dunes (i.e. larger sand waves than ripples); (3) plane bed, where there are no elevations or depressions on the bed; (4) antidunes or standing waves with their flow moving downstream and bed sand waves and surface waves in phase moving upstream; (5) chutes and pools which are large and long accumulations of sediment; and (6) bars

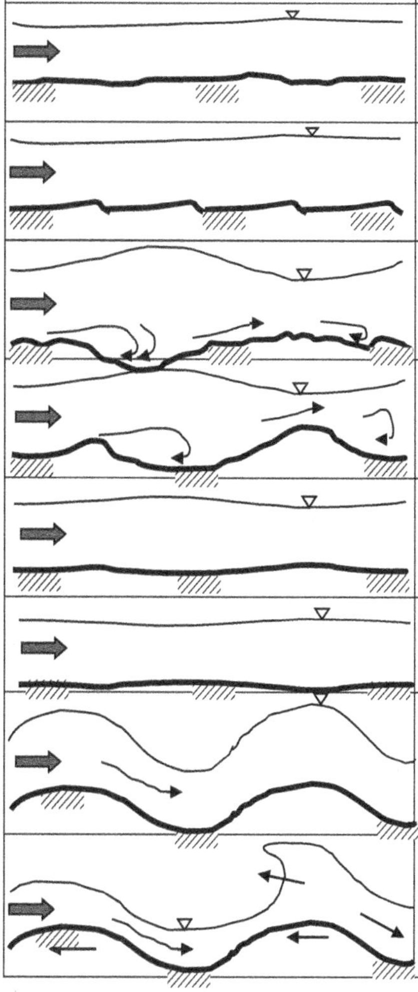

Schematic view	Bed form	Regime	$\tau_o V$	V	Fr	*Res.*
	Plane bed without sediment transport	Lower flow regime	+	+	+	+
	Ripples	Lower flow regime	+	+	+	+
	Ripples on dunes	Lower flow regime	+	+	+	+
	Dunes	Lower flow regime	+	+	+	+
	Washed-out dunes	Transition	+	+	+	–
	Plane bed with sediment transport	Upper flow regime	+	+	+	±
	Antidunes – standing waves	Upper flow regime	+	+	+	+
	Antidunes – breaking waves	Upper flow regime	+	+	+	+

Note: (+) Increases, (–) Decreases, (±) Nearly unchanged or slightly increasing

Figure 8.3 Bed forms in alluvial streams.

which are sediment accumulations having dimensions comparable to the channel width and mean depth. When the material is medium sand and finer (D_{50} < 0.6 mm), the bed forms are created by the flow in the following order as the stream power (defined as $\tau_o V$ with τ_o taken from Equation 8.14) increases: plane bed without sediment transport, ripples, ripples superimposed on dunes, dunes, washed-out dunes, plane bed with sediment transport, antidune standing wave, antidune breaking wave, chutes and pools. Figure 8.3 depicts this. For coarser bed material (D_{50} > 0.6 mm), the first bed form at small stream power is dunes (no ripples are formed). As Figure 8.3 shows, similar to the stream power increase, as we move from ripples to antidunes, the velocity and the Froude number (*Fr*) increase while the resistance to flow generally increases from a small value equivalent to the skin friction value (e.g. estimated by Equations 8.16) provided by the sediment grains (plain bed – no sediment transport) to the largest value when dunes are created. Then, flow resistance falls to values comparable to the skin friction roughness for plain bed with sediment transport, and then

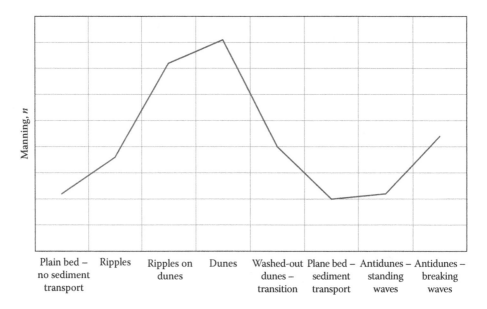

Figure 8.4 Bed forms and qualitative flow resistance.

again it increases when antidunes are the prevailing bed forms. A qualitative schematic diagram for this change is shown in Figure 8.4.

Generally, ripples and dunes are associated with the lower flow regime for subcritical flows (small stream power), while plane bed with sediment transport, antidunes, and chutes and pools occur in the upper flow regime for supercritical flows (larger stream power). There is also the transition regime where the bed configuration changes from dunes to plane bed and the flow is close to critical. As mentioned, the lower flow regime is characterized by increased resistance and low sediment transport. The opposite (small resistance, significant sediment transport) is true for the upper flow regime.

Which bed form develops depends on various parameters, including the (Yang, 1996) (1) flow depth and mostly the relative roughness (i.e. ratio of roughness element dimension to flow depth), (2) channel slope, (3) sediment–water mixture density, (4) bed material gradation, (5) fall velocity, (6) temperature and (7) shape of the channel cross section. To predict the bed form, several methods, mostly graphical, have been developed, relating various flow and sediment parameters. An evaluation of these methods is presented by Yang (1996). Common methods include that by Engelund and Hansen (1966), who used the mean velocity, the shear velocity (Equation 8.13) and the Froude number in a graph to predict bed forms. Similarly, Simon and Richardson (1966) used a graphical method based on the stream power (Equation 8.14) and the median fall diameter to map areas where each bed form develops (Figure 8.5).

As mentioned, bed forms significantly affect resistance to flow (Figure 8.4) by exerting bed form drag in addition to grain skin friction. One way to address this problem is by computing separately the resistance due to the grains and due to the bed forms, and adding the two together. Therefore, the total shear stress (Equation 8.14) can be written as

$$\tau_o = \gamma R S = \tau_o' + \tau_o'' = \gamma R' S + \gamma R'' S \tag{8.17}$$

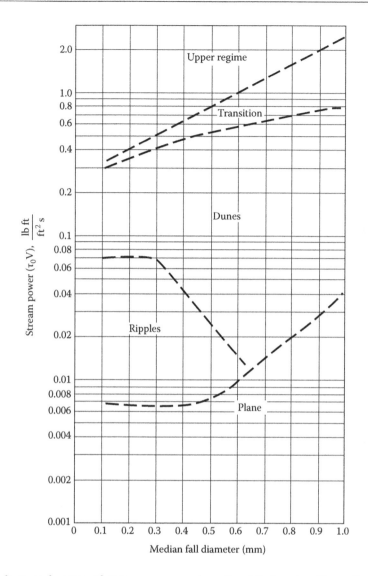

Figure 8.5 Bed form as function of stream power and median fall diameter. (From Simons, D.B. and Richardson, E.V., *Resistance to Flow in Alluvial Channels*, U.S. Geological Survey Professional Paper 422-J, 1966.)

where τ'_o and τ''_o are the shear stresses due to skin friction and bed form roughness, respectively, and R' and R'' are the parts of the hydraulic radius R ($R = R' + R''$) which are related to skin and form resistance, respectively. Using the same approach, the roughness coefficient (either Darcy–Weisbach f or Chézy C or Manning n) can be separated into two resistance coefficient parts. For example, for Manning n

$$n = n' + n'' \tag{8.18}$$

As mentioned, values for n' can be estimated using either Equation 8.16 or the various tables presented (e.g. Table 8.3). Figure 8.6 presents typical values for total n (Equation 8.18)

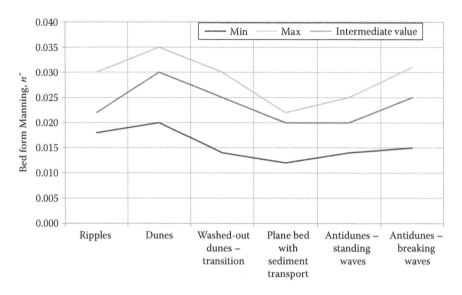

Figure 8.6 Bed form Manning n'' values for fine-to-medium sand channels.

for various bed forms in a fine-to-medium sand channel. One can relate these values to the qualitative Manning n variation presented in Figure 8.4. The graph presents a range (minimum to maximum) of n values. For flood control design in alluvial streams, the maximum curve is recommended because it gives conservative depths, while for sediment transport and scour design, the minimum or intermediate value curves can be used to compute conservative velocities.

Several methods have been developed, based on the approach to separate grain and bed form roughness, to compute total resistance in alluvial channels. The well-known methods include those by Einstein (1950), Einstein and Barbarossa (1952), Engelund and Hansen (1966), Alam and Kennedy (1969), Richardson and Simons (1967), and Shen (1962), among others. We present here Einstein's method, while details on the others can be found in specialized sediment transport textbooks (e.g. Simons and Sentürk, 1992).

According to Einstein (1950), the mean flow velocity V is computed based on the hydraulic radius R' due to grain roughness as follows:

$$\frac{V}{V'_*} = 5.75 \log\left(12.75\frac{R'x}{k_s}\right) \tag{8.19}$$

where

V'_* is the shear velocity due to grain roughness (calculated by Equation 8.8 using R') (m/s)

k_s is the equivalent grain roughness dimension (m) (Einstein set $k_s = D_{65}$)

x (–) is a parameter depending on the ratio $k_s/\delta = D_{65}/\delta$, where δ (m) is the laminar boundary sub-layer thickness given by

$$\delta = \frac{11.6\upsilon}{V'_*} \tag{8.20}$$

where υ is the kinematic viscosity (m²/s). The parameter x is given in Figure 8.7 as function of $k_s/\delta = D_{65}/\delta$. Rao and Kumar (2009) developed the following equation to predict

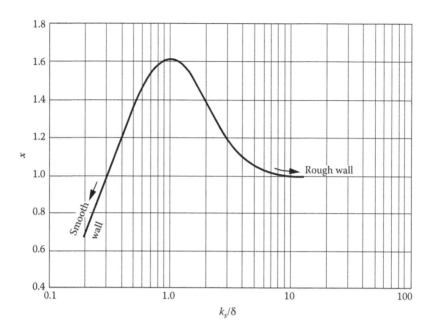

Figure 8.7 Values of parameter x in Equation 8.19. (From Einstein, H.A., The bed load function for sediment transportation in open channel flows, Technical Bulletin No. 1026, U.S. Department of Agriculture, Washington, DC, 1950.)

Einstein's parameter x, which was tested with Nikuradse's data for pipes and approximates the graph of Figure 8.7; so instead of the graph of Figure 8.7, the following equation can be used:

$$x = \frac{1}{7.4} \frac{\left\{ 11.6 \frac{k_s}{\delta} \right\}}{\left(0.444 + 0.135 \left\{ 11.6 \frac{k_s}{\delta} \right\} \right) \left[1 - 0.55 e^{-0.33 \left| \ln \left[\frac{\left\{ 11.6 \frac{k_s}{\delta} \right\}}{6.5} \right] \right|^2} \right]} \tag{8.21}$$

Similar to the grain roughness, Einstein (1950) proposed the following function to compute the mean flow velocity V based on the hydraulic radius R'' due to bed form roughness:

$$\frac{V}{V_*''} = f(\Psi') = f\left(\frac{\gamma_s - \gamma}{\gamma} \frac{D_{35}}{S R'} \right) = f\left([S_g - 1] \frac{D_{35}}{S R'} \right) \tag{8.22}$$

where
  f implies 'function of'
  S_g is the specific gravity (–)
  S is the slope (–)

The function f was developed by Einstein and Barbarossa (1952) based on field data, and is presented in Figure 8.8.

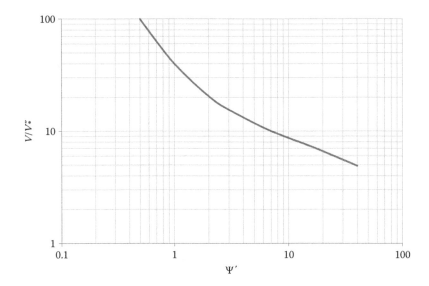

Figure 8.8 Einstein's function f (Equation 8.22). (From Einstein, H.A. and Barbarossa, N.L., *Trans. ASCE*, 117, 1121, 1952.)

The Einstein method can be used based on the following steps suggested by Einstein and Barbarossa (1952) to relate discharge and total hydraulic radius of a given alluvial stream and flow condition. If the discharge Q is given, perform the following:

1. Collect data on slope S, sediment sizes (D_{65}, D_{35}) and cross-sectional dimensions. Cross-sectional dimension data can be used to produce a relationship (i.e. a rating curve) between flow area A and total hydraulic radius R, i.e. $A = f_1(R)$ (where f_1 implies 'function of').
2. Assume a value for the grain roughness part of the hydraulic radius R'.
3. Compute the shear velocity V'_* based on Equations 8.13 and 8.14: $V'_* = \sqrt{g R' S}$.
4. Use either Figure 8.7 or Equation 8.21 to determine parameter x.
5. Determine the mean flow velocity V using Equation 8.19 setting $k_s = D_{65}$.
6. Compute from Equation 8.22 Einstein's parameter: $\Psi' = [S_g - 1]\dfrac{D_{35}}{S R'}$.
7. Use the computed Ψ' value to determine the ratio V/V''_* from Figure 8.8.
8. Use the mean flow velocity V value computed in Step 5 to determine V''_*.
9. Use Equations 8.13 and 8.14 ($V''_* = \sqrt{g R'' S}$) to determine R''.
10. Compute $R = R' + R''$.
11. Use the relationship $A = f_1(R)$ from Step 1 to compute the flow area A.
12. Compute the discharge: $Q = A\, V$.
13. If the computed Q in Step 12 is equal to the given one, then the procedure is finished. Otherwise, go to Step 2 and assume a new value for R' and repeat the procedure until it converges.

If the total hydraulic radius R is now given instead of the discharge Q, then follow the same Steps 1–10 and check if the computed value R (Step 10) agrees with the given one. If it does, the procedure is finished; otherwise, assume a new value for R' and repeat the procedure. Then, follow Steps 11 through 12 to compute A and Q.

8.4 INCIPIENT MOTION

Incipient motion is the commonly used term (other terms include beginning of motion, initiation of motion, initial motion and critical movement) for describing the flow conditions leading to the movement of a sediment particle lying on the stream bed. The theory presented here is important, because, in addition to sediment transport and channel degradation processes, it can be used to design a stable channel or to size riprap for bank protection.

Several methods have been proposed to access incipient motion. Here we will present some common methods while several others can be found in specialized textbooks. One well-known method has been developed by Shields (1936) based on similarity analysis. Although it has received criticism, it still finds application. The method uses two parameters in a graph (Figure 8.9). The vertical axis contains the dimensionless Shields parameter, τ_*, defined by

$$\tau_* = \frac{\tau_c}{(\gamma_s - \gamma)D_s} = \frac{\tau_c}{\gamma\,(S_g - 1)D_s} \tag{8.23}$$

where

τ_c is the critical shear stress for motion (N/m^2)
γ_s and γ are the sediment and water-specific weights
D_s is the mean sediment size

The horizontal axis contains the critical boundary Reynolds number, R_*, defined by

$$R_* = \frac{V_* D_s}{\nu} \tag{8.24}$$

where ν is the kinematic viscosity (m^2/s) and the other parameters are already defined. The original Shields graph contained only experimental data. The line through the data was added later by Rouse (1939). The graph of Figure 8.9 also contains a family of parallel lines,

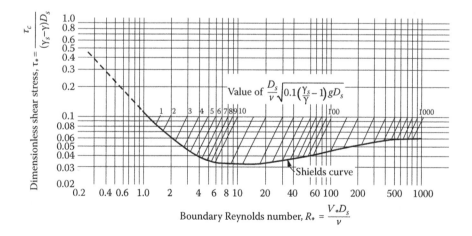

Figure 8.9 Shields graph of incipient motion. (Adapted from Vanoni, V., Measurements of critical shear stress for entraining fine sediments in a boundary layer, Report KH-R-7, W.M. Keck Laboratory of Hydraulics and Water Resources, California Institute of Technology, Pasadena, CA, 1964; Vanoni, V.A., ed., *Sedimentation Engineering*, ASCE Manuals and Reports on Engineering Practice No. 54, American Society of Civil Engineers, Reston, VA, 2006.)

added later by Vanoni (1964) and not contained in the original graph (Shields, 1936), representing the following dimensionless parameter (Equation 8.25). The use of this parameter makes the use of the graph easier avoiding necessary iterations because the parameter D_s is contained in both axes:

$$\frac{D_s}{\nu}\sqrt{0.1\,(S_g - 1)\,g\,D_s} \tag{8.25}$$

It is noticed that to the right (fully developed turbulent flow), τ_* tends to the constant 0.060, while Meyer–Peter and Müller (1948) suggested the value 0.047 instead, a value also suggested by several other researchers. For $R_* = 10$, the lowest value of τ_* occurs, which is about 0.033. Thus, it has been suggested to use this value, irrespective of the R_* value, as a conservative criterion defining motion or no motion (Shen and Julien, 1993). Alternatively, a safety factor (usually 2–4) can be used when flow is turbulent (Shen and Julien, 1993).

The Shields method applies for nearly uniform bed materials of size D_s. When the material is non-uniform, the phenomenon becomes more complex. The factors affecting it include, among others, armouring of the bed (i.e. larger particles protect finer particles from moving by the flow), and considering the fact that larger particles protrude more than finer particles, thus, larger particles may move first. The following criterion has been proposed in these cases to define stability (Shen and Julien, 1993):

$$\frac{\tau_c}{\gamma\,(S_g - 1)\,D_{30}} < 0.028 \tag{8.26}$$

Another method for incipient motion determination was developed by Yang (1973) based on the balance of forces acting on the sediment particle. The method computes a critical velocity V_c at incipient motion:

$$\frac{V_c}{V_f} = \begin{cases} \dfrac{2.5}{\log R_* - 0.06} + 0.66 & \text{for } 1.2 < R_* < 70 \\[2mm] 2.05 & \text{for } R_* \geq 70 \end{cases} \tag{8.27}$$

where
   R_* is the critical boundary Reynolds number (Equation 8.24)
   V_f is the fall velocity (m/s)

Finally, Chow (1959) presents permissible velocity, i.e. maximum non-erosive mean velocity (originally developed by Fortier and Scobey, 1926) and critical unit tractive force values (Table 8.6), while the U.S. Bureau of Reclamation presents the graph of Figure 8.10 to compute critical unit tractive force as function of the sediment size.

The aforementioned methods (values of permissible velocities or shear stresses computed from the Shields graph or values of critical unit tractive forces) can be used to design stable channels. Regarding the shear stress at a channel section, this is not constant throughout the cross section. The maximum shear stress is computed by Equation 8.14 when the channel is wide (width-to-depth ratio greater than 6.0). In other cases, the unit tractive force is not uniformly distributed in the channel section but presents different maximum values on the sides and the bed. Then, Equation 8.14 reads

$$\tau_o = a_{max}\,\gamma RS \tag{8.28}$$

Table 8.6 Permissible velocities and critical tractive forces recommended by Fortier and Scobey (1926)

Material	Manning n	Clear water		Water transporting colloidal silts	
		V_p (m/s)	τ_c (N/m²)	V_p (m/s)	τ_c (N/m²)
Fine sand, colloidal	0.020	0.46	1.21	0.76	3.37
Sandy loam, non-colloidal	0.020	0.53	1.66	0.76	3.37
Silt loam, non-colloidal	0.020	0.61	2.15	0.91	4.94
Alluvial silts, non-colloidal	0.020	0.61	2.15	1.07	6.73
Ordinary firm loam	0.020	0.76	3.37	1.07	6.73
Volcanic ash	0.020	0.76	3.37	1.07	6.73
Stiff clay, very colloidal	0.025	1.14	11.67	1.52	20.64
Alluvial silts, colloidal	0.025	1.14	11.67	1.52	20.64
Shales and hardpans	0.025	1.83	30.07	1.83	30.07
Fine gravel	0.020	0.76	3.37	1.52	14.36
Graded loam to cobbles when non-colloidal	0.030	1.14	17.05	1.52	29.62
Graded silts to cobbles when non-colloidal	0.030	1.22	19.30	1.68	35.90
Coarse gravel, non-colloidal	0.025	1.22	13.46	1.83	30.07
Cobbles and shingles	0.035	1.52	40.84	1.68	49.37

Source: Chow, V.T., *Open-Channel Hydraulics*, McGraw-Hill Inc., New York, 1959.

where α_{max} is a coefficient less than one. Values for α_{max} were computed theoretically for various channel sections (i.e. trapezoidal and rectangular) and are given in the graph of Figure 8.11 adopted from the U.S. Bureau of Reclamation (Chow, 1959) as a function of the ratio bed width b to flow depth d. As an example, maximum values for α_{max} are 0.97 and 0.75 for the bed and the bank, respectively, of a trapezoidal canal for $b/d = 4$ and side slope of the bank 1.5–1 (horizontal to vertical). The aforementioned methods can also be used to design appropriate riprap size for channel protection.

8.5 SEDIMENT TRANSPORT FORMULAS

The sediment transport formulas are used to compute the sediment transport capacity of a given stream cross section or reach under specific hydraulic conditions and bed characteristics. Several formulas have been developed and are presented in various textbooks. Here, only typical classic and well-known formulas will be presented and their use will be explained. We will utilize the following commonly used terms:

- *Bed layer*: The flow layer adjacent and immediately above the bed, which is usually considered about two grain diameters thick.
- *Bed material*: The mixture of all sediment sizes found in the stream bed.
- *Sediment load*: The sediment transport rate expressed as either weight per unit time or volume per unit time.
- *Bed load* (BL): The sediment quantity that moves within the bed layer. The mechanisms of movement include saltation (i.e. picking from and dropping on the bed of a particle by turbulence), rolling and sliding of sediment particles.
- *Suspended load* (SL): The sediment quantity that stays due to turbulence in suspension in the water column (i.e. between the top of the bed layer and the water surface) and is transported by the streamflow.

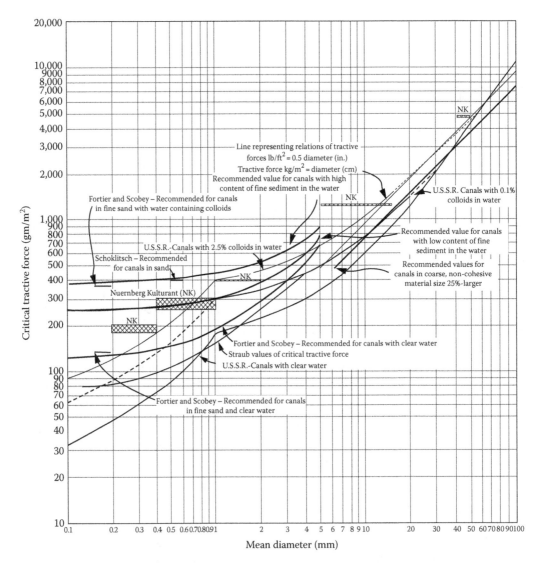

Figure 8.10 Critical unit tractive force as function of sediment size. (From Lane, E.W., *Proceedings of the ASCE*, 79, 1953; U.S. Bureau of Reclamation, *Design of Small Dams*, U.S. Department of the Interior, Bureau of Reclamation, Washington, DC, 816pp., 1977.)

- *Total sediment load* (TSL): The total sediment quantity in the stream, which is the sum of BL and SL. Usually, the BL varies between 5% and 25% of the SL.
- *Wash load* (WL): The part of the total load comprising particles finer than those existing in the bed material mixture. Most part of this load originates from erosion in the upper watershed and the banks. The WL makes one part of the SL, the other part being the suspended bed material load (*SBML*; i.e. coarser particles in suspension also existing in the bed material).
- *Bed material load* (BML): The part of the total load comprising sizes found in the bed material. Most times, the BML is considered to have sizes greater than 0.0625 mm (i.e. the limit between sand and silt; see Table 8.1). The SBML and the BL make the BML.

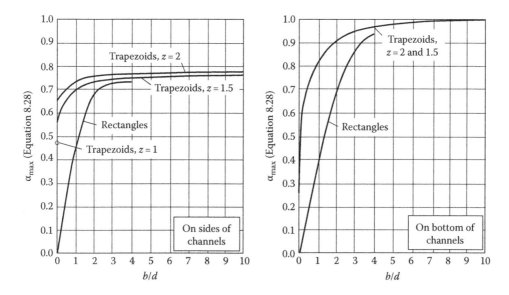

Figure 8.11 Ratio of a maximum unit tractive force divided by shear stress (Equation 8.28) in a channel section as function of ratio *b/d*. (From Chow, V.T., *Open-Channel Hydraulics*, McGraw-Hill Inc., New York, 1959.)

- *Sediment concentration*: The quantity of sediment in a given quantity of fluvial water transporting the sediment. The sediment and/or water quantity can be measured by weight or by volume. Usually, sediment concentration is expressed as the ratio of sediment transport rate (mass per time) to discharge (volume per time), in units mg/L or kg/m^3. Alternatively, concentration has also been expressed as the ratio of weight of sediment to total (water + sediment) weight or volume of sediment to total volume. The sediment concentration has a vertical distribution in the water column, which generally differs from that of the velocity. Typical qualitative velocity and sediment concentration profiles are shown in Figure 8.12.
- *Sediment discharge*: The quantity of sediment passing a given stream cross section in unit time. We have bed material, suspended and total sediment discharges, respectively.

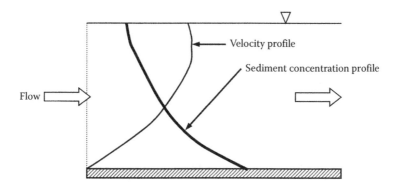

Figure 8.12 Typical qualitative velocity and sediment concentration profiles in a river.

The relation between the various types of sediment load is presented in the following equations:

$$TSL = BL + SL = \{BML - SBML\} + \{WL + SBML\} = BML + WL \tag{8.29}$$

where

$$BML = SBML + BL \tag{8.30}$$

$$SL = WL + SBML \tag{8.31}$$

8.5.1 Bed load

BL is the quantity of transported bed material; transport occurs when critical conditions for motion of the bed material (see previous section on incipient motion) prevail. Several formulas have been proposed by various researchers, as summarized in various specialized textbooks. Here, we indicatively present two representative formulas: the Meyer–Peter and Müller (1948) formula and the Einstein (1950) formula.

8.5.1.1 Meyer–Peter and Müller method

The Meyer–Peter and Müller (1948) formula, which is one of the most widely and diachronically used and accepted formulas, reads

$$q_{Bm} = \frac{8}{g\sqrt{\rho}} \left(\frac{1}{S_g - 1} \right) (\tau_o - \tau_c)^{1.5} \tag{8.32}$$

where
q_{Bm} is the BL transport rate by dry weight (mass) per unit width of channel (kg/s/m) (to express the transport by BL volume q_{Bv} (m³/s/m), one has to divide the right-hand side of the Equation 8.32 by the specific weight γ_s)
τ_o is the boundary shear stress (unit tractive force) given by Equation 8.14 (N/m²)
τ_c (N/m²) is the critical shear stress, which, for fully developed turbulent flow, is given by (see also the Shields diagram; Figure 8.9)

$$\tau_c = 0.047 \gamma (S_g - 1) D_s \tag{8.33}$$

A commonly used form of Equation 8.32 (transport by volume), which assumes $S_g = 2.65$ (i.e. quartz), is

$$q_{Bm} = \frac{12.85}{\gamma_s \sqrt{\rho}} (\tau_o - \tau_c)^{1.5} \tag{8.34}$$

In computing τ_o, the hydraulic radius due to grain friction R' and the corresponding Manning n' are used. We note that, unlike Einstein's approach for grain and bed form roughness (i.e. Equations 8.17 and 8.18), Meyer–Peter and Müller (1948) assumed that

$$\frac{R'}{R} = \frac{S'}{S} = \frac{n'}{n} \tag{8.35}$$

where

R' and R are the respective hydraulic radii due to grain friction and the total (m)
S' and S are the respective energy slopes due to grain friction and total
n' and n are the respective Manning n values due to grain friction and total

To compute Manning n', Equation 8.16c (Table 8.4) is used with $D_x = D_{90}$ (m) and $b = 26$.

Equation 8.32 can also be used to compute the transport of material containing certain different sizes. In this case, a mean sediment sample diameter is used instead of D_s, as follows:

$$D_s = \sum_{i=1}^{N} (p_i D_{si})$$ (8.36)

where p_i is the percentage by weight of a certain D_{si} in the sediment sample.

8.5.1.2 Einstein method

Einstein's (1950) formula is based on the most comprehensive theory among all formulas. Furthermore, it has been demonstrated that several other proposed BL formulas, following mathematical manipulation, can be converted to Einstein's (Shen and Julien, 1992). Einstein's (1950) BL formula relates two dimensionless parameters: one is the dimensionless transport rate, Φ, defined by

$$\Phi = \frac{q_{Bw}}{\gamma_s} \left[\frac{1}{(S_g - 1) g D_s^3} \right]^{1/2}$$ (8.37)

and the other is the dimensionless flow intensity, Ψ', already defined in Equation 8.22

$$\Psi' = (S_g - 1) \frac{D_{35}}{R'S}$$ (8.38)

where q_{Bw} is the weight of transported BL per unit time and channel unit width (N/s/m) and the other parameters have already been defined. The two parameters Φ and Ψ' of Equations 8.37 and 8.38 can be used to predict the transport of nearly uniform bed material, in which case D_{35} is used as the representative sediment size. Einstein (1950) proposed for this the function relating Φ and Ψ', which is based on field data and is presented in Figure 8.13. According to Shen and Julien (1992), several other BL formulas (e.g. those by Engelund–Fredsoe, Yalin, Meyer–Peter, Ackers and White and Bagnold) can be converted to fall around or on this curve, showing the general validity of Einstein's procedure.

In the case of non-uniform bed material, Einstein (1950) proposed a separate computation of BL transport of each size class. In this case, Equation 8.37 is also used to compute the dimensionless transport rate Φ_i^* for class i, i.e. in Equation 8.37 q_{Bw} is replaced by q_{Bwi}, which refers to the transport of the mean grain size D_{Si} of class i whose percentage by weight in the bed sample is p_{Bi} (thus, D_s in Equation 8.37 is replaced by D_{si}). The dimensionless flow intensity Ψ_i' of Equation 8.38 also refers to D_{si} (therefore, D_{35} is

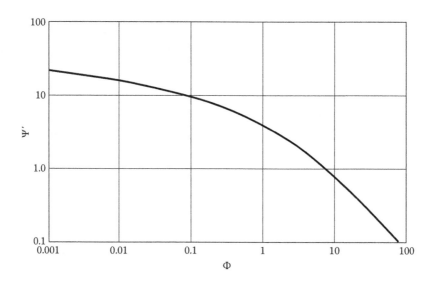

Figure 8.13 Function relating Ψ' and Φ in Einstein's (1950) bed load formula.

replaced by D_{si}). Furthermore, the corrected dimensionless flow intensity Ψ_i^* is computed in this case, which is given by

$$\Psi_i^* = \xi_i \, Y \left[\frac{\log(10.6)}{\log(10.6 \, (x \, X/D_{65}))} \right]^2 \Psi_i' \tag{8.39}$$

where
  ξ_i is the hiding correction factor, accounting for hiding of small grains with sizes less than size X by larger grains (Einstein's Figure 8.14 is used to estimate ξ_i for a given size D_{si} as function of the ratio D_{si}/X)
  Y is the pressure correction factor, accounting for the different resistance to flow provided by grains of different sizes (Einstein's Figure 8.15 is used to estimate Y as function of the ratio D_{65}/δ, where δ is the laminar boundary sub-layer thickness given by Equation 8.20)
  x is the parameter already presented in Equation 8.19, which is given as function of the laminar boundary sub-layer thickness δ in Figure 8.7 (or, alternatively, it can be computed by Equation 8.21)

Finally, the grain size X is computed by the following formulas:

$$X = \begin{cases} 0.77 \dfrac{D_{65}}{x} & \text{for } \dfrac{D_{65}}{x\delta} > 1.8 \\[2mm] 1.398\delta & \text{for } \dfrac{D_{65}}{x\delta} < 1.8 \end{cases} \tag{8.40}$$

According to the previous discussion, the procedure to estimate BL transport is as follows: Compute δ from Equation 8.20. Compute x from Figure 8.7 or Equation 8.21. Compute X from Equation 8.40. Compute ξ_i for each sediment class from Figure 8.14. Compute Y from Figure 8.15. Compute Ψ_i' from Equation 8.38 for each sediment class. Compute Ψ_i^* from

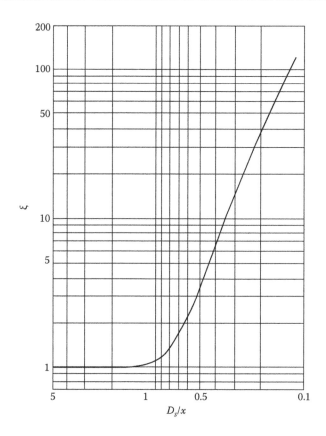

Figure 8.14 Estimate of hiding factor ξ. (From Einstein, H.A., The bed load function for sediment transportation in open channel flows, Technical Bulletin No. 1026, U.S. Department of Agriculture, Washington, DC, 1950.)

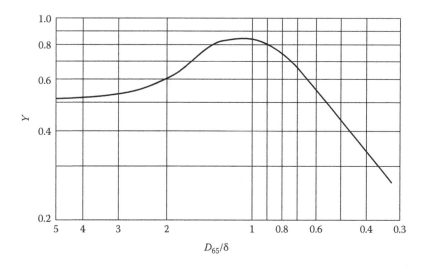

Figure 8.15 Estimate of pressure correction factor Y. (From Einstein, H.A., The bed load function for sediment transportation in open channel flows, Technical Bulletin No. 1026, U.S. Department of Agriculture, Washington, DC, 1950.)

Equation 8.39. Compute Φ_i^* using Figure 8.13 for each sediment class. Then, the BL transport rate q_{Bw} is computed following Equation 8.37 as follows:

$$q_{Bw} = \gamma_s \left[(S_g - 1)g\right]^{1/2} \sum_{i=1}^{N} \left(p_{Bi} \Phi_i^* D_{si}^{3/2}\right) \tag{8.41}$$

8.5.1.3 Einstein–Brown method

This method was based on Einstein (1942) formula and was presented by Brown (1950). It is based on the functional relation of Φ and Ψ which is presented in Figure 8.16, where

$$\Phi = \frac{q_{Bm}}{\lambda \gamma_s} \left[\frac{1}{(S_g - 1)g D_s^3}\right]^{1/2} \tag{8.42}$$

is a dimensionless transport rate by mass (see Equation 8.37) and Ψ is the inverse of the Shields parameter τ_* (Equation 8.23):

$$\frac{1}{\Psi} = \tau_* = \frac{\tau_o}{(\gamma_s - \gamma)D_s} = \frac{\tau_o}{\gamma\,(S_g - 1)D_s} \tag{8.43}$$

Finally, λ is a term we have also seen in Rubey's formula for fall velocity (Equation 8.4):

$$\lambda = \left[\frac{2}{3} + \frac{36v^2}{gD_s^3\,(S_g - 1)}\right]^{1/2} - \left[\frac{36v^2}{gD_s^3\,(S_g - 1)}\right]^{1/2} \tag{8.44}$$

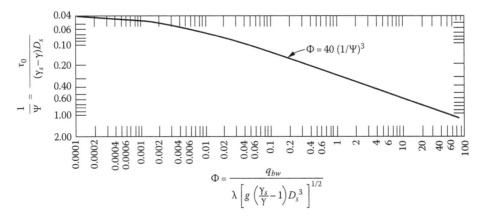

Figure 8.16 Einstein–Brown function $\Phi - \left(\dfrac{1}{\Psi}\right)$. (From Brown, C.B.: Sediment transportation, in *Engineering Hydraulics*, H. Rouse, ed. 1950. Copyright Wiley-VCH Verlag GmbH & Co. KGaA.)

As seen in Figure 8.16, when $\dfrac{1}{\Psi} > 0.09$, Equation 8.42 reads

$$\Phi = 40\left(\frac{1}{\Psi}\right)^3 \tag{8.45}$$

8.5.2 Suspended load

As previously mentioned and presented in Figure 8.12, the vertical distribution of the suspended sediment concentration in the water column has a special profile, showing low values at the surface and increased values at the bed. This is a result of transported sediment particles falling due to gravity towards the bed and other particles brought into suspension by turbulence from the bed. Under equilibrium conditions, Rouse (1937) derived the following equation for the sediment concentration distribution with distance above the bed:

$$\frac{C(y)}{C_a} = \left[\frac{d-y}{y}\frac{a}{d-a}\right]^z \tag{8.46}$$

where
$C(y)$ is the suspended sediment concentration at distance y above the bed (m)
C_a is the sediment concentration at a reference level a above the bed

The level a is that below which we have BL transport and is usually taken as $a = 2D_s$; d is the water depth (m) (note that for a wide channel $R' \approx d$); and the exponent z is given by

$$z = \frac{V_f}{\kappa V_*} \tag{8.47}$$

where
V_f is the fall velocity (m/s) (Rubey's formula–Equations 8.3 and 8.4, can be used to calculate it)
V_* is the shear velocity (m/s)
κ is the von Karman constant (≈ 0.4 for clear water, but it decreases with increasing sediment concentration, as presented by Einstein and Chien, 1953, 1954)

Einstein (1950) replaced V_f with V_f' in Equation 8.47. To compute the suspended sediment load discharge for a specific sediment size class, he integrated in the direction perpendicular to the bed above the reference level a:

$$q_{Swi} = \gamma_s \int_a^d V(y)C(y)\,dy \tag{8.48}$$

where
q_{swi} is the SL discharge per unit channel width by weight (N/S/m)
$V(y)$ is the velocity at distance y from the bed (m/s)
$C(y)$ the sediment concentration by volume at distance y (m³/m³)

For the velocity $V(y)$, the following formula was used:

$$V(y) = 5.75 V_*' \log\left(30.2 \frac{yx}{D_{65}}\right) \tag{8.49}$$

where x is the parameter of Equation 8.19 and Figure 8.7. Note that this equation was used after integration to derive Equation 8.19 for the mean velocity. Introducing Equations 8.47 and 8.49 into Equation 8.48, Einstein (1950) got

$$q_{Swi} = 11.6 V_*' C_{ai} a_i \left[2.303\log\left(\frac{30.2 R'x}{D_{65}}\right) I_{1i} + I_{2i}\right] \tag{8.50}$$

where I_{1i} and I_{2i} are Einstein's integral functions (Yang, 1982; Dimons and Sentürk, 1992; Vanoni, 2006), defined as

$$I_{1i} = 0.216 \frac{E^{z-1}}{(1-E)^z} \int_E^1 \left(\frac{1-y}{y}\right)^z dy \tag{8.51}$$

and

$$I_{2i} = 0.216 \frac{E^{z-1}}{(1-E)^z} \int_E^1 \left(\frac{1-y}{y}\right)^z \ln y \, dy \tag{8.52}$$

where
  E is the ratio $E = a_i/d$
  d is the depth of flow (m)

The numerical value of the two integrals can be evaluated using Figures 8.17 and 8.18 as function of parameters z and E. As mentioned, $a_i = 2D_{si}$. The concentration C_{ai} is computed by the following equation which was derived based on Einstein's (1950) experimental data:

$$C_{ai} = \frac{1}{11.6} \frac{p_{Bi} q_{Bwi}}{a V_*'} \tag{8.53}$$

This leads to the following equation for SL transport, relating it to BL transport q_{Bwi}:

$$q_{Swi} = p_{Bi} q_{Bwi} \left[2.303\log\left(\frac{30.2 dx}{D_{65}}\right) I_{1i} + I_{2i}\right] \tag{8.54}$$

And the TSL is computed by addition of that of each size class:

$$q_{Sw} = \sum_{i=1}^{N} \left\{ p_{Bi} q_{Bwi} \left[2.303\log\left(\frac{30.2 dx}{D_{65}}\right) I_{1i} + I_{2i}\right]\right\} \tag{8.55}$$

One should notice that in Equations 8.54 and 8.55, Einstein's BL equation (Equation 8.41) or any other BL equation (e.g. Meyer–Peter and Müller's) can be used.

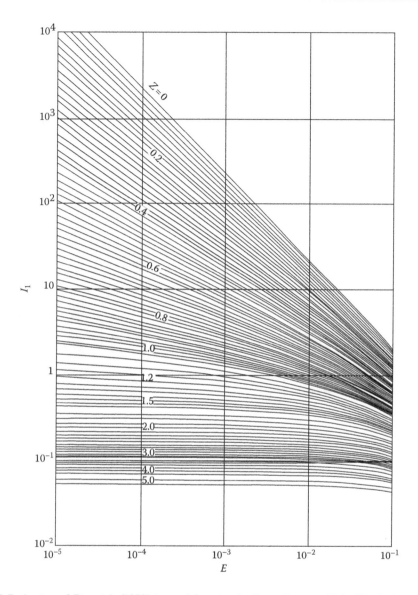

Figure 8.17 Evaluation of Einstein's (1950) integral function I_1. (From Einstein, H.A., The bed load function for sediment transportation in open channel flows, Technical Bulletin No. 1026, U.S. Department of Agriculture, Washington, DC, 1950.)

8.5.3 Total sediment load

8.5.3.1 Einstein method

Taking into account Equation 8.29, the TSL can be calculated from Equations 8.41 and 8.54 as

$$q_{Tm} = q_{Bmi} + q_{Smi} = \sum_{i=1}^{N} q_{Bmi} \left\{ 1 + p_{Bi} \left[2.303 \log \left(\frac{30.2 \, dx}{D_{65}} \right) I_{1i} + I_{2i} \right] \right\}$$ (8.56)

where q_T is the TSL rate per unit width of the stream (kg/s/m).

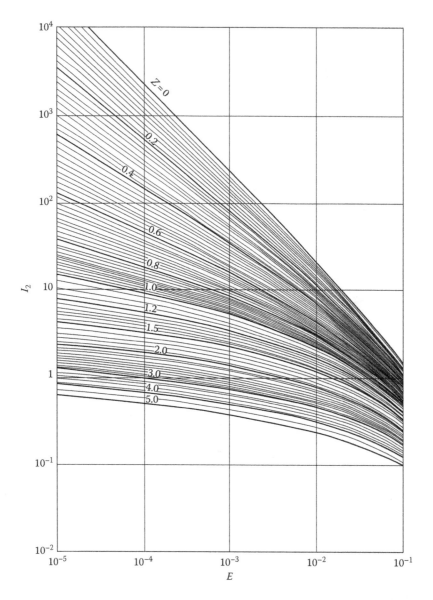

Figure 8.18 Evaluation of Einstein's (1950) integral function I_2. (From Einstein, H.A., The bed load function for sediment transportation in open channel flows, Technical Bulletin No. 1026, U.S. Department of Agriculture, Washington, DC, 1950.)

8.5.3.2 Ackers and White method

The method is based on computing the mean sediment concentration C_m by weight (kg/kg) based on the following empirical equation determined from a large number of experimental data, which contains four empirical coefficients a_1, a_2, a_3, a_4:

$$C_m = \left(\frac{V}{V_*}\right)^{a_1} \frac{S_g D_s}{d} a_2 \left(\frac{F}{a_3} - 1\right)^{a_4} \tag{8.57}$$

where

V is the mean flow velocity (m/s)

F is given by the following equation

$$F = \frac{V_*^{a_1}}{\sqrt{gd\ (S_g - 1)}} \left[\frac{V}{\sqrt{32}\ \log(10\ (d/D_s))} \right]^{1-a_1} \tag{8.58}$$

The empirical coefficients a_1, a_2, a_3 and a_4 depend on the dimensionless grain diameter D_*:

$$D_* = D_s \left[\frac{g(S_g - 1)}{v^2} \right]^{1/3} \tag{8.59}$$

For $D_* \geq 60$, they are constants assuming the following values: $a_1 = 0$, $a_2 = 0.025$, $a_3 = 0.17$ and $a_4 = 1.5$, and for $1.0 < D_* \leq 60$, they are given by the following equations:

$$a_1 = 1.00 - 0.56\ \log D_* \tag{8.60}$$

$$a_2 = 10^{(2.86\ \log D_* - (\log D_*)^2 - 3.53)} \tag{8.61}$$

$$a_3 = \frac{0.23}{\sqrt{D_*}} + 0.14 \tag{8.62}$$

$$a_4 = \frac{9.66}{D_*} + 1.34 \tag{8.63}$$

As mentioned, the sediment concentration C_m is by weight defined as mass of sediment divided by total mass of water and sediment or

$$C_m = \frac{\rho_s\ q_{Tv}}{\rho q + \rho_s\ q_{Tv}} = \frac{S_g\ q_{Tv}}{q + S_g\ q_{Tv}} \tag{8.64}$$

where

q is the water discharge per unit channel width (m³/s/m)

q_{Tv} is the sediment discharge per unit channel width (m³/s/m)

This equation can be used to express q_{Tv} as function of a known discharge q as follows:

$$q_{Tv} = \frac{C_m q}{S_g (1 - C_m)} \tag{8.65}$$

The total BL discharge per unit channel width by weight q_{Tm} (kg/s/m) can be determined from this equation by multiplying the sediment density ρ_s (kg/m³). The steps to apply the method are the following: Compute D_* from Equation 8.59 for a given grain size (usually D_{50}). Determine the coefficients a_i from Equations 8.60 through 8.63. Compute parameter F from

Equation 8.58 and the flow parameters. Compute the mean sediment concentration C_m from Equation 8.57. Use Equation 8.65 to compute the unit TSL discharge (by volume or by weight).

8.5.3.3 Yang's method

Yang (1972) derived an empirical equation to predict the mean sediment concentration C_m based on regression analysis of laboratory flume data, relating TSL transport of sand to the unit stream power (i.e. product of mean velocity V and energy slope S):

$$\log C_m = 5.435 - 0.286 \log\left(\frac{V_f D_s}{\nu}\right) - 0.457 \log\left(\frac{V_*}{V_f}\right)$$

$$+ \left[1.799 - 0.409 \log\left(\frac{V_f D_s}{\nu}\right) - 0.314 \log\left(\frac{V_*}{V_f}\right)\right] \log\left(\frac{VS}{V_f} - \frac{V_{cr}S}{V_f}\right) \tag{8.66}$$

where C_m is the mean sediment concentration by weight (ppm). To convert it to kg/kg, the value predicted by Equation 8.66 should be multiplied by the unit conversion factor 10^{-6}; V_{cr} is the critical velocity for incipient motion (m/s), which has also been presented (Equation 8.27), and the other parameters have already been defined. After computing C_m based on Equation 8.66, Equation 8.65 can be used to compute the total BL either by volume or by weight.

8.6 LAND EROSION AND WATERSHED SEDIMENT YIELD

8.6.1 Introduction

The soil erosion from natural, agricultural or other lands is a very important process affecting both the morphology of the eroding land itself and the erosion and sedimentation processes in the receiving streams where the land drains and supplies sediments. Extensive land erosion can result, among others, in the loss of topsoil in agricultural areas, leading to lower productivity, and rilling and gullying of the land surface, creating land use problems. Transported by streams, eroded sediments can cause reservoir silting, while sediments deposited in streams can lead to increased flood stages. On the other hand, measures to control land erosion may lead to sediment starvation in the receiving stream and excess bed degradation and/or bank instability. These examples demonstrate the need for accurate estimation of soil erosion from land and watershed sediment yields and the importance of quantifying these processes in sedimentation studies.

In this section, we present the basics of erosion, while comprehensive treatments of the subject can be found in specialized textbooks (e.g. Kirkby and Morgan, 1980; Goldman et al., 1986; Simons and Sentürk, 1992).

The erosion process comprises two parts: (1) loosening of soil grains by rainfall impact, something aided by the cycles of freezing–thawing and wetting–drying, and (2) transportation of loose soil grains by flowing rainwater (runoff). There are several types of soil erosion from land (Goldman et al., 1986):

- *Splash erosion*: This type occurs on unvegetated bare soil and is due to raindrop impact on soil and splashing away of soil grains. It is more significant in heavy intense rainfalls of large raindrop size and on sloping ground.

- *Sheet erosion*: This type occurs on relatively flat sloping ground under thin sheet flow (overland flow) which transports soil grains which were earlier detached due to rainfall impact.
- *Rill flow*: Due to ground irregularities, sheet flow starts concentrating in deeper parts. As flow concentrates there, velocity and turbulence increase, soil grains are detached and transported more easily, and small well-defined channels (rills) are formed. Rill dimensions are of a few centimetres.
- *Gully erosion*: Gullies are larger channels formed either when rills are cut deeper due to heavy rainfall runoff or when rills confluence to form deeper and wider channels.

There are several factors affecting soil erosion. The most important ones include climatic and soil characteristics, topography and ground cover (Kirkby and Morgan, 1980; Goldman et al., 1986). Climate is the main factor that affects erosion since erosion is a rain-driven process. The main rainfall characteristics affecting erosion is rainfall intensity (with short duration highly intense rainfall being more erosive) and rainfall droplet size (the larger the size, the more erosive the rainfall). Climate indirectly affects erosion by affecting the growth of vegetation, which is the most effective shield to erosion. As a result, erosion rates are lower in mild temperature and rainy climates, where vegetation grows fast and provides a dense ground cover; on the other hand, in cold and in dry climates, vegetation is sparse and, thus, erosion rates are higher.

Soil characteristics influencing erosion include texture, organic matter content, structure and permeability. With respect to texture (see Figure 7.6), the most erodible soils are those with high content of fine sand and silt and low organic matter and clay content. Sandy soils and clay soils are less erodible, the former because of the high permeability and the latter because clay particles tend to stick together and coagulate. Organic matter tends to improve soil structure and increase permeability, thus making the soil less erodible. With respect to structure, i.e. the tendency of soil particles to aggregate, a loose granular structure enhances infiltration leading to less runoff and erosion. Similarly, the more permeable the soil, the less the runoff and the erosion.

The topography of the area is also a key factor, with slope length and steepness being important parameters, because they influence runoff velocity. As both length and slope increase, so does flow velocity and erosion potential. Finally, ground cover (i.e. vegetation or protection measures) is important in reducing runoff. Vegetation reduces raindrop impact, decreases runoff velocity, reinforces the soil through the plant roots which hold particles together, and enhances infiltration.

8.6.2 Universal soil loss equation (USLE)

The Universal Soil Loss Equation (USLE) (Smith and Wischmeier, 1957; Meyer and Monke, 1965; Wischmeier and Smith, 1965, 1978; Wischmeier et al., 1971; Wischmeier, 1972, 1973) is the most commonly used empirical model to predict land erosion. It has been developed by the Soil Conservation Service (now Natural Resources Conservation Service) of the U.S. Department of Agriculture based on monitoring of experimental plots throughout the United States, which started in the 1930s and lasted for several decades. Detailed presentations of the USLE can be found in specialized textbooks (e.g. Kirkby and Morgan, 1980; Goldman et al., 1986; Simons and Sentürk, 1992).

The form of the USLE equation is

$$A = \beta R K LSC P \tag{8.67}$$

where

A is the annual or for a specific time period soil loss (U.S. tons/acre or metric tonnes/ha)

R is the rainfall erosivity factor (or rainfall erosion index) (–), which expresses the effectiveness of rainfall in causing erosion

K is the soil erodibility factor (tons/acre), which expresses in one parameter the soil characteristics relevant to erosion

LS is the slope length and steepness factor (–), which accounts for slope length and steepness in erosion potential

C is the vegetative cover and cropping-management factor (–), which expresses the effect of vegetation or mulch on reducing erosion and can be variable during the year related to vegetation canopy growth

P is the erosion control practices factor, i.e. it describes how human interventions, such as contouring and terracing, control erosion (–)

Finally, β is a unit conversion factor which is equal to 1.0 when parameter A is in tons/acre or 2.241 when A is in metric tonnes/ha.

8.6.2.1 Rainfall erosivity factor R

The rainfall erosivity index R is the product of rainfall kinetic energy at the ground and maximum intensity for 30 min duration (I_{30}) and can be evaluated for either one rainfall event or as a mean annual value for a series of years. According to Kirkby and Morgan (1980), R for a single storm can be computed by the following equation:

$$R = \frac{\left[\sum_{j=1}^{n} (1.213 + 0.890 \log I_j)\, (I_j\, \Delta t_j) \right] I_{30}}{173.6} \tag{8.68}$$

where

R is the rainfall erosivity factor (–)

I_j is the rainfall intensity for a specific time increment of the storm (mm/h)

Δt_j is the corresponding time increment (h)

I_{30} is the maximum 30 min intensity (mm/h)

j is the time increment

n is the number of time increments in the storm

R, although it has units from its definition, for simplicity, is considered non-dimensional and the dimensions are given to the following soil erodibility factor K (same as for A). This way, conversion between English and SI units is easy and only based on the value of β (Equation 8.67). Specifically for the United States, Wischmeier and Smith (1978) have produced the map of Figure 8.19 presenting average annual R values.

8.6.2.2 Soil erodibility factor K

The soil erodibility factor K can be predicted by the nomograph of Figure 8.20 given by Wischmeier and Smith (1978) or by simply using Table 8.7. Percent organic matter should be measured. Permeability class descriptions are given by Wischmeier et al. (1971). Table 8.7 gives K values for the main soil texture classes of Figure 7.6 and others.

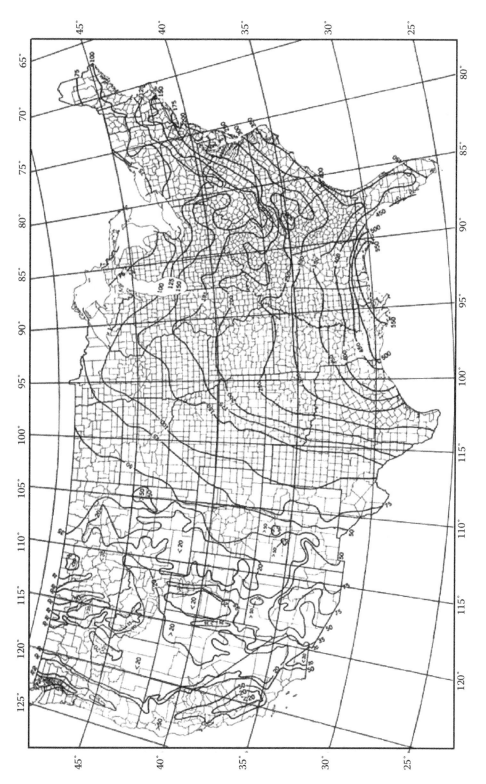

Figure 8.19 Average annual values for rainfall erosivity factor R for the United States. (From Wischmeier, W.H. and Smith, D.D., *Predicting Rainfall Erosion Losses – A Guide to Conservation Planning*, Agricultural Handbook No. 537, USDA, 1978.)

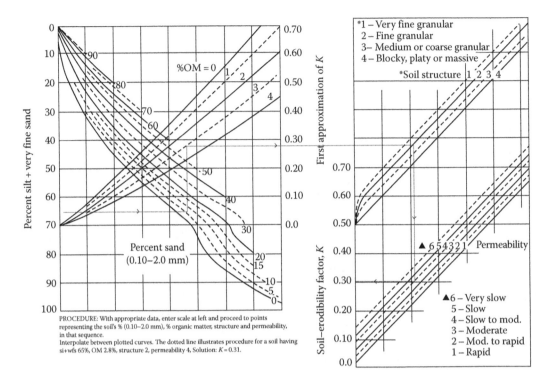

Figure 8.20 Nomograph to estimate the soil erodibility factor *K*. (From Wischmeier, W.H. et al., *J. Soil Water Conserv.*, 26(5), 189, 1971.)

8.6.2.3 Slope length and steepness factor LS

The slope length and steepness factor *LS* is given by the following equation (Smith and Wischmeier, 1957; Wischmeier and Smith, 1965):

$$LS = \left(\frac{l}{22.13}\right)^{m} (0.065 + 4.56\sin\theta + 65.41\sin^2\theta) \tag{8.69}$$

where

l is the slope length from the point where overland (sheet) flow starts to the point where slope is either significantly milder to allow deposition or to the point where a structure is set that interrupts runoff (e.g. a drainage pipe across the slope) (m)

θ is the angle of the slope which is defined according to

$$\sin\theta = \frac{s}{\sqrt{10^4 + s^2}} \tag{8.70}$$

where

s is the slope gradient (%)

m is an exponent depending on slope (Wischmeier and Smith, 1978), as shown in Table 8.8

Table 8.7 Soil erodibility factor K values

	Organic matter content (%)		
	<0.5%	2%	4%
		K	
Sand	0.05	0.03	0.02
Fine sand	0.16	0.14	0.10
Very fine sand	0.42	0.36	0.28
Loamy sand	0.12	0.10	0.08
Loamy fine sand	0.24	0.20	0.16
Very loamy fine sand	0.44	0.38	0.30
Sandy loam	0.27	0.24	0.19
Fine sandy loam	0.35	0.30	0.24
Very fine sandy loam	0.47	0.41	0.33
Loam	0.38	0.34	0.29
Silty loam	0.48	0.42	0.33
Silt	0.60	0.52	0.42
Sandy clay loam	0.27	0.25	0.,21
Clay loam	0.28	0.25	0.21
Silty clay loam	0.37	0.32	0.26
Sandy clay	0.14	0.13	0.12
Silty clay	0.25	0.23	0.19
Clay	0.13–0.29		

Sources: Wischmeier, W.H. et al., *J. Soil Water Conserv.*, 26(5), 189, 1971; Kirkby, M.J. and Morgan, R.P.C., *Soil Erosion*, John Wiley & Sons, Chichester, U.K.

Table 8.8 Values of the exponent m of Equation 8.69

s (%)	M
<1	0.2
1–3	0.3
3–5	0.4
>5	0.5

8.6.2.4 Vegetative cover and cropping management factor C

The cropping-management factor C shows the effect of crops and/or mulch on reducing erosion compared to cultivated bare soil. The C factor can also be used in construction sites to evaluate the effect of seeding of grass or mulch in controlling erosion during the construction period. In other words, this factor shows the effect of measures to reduce the raindrop impact. Comprehensive tables with C values are given by Wischmeier and Smith (1978) for specific crops and mulch application. General values can be found in Table 8.9 for pasture, range and idle land, Table 8.10 for forest land and Table 8.11 for mulch application in construction sites. The last table also presents mulch rate application for various mulches, slope steepness range and maximum slope length for which the C values are applicable.

Table 8.9 Factor C values for permanent pasture, range and idle land

Vegetative canopy			Cover that contacts the soil surface					
			Percent ground cover					
Type and height	Percent cover	Type	0	20	40	60	80	95+
No appreciable canopy		G	0.45	0.20	0.1	0.042	0.013	0.003
		W	0.45	0.24	0.15	0.091	0.043	0.011
Tall weeds or short brush with average drop fall height of 50 cm	25	G	0.36	0.17	0.09	0.038	0.013	0.003
		W	0.36	0.20	0.13	0.083	0.041	0.011
	50	G	0.26	0.13	0.07	0.035	0.012	0.003
		W	0.26	0.16	0.11	0.076	0.039	0.011
	75	G	0.17	0.10	0.06	0.032	0.011	0.003
		W	0.17	0.12	0.09	0.068	0.038	0.011
Appreciable brush or bushes, with average drop fall height of 2 m	25	G	0.4	0.18	0.09	0.04	0.013	0.003
		W	0.4	0.22	0.14	0.087	0.042	0.011
	50	G	0.34	0.16	0.08	0.038	0.012	0.003
		W	0.34	0.19	0.13	0.082	0.041	0.011
	75	G	0.28	0.14	0.08	0.036	0.012	0.003
		W	0.28	0.17	0.12	0.078	0.04	0.011
Trees, but no appreciable low brush, with average drop fall height of 4 m	25	G	0.42	0.19	0.10	0.041	0.013	0.003
		W	0.42	0.23	0.14	0.089	0.042	0.011
	50	G	0.39	0.18	0.09	0.04	0.013	0.003
		W	0.39	0.21	0.14	0.087	0.042	0.011
	75	G	0.36	0.17	0.09	0.039	0.012	0.003
		W	0.36	0.20	0.13	0.084	0.041	0.011

Source: Wischmeier, W.H. and Smith, D.D., *Predicting Rainfall Erosion Losses – A Guide to Conservation Planning*, Agricultural Handbook No. 537, USDA, 1978.

Notes: G – The slope cover is grass or grass-like plants, decaying compacted duff or litter 50 mm deep. W – The slope cover is broadleaf herbaceous plants with little lateral root network near the surface or undecayed residues or both.

Table 8.10 Factor C values for undisturbed forest land

Percent of area covered by canopy of trees and undergrowth	Percent of area covered by decaying forest duff (litter) at least 5 cm thick	C value
100–75	100–90	0.0001–0.001
70–45	85–75	0.002–0.004
40–20	70–40	0.003–0.009

Source: Wischmeier, W.H. and Smith, D.D., *Predicting Rainfall Erosion Losses – A Guide to Conservation Planning*, Agricultural Handbook No. 537, USDA, 1978.

8.6.2.5 Erosion control practices factor P

The value of the erosion control practices factor *P* quantifies the effect of tillage and plough-ing practices which reduce the velocity of runoff and affect its tendency to flow downhill. The no-practice (baseline) condition is ploughing directly up and down the slope for which *P* is set equal to 1. Soil conservation practices include contouring, terracing, crop rotations and retention of residues on the surface. Wischmeier and Smith (1978) present detailed descriptions of the various practices and Table 8.12 summarizes *P* values.

Table 8.11 Factor *C* values, length limits and slope for construction sites

Type of mulch	Mulch rate (tons/ha)	Land slope (%)	Factor C	Length limit (m)
None	0.0	All	1.00	
Straw or hay tied down by anchoring and tacking equipment	2.2	1–5	0.20	61
	2.2	6–10	0.20	30
	3.4	1–5	0.12	91
	3.4	6–10	0.12	46
	4.5	1–5	0.06	122
	4.5	6–10	0.06	61
	4.5	11–15	0.07	46
	4.5	16–20	0.11	30
	4.5	21–25	0.14	23
	4.5	26–33	0.17	15
	4.5	34–50	0.20	11
Crushed stone, 1/4–1 1/2 in.	303	<16	0.05	61
	303	16–20	0.05	46
	303	21–33	0.05	30
	303	34–50	0.05	23
	538	<21	0.02	91
	538	21–33	0.02	61
	538	34–50	0.02	46
Wood chips	16	<16	0.08	23
	16	16–20	0.08	15
	27	<16	0.05	46
	27	16–20	0.05	30
	27	21–33	0.05	23
	56	<16	0.02	61
	56	16–20	0.02	46
	56	21–33	0.02	30
	56	34–50	0.02	23

Source: Wischmeier, W.H. and Smith, D.D., *Predicting Rainfall Erosion Losses – A Guide to Conservation Planning*, Agricultural Handbook No. 537, USDA, 1978.

Table 8.12 Values of the erosion control practices factor *P*

Land slope (%)	Contouring	Strip cropping	Terracing
1–2	0.60	0.30	0.12
3–5	0.50	0.25	0.10
6–8	0.50	0.25	0.10
9–12	0.60	0.30	0.12
13–16	0.70	0.35	0.14
17–20	0.80	0.40	0.16
21–25	0.90	0.45	0.18

8.6.3 Modified universal soil loss equation

The main limitation of the USLE is that it computes annual average soil loss due to sheet and rill erosion in farmland and construction sites. The application of USLE has limitations in arid areas (e.g. southwestern United States) which receive a significant portion of annual rainfall in the form of short-duration intense storms. To address this problem, Williams and Berndt (1972) modified the USLE and developed a procedure to compute sediment yields from watersheds produced from single-storm events. In the produced Modified Universal Soil Loss Equation (MUSLE), the rainfall erosivity factor was replaced by a runoff term considering volume and peak of runoff.

The form of the MUSLE equation is

$$Y_s = \beta' 95 (Q_V q_p)^{0.56} KLSCP \qquad (8.71)$$

where

Y_s is the sediment yield for a specific storm (U.S. tons or tonnes)
Q_V is the runoff volume (acre-feet or m^3)
q_p is the peak flow rate of runoff (cfs or m^3/s)

All other parameters (i.e. K, LS, C and P) are as defined for USLE. Finally, β' is a unit conversion factor which is equal to 1.0 when Y_s is computed in U.S. tons or 124 when Y_s is computed in metric tonnes. The coefficient 95 and the exponent 0.56 were based on calibration of the method in watersheds in Texas and Nebraska.

8.6.4 Erosion control measures

Erosion control is desirable due to the mentioned adverse impacts of sediments. There are two general types of measures to adopt to control erosion: (1) measures to reduce erosion at the locations where it occurs, i.e. where sediments are produced (control at the source), and (2) structural measures to trap eroded and transported sediments.

8.6.4.1 Erosion control at the source

Such measures are usually taken at the upper sloping parts of the watershed. One can get a good feeling about what kind of measures they should be by analysing the terms of the USLE.

8.6.4.1.1 Changing the R and K factors

Factors R (rainfall erosivity index) and K (soil erodibility index) cannot be affected much since the first one depends on rainfall characteristics of the area and the other on local soil texture, organic content and permeability (Figure 8.20). Attempts to add organic matter (decomposed vegetation residues) could result in lower K values and erosion rates.

8.6.4.1.2 Changing the LS factor

As mentioned, the length of the slope is an important factor in erosion production. Reducing the length can reduce erosion rates. Possible ways to reduce length is by providing better drainage of the slope and this can be done by the addition of open side drains (nearly parallel to the contours and perpendicular to the main slope direction) which confluence

Figure 8.21 Side drains and downdrains used to reduce slope length. (Photo by V.A. Tsihrintzis.)

to downdrains (along the slope direction). Such a method is shown in Figure 8.21, where practically the slope length between side drains is reduced to one-third of the entire slope length. The vegetation development between side drains can further, and more effectively, reduce erosion as discussed in the next paragraph.

8.6.4.1.3 Changing the C factor

Changing the C factor is the best way to reduce erosion, and this can be done (as seen in Tables 8.9 and 8.11) by either enhancing the vegetation cover or using mulch. Both measures significantly reduce the value of the C factor, particularly vegetation. Details on the type of vegetation and site preparation are given by Goldman et al. (1986).

8.6.4.1.4 Changing the P factor

As Table 8.12 shows, contouring (and minimum tillage), strip cropping and terracing are effective erosion control practices, significantly reducing the value of P factor. Terraces (Figure 8.22) reduce both slope steepness and length and also enhance infiltration, thus reducing runoff. Terraces can also be considered as a structural measure; however, it is presented here since it can control sediments at the source.

8.6.4.1.5 Measures during construction

Construction is a major activity producing sediments. Temporary measures should be taken during construction to minimize soil losses. Measures to change the LS, C and P factors are effective. Examples are shown in Figures 8.23 and 8.24. In Figure 8.23, the slope, in the rainy season during construction, is covered with plastic, and sandbags are used to keep plastic in place. Figure 8.24 demonstrates poor construction practices in a small urban project.

Figure 8.22 Terracing to reduce slope steepness. (Photo by V.A. Tsihrintzis.)

Figure 8.23 Plastic used to cover slope during construction. (Photo by V.A. Tsihrintzis.)

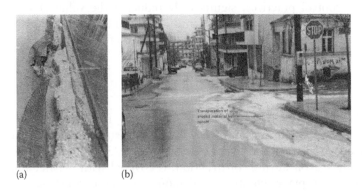

(a)   (b)

Figure 8.24 (a) Poor construction practices leave trench unprotected from rainfall impact to erode; (b) eroded material from the trench is transported downslope. (Photos by V.A. Tsihrintzis.)

8.6.4.2 Structural measures to control erosion

These can be of three types: (1) indirect, which involves installation of water conveyance and energy dissipation structures—such structures provide conveyance means for water to flow in, safely minimizing erosion potential; (2) channel protection from degradation (channel stabilization); and (3) direct, which involves building special structures whose purpose is to trap eroded sediments (sediment retention structures such as debris basins [Figure 8.25]). We will discuss them briefly in Table 8.13, since there are specialized textbooks exhausting the issue (e.g. Goldman et al. 1986; Simons and Sentürk, 1992).

(a)

(b)

Figure 8.25 Debris basin: (a) Looking from the watershed downslope, the upstream slope of the earthen dam, which is covered with concrete, is shown as well as the riser and the emergency spillway. (b) Looking upslope, the downstream slope of the dam and the emergency spillway are also shown. During construction, erosion measures are installed by covering the face of the earthen dam with plastic. The right side of the picture shows some erosion in the uncovered face of the dam. (Photos by V.A. Tsihrintzis.)

Table 8.13 Structural measures to control erosion

Type of measure	Structure	Short description
Water conveyance structures – outlet energy dissipation	Dikes	Dikes are small temporary non-engineered compacted soil embankments placed on the head of the slope to prevent or minimize water flowing down the slope.
	Swales	Swales are small non-engineered channels collecting and conveying drainage. They are earthen or grassed.
	Side drains and downdrains	They are placed on the slope along the contours and in the main slope direction, as presented in Figure 8.21. They can either be corrugated metal or paved with concrete or gunite.
	Pipe slope drains	They are plastic pipes used to collect and convey runoff down a slope preventing erosion. They are also effective when installed inside gullies to minimize runoff and, thus, gully erosion.
	Culverts	They are placed underneath roads at stream or gully crossings to provide drainage conveyance and protect road from erosion and failure.
	Outlet protection	All the aforementioned structures require protection at the outlet, to either resist or reduce the energy of the flow and prevent local scour. Either riprap or small energy dissipators are used.
Channel stabilization	Channel linings	Several materials can be used to line a channel and make it erosion resistant. These include riprap (i.e. gravel and rock), grouted riprap, gabions, reinforced concrete and grass, among others.
	Bed stabilization	Specialized stabilization or grade control structures (concrete or riprap) can be used at point where major head cutting and bed degradation is expected. Such structures can be designed as retaining walls across the channel to make upstream slope milder and retain sediments.
	Local protection	Specialized structures can be used at places in the channel where local scour occurs. Examples include bank stabilization methods (e.g. jetties, groynes) and bridge pier protection from local scour and undermining.
Sediment retention	Sediment or debris basins	These do not stop erosion but trap eroded sediments from moving downstream with runoff. They essentially comprise a small dam with a riser and an emergency spillway (Figure 8.25). The riser is a vertical, usually perforated corrugated metal pipe, which acts as a 'strain' allowing runoff to pass through but retaining sediments. The sediments then accumulate behind. Obviously, coarser materials are trapped and finer pass through. Equations like the USLE (Equation 8.67) or MUSLE (Equation 8.71) can be used to compute the volume of the accumulated sediments with time and size the debris basin. Access should be provided for clean up.
	Sediment barriers	These are temporary non-engineered structures which retain sediments. They can be either permeable fences along the contours made of permeable fabric or straw bales anchored at the base of the slope. They are both usually attached to a wire fence.

Source: Adapted from Goldman et al., 1986.

REFERENCES

Ackers, P. and White, W.R., 1973, Sediment transport: New approach and analysis, *Journal of Hydraulic Engineering*, ASCE, 99(11), 2041–2060.

Alam, M.S. and Kennedy, J.F., 1969, Friction factors for flow in sand-bed channels, *Journal of the Hydraulics Division*, ASCE, 95(HY6), 1973–1992.

Arcement, G.J. and Schneider, V.R., 1989, Guide for selecting Manning's roughness coefficients for natural channels and flood plains, U.S. Geological Survey, Water-Supply Paper 2339, Denver, CO.

Barnes, H.H., 1967, Roughness characteristics of national channels, U.S. Geological Survey, Water Supply Paper 1849, Washington, DC.

Brown, C.B., 1950, Sediment transportation, in *Engineering Hydraulics*, H. Rouse (ed.), John Wiley & Sons, New York, pp. 769–858.

Chaudhry, M.H., 1993, *Open-Channel Flow*, Prentice Hall, Englewood Cliffs, NJ.

Chow, V.T., 1959, *Open-Channel Hydraulics*, McGraw-Hill Inc., New York.

Cowan, W.L., 1956, Estimating hydraulic roughness coefficient, *Agricultural Engineering*, 37(7), 473–475.

Coyle, J.J., 1975, *Grassed Waterways and Outlets*, Engineering Field Manual, U.S. Department of Agriculture, Soil Conservation Service, Washington, DC, pp. 7.1–7.43.

Einstein, H.A., 1942, Formula for the transportation of bed-load, *Transactions of the ASCE*, 107, 561–573.

Einstein, H.A., 1950, The bed load function for sediment transportation in open channel flows, Technical Bulletin No. 1026, U.S. Department of Agriculture, Washington, DC.

Einstein, H.A. and Barbarossa, N.L., 1952, River channel roughness, *Transactions of the ASCE*, 117, 1121–1132.

Einstein, H.A. and Chien, N., 1953, Transport of sediment mixtures with large range of grain sizes, MRD Series Report No. 2, Missouri River Division, U.S. Army Corps of Engineers, Omaha, NE.

Einstein, H.A. and Chien, N., 1954, Second approximation to the solution of suspended-load theory, MRD Series Report No. 2, University of California and Missouri River Division, U.S. Army Corps of Engineers, Institute of Engineering Research, Omaha, Nebraska, June.

Engelund, F. and Hansen, E., 1966, Investigations of flow in alluvial streams. *Acta Polytechnica Scandinavica*, Civil Engineering and Building Construction Series, Vol. 35. Finnish Academy of Technology, Copenhagen, Denmark, 100pp.

Fortier, S. and Scobey, F.C., 1926, Permissible canal velocities, *Transaction of ASCE*, 89, 940–984, Paper No. 1588.

French, R.H., 1985, *Open Channel Hydraulics*, McGraw-Hill Inc., New York.

Goldman, S.J., Jackson, K. and Bursztynsky, T.A., 1986, *Erosion and Sediment Control Handbook*, McGraw-Hill Inc., New York.

Guillen, D., 1996, *Determination of Roughness Coefficients for Streams in West-Central Florida*, U.S. Geological Survey, in cooperation with the Southwest Florida Water Management District, Report 96-226, pp. 1–93.

Henderson, F.M., 1966, *Open Channel Flow*, Macmillan Publishing Co., Inc., New York.

Kirkby, M.J. and Morgan, R.P.C., 1980, *Soil Erosion*, John Wiley & Sons, Chichester, U.K.

Lane, E.W., 1947, Report of the subcommittee on sediment terminology, *Transactions of the American Geophysical Union*, 28(6), 936–938.

Lane, E.W., 1953, Progress report on studies on the design of stable channels of the Bureau of Reclamation, *Proceedings of the ASCE*, 79, Separate No. 280, pp. 1–31.

Meyer, L.D. and Monke, E.J., 1965, Mechanics of soil erosion by rainfall and overland flow, *Transactions of the ASCE*, 580, 572–577.

Meyer-Peter, E. and Müller, R., 1948, Formula for bed-load transport, *Second Meeting International Association for Hydraulic Research*, Stockholm, Sweden, June 1948.

Rao, A.R. and Kumar, B., 2009, Analytical formulation of the correction factor applied in Einstein and Barbarossa equation (1952), *Journal of Hydrology and Hydromechanics*, 57(1), 40–44.

Richardson, E.V. and Simons, D.B., 1967, Resistance to flow in sand channels, *Proceedings of the 12th Congress of the IAHR*, Vol. 1, Fort Collins, CO, pp. 141–150.

Rouse, H., 1937, Modern conceptions of the mechanics of turbulence, *Transactions of ASCE*, 102 (1), 463–543.

Rouse, H., 1939, An analysis of sediment transportation in the light of fluid turbulence, Soil Conservation Service Report No. SCS-TP-25, U.S. Department of Agriculture, Washington, DC.

Rubey, W.W., 1933, Settling velocities of gravel, sand and silt particles, *American Journal of Science*, 25, 325–338.

Shen, H.W., 1962, Development of bed roughness in alluvial channels, *Journal of the Hydraulics Division*, ASCE, 88(HY3), 45–58.

Shen, H.W. and Julien, P.Y., 1992, Erosion and sediment transport, in *Handbook of Hydrology*, D.R. Maidment (ed.), McGraw-Hill Inc., New York, pp. 12.1–12.61.

Shields, A., 1936, *Application of Similarity Principles, and Turbulence Research to Bed-load Movement*, Soil Conservation Service, Cooperative Laboratory, California Institute of Technology, Pasadena, CA (translated by W.P. Ott and J.C. van Uchelen from "Anwendung der Aehnlichkeitsmechanilr und der Turbulenzforschung ad die Geschiebe-bewegung," Mitteilungen der Preussischen Versuchsanstalt fur Nassexbau nd Schiffbau, Berlin, Germany, 1936).

Simons, D.B. and Richardson, E.V., 1966, Resistance to flow in alluvial channels, U.S. Geological Survey Professional Paper 422-J.

Simons, D.B. and Senturk, F., 1992, *Sediment Transport Technology – Water and Sediment Dynamics*, Water Resources Publications, Littleton, CO.

Smith, D.D. and Wischmeier, W.H., 1957, Factors affecting sheet and rill erosion, *Transactions American Geophysical Union*, 38, 889–896.

Stokes, G.G., 1851, On the effect of internal friction of fluids on the motion of pendulums, *Transaction of the Cambridge Philosophical Society*, 9(2), 8–106.

Toffaletti, F.B., 1969, Definite computation of sand discharges in rivers, *Journal of Hydraulics Division*, ASCE, 95(HY1), 225–248.

Tsihrintzis, V.A. and Madiedo, E.E., 2000, Hydraulic resistance determination in marsh wetlands, *Water Resources Management*, 14(4), 285–309.

U.S. Bureau of Reclamation, 1977, *Design of Small Dams*, U.S. Department of the Interior, Bureau of Reclamation, Washington, D.C., 816pp.

Vanoni, V., 1964, Measurements of critical shear stress for entraining fine sediments in a boundary layer, Report KH-R-7, W.M. Keck Laboratory of Hydraulics and Water Resources, California Institute of Technology, Pasadena, CA.

Vanoni, V.A. (ed.), 2006, *Sedimentation Engineering*, ASCE Manuals and Reports on Engineering Practice No. 54, American Society of Civil Engineers, Reston, VA.

Williams, J.R. and Berndt, H.D., 1972, Sediment yield computed with universal equation, *Journal of the Hydraulics Division*, ASCE, 98(HY12), 2087–2098.

Wischmeier, W.H., 1972, Estimating the cover and management factor for undisturbed land, Sediment Yield Workshop, USDA Sedimentation Laboratory, Oxford, MS.

Wischmeier, W.H., 1973, Upslope erosion analysis, in *Environmental Impact on Rivers*, H.E. Shen (ed.), Chapter 15, Colorado State University, Fort Collins, CO.

Wischmeier, W.H., Johnson, C.B. and Cross, B.V., 1971, A soil-erodibility nomograph for farmland and construction sites, *Journal of Soil and Water Conservation*, 26(5), 189–193.

Wischmeier, W.H. and Smith, D.D., 1965, *Predicting Rainfall-Erosion Losses from Cropland East of the Rocky Mountains*, Agriculture Handbook No. 282, USDA, Agricultural Research Service, Washington, DC.

Wischmeier, W.H. and Smith, D.D., 1978, *Predicting Rainfall Erosion Losses – A Guide to Conservation Planning*, Agricultural Handbook No. 537, USDA.

Yang, C.T., 1972, Unit stream power and sediment transport, *Journal of the Hydraulics Division*, ASCE, 98(HY10), 1805–1826.

Yang, C.T., 1973, Incipient motion and sediment transport, *Journal of the Hydraulics Division*, ASCE, 99(HY10), 1679–1704.

Yang, C.T., 1996, *Sediment Transport – Theory and Practice*, McGraw-Hill Inc., New York.

Yen, B.C., 1992, Hydraulic resistance in open channels, in *Channel Flow Resistance: Centennial of Manning's Formula*, B.C. Yen (ed.), Water Resources Publications, Littleton, CO, pp. 1–135.

Index